BIOREMEDIATION AND PHYTOREMEDIATION TECHNOLOGIES IN SUSTAINABLE SOIL MANAGEMENT

Volume 1

Fundamental Aspects and Contaminated Sites

Bioremediation and Phytoremediation Technologies in Sustainable Soil Management, 4 volume set:
ISBN: 978-1-77463-720-3 (hbk)
ISBN: 978-1-77491-100-6 (pbk)
ISBN: 978-1-00328-133-7 (ebk)

Volume 1: Fundamental Aspects and Contaminated Sites
ISBN: 978-1-77463-718-0 (hbk)
ISBN: 978-1-77463-958-0 (pbk)
ISBN: 978-1-00328-065-1 (ebk)

Volume 2: Microbial Approaches and Recent Trends
ISBN: 978-1-77463-719-7 (hbk)
ISBN: 978-1-77463-892-7 (pbk)
ISBN: 978-1-00328-068-2 (ebk)

Volume 3: Inventive Techniques, Research Methods, and Case Studies
ISBN: 978-1-77463-987-0 (hbk)
ISBN: 978-1-77463-988-7 (pbk)
ISBN: 978-1-00328-117-7 (ebk)

Volume 4: Degradation of Pesticides and Polychlorinated Biphenyls
ISBN: 978-1-77491-038-2 (hbk)
ISBN: 978-1-77491-039-9 (pbk)
ISBN: 978-1-00328-120-7 (ebk)

Innovations in Agricultural and Biological Engineering

BIOREMEDIATION AND PHYTOREMEDIATION TECHNOLOGIES IN SUSTAINABLE SOIL MANAGEMENT

Volume 1

Fundamental Aspects and Contaminated Sites

Edited by

Junaid Ahmad Malik, PhD
Megh R. Goyal, PhD, PE
Khursheed Ahmad Wani, PhD

AAP | APPLE ACADEMIC PRESS

First edition published 2023

Apple Academic Press Inc.
1265 Goldenrod Circle, NE,
Palm Bay, FL 32905 USA

4164 Lakeshore Road, Burlington,
ON, L7L 1A4 Canada

CRC Press
6000 Broken Sound Parkway NW,
Suite 300, Boca Raton, FL 33487-2742 USA

4 Park Square, Milton Park,
Abingdon, Oxon, OX14 4RN UK

Apple Academic Press exclusively co-publishes with CRC Press, an imprint of Taylor & Francis Group, LLC

Library and Archives Canada Cataloguing in Publication

Title: Bioremediation and phytoremediation technologies in sustainable soil management / edited by Junaid Ahmad Malik, PhD, Megh R. Goyal, PhD, PE, Khursheed Ahmad Wani, PhD.
Names: Malik, Junaid Ahmad, 1987- editor. | Goyal, Megh R., editor. | Wani, Khursheed Ahmad, editor.
Series: Innovations in agricultural and biological engineering.
Description: First edition. | Series statement: Innovations in agricultural and biological engineering | Includes bibliographical references and indexes. | Content: Volume 1: Fundamental Aspects and Contaminated Sites.
Identifiers: Canadiana (print) 20220155194 | Canadiana (ebook) 20220155224 | ISBN 9781774637180 (v. 1 ; hardcover) | ISBN 9781774639580 (v. 1 ; softcover) | ISBN 9781003280651 (v. 1 ; ebook)
Subjects: LCSH: Soil remediation. | LCSH: Phytoremediation. | LCSH: Bioremediation.
Classification: LCC TD878.48 .B56 2022 | DDC 628.5/5—dc23

Library of Congress Cataloging-in-Publication Data

CIP data on file with US Library of Congress

ISBN: 978-1-77463-718-0 (hbk)
ISBN: 978-1-77463-958-0 (pbk)
ISBN: 978-1-00328-065-1 (ebk)

ABOUT THE BOOK SERIES: INNOVATIONS IN AGRICULTURAL AND BIOLOGICAL ENGINEERING

Under this book series, Apple Academic Press Inc. is publishing book volumes over a span of 8–10 years in the specialty areas defined by the American Society of Agricultural and Biological Engineers (www.asabe.org). Apple Academic Press Inc. aims to be a principal source of books in agricultural and biological engineering. We welcome book proposals from readers in areas of their expertise.

The mission of this series is to provide knowledge and techniques for agricultural and biological engineers (ABEs). The book series offers high-quality reference and academic content on agricultural and biological engineering (ABE) that is accessible to academicians, researchers, scientists, university faculty and university-level students, and professionals around the world.

Agricultural and biological engineers ensure that the world has the necessities of life, including safe and plentiful food, clean air and water, renewable fuel and energy, safe working conditions, and a healthy environment by employing knowledge and expertise of the sciences, both pure and applied, and engineering principles. Biological engineering applies engineering practices to problems and opportunities presented by living things and the natural environment in agriculture.

ABE embraces a variety of the following specialty areas (www.asabe.org): aquaculture engineering, biological engineering, energy, farm machinery and power engineering, food, and process engineering, forest engineering, information, and electrical technologies, soil, and water conservation engineering, natural resources engineering, nursery, and greenhouse engineering, safety, and health, and structures and environment.

For this book series, we welcome chapters on the following specialty areas (but not limited to):

1. Academia to industry to end-user loop in agricultural engineering.
2. Agricultural mechanization.
3. Aquaculture engineering.
4. Biological engineering in agriculture.

5. Biotechnology applications in agricultural engineering.
6. Energy source engineering.
7. Farm to fork technologies in agriculture.
8. Food and bioprocess engineering.
9. Forest engineering.
10. GPS and remote sensing potential in agricultural engineering.
11. Hill land agriculture.
12. Human factors in engineering.
13. Impact of global warming and climatic change on agriculture economy.
14. Information and electrical technologies.
15. Irrigation and drainage engineering.
16. Micro-irrigation engineering.
17. Milk Engineering.
18. Nanotechnology applications in agricultural engineering.
19. Natural resources engineering.
20. Nursery and greenhouse engineering.
21. Potential of phytochemicals from agricultural and wild plants for human health.
22. Power systems and machinery design.
23. Robot engineering and drones in agriculture.
24. Rural electrification.
25. Sanitary engineering.
26. Simulation and computer modeling.
27. Smart engineering applications in agriculture.
28. Soil and water engineering.
29. Structures and environment engineering.
30. Waste management and recycling.
31. Any other focus areas.

Books in the Innovations in Agricultural & Biological Engineering Series

- Advances in Food Process Engineering: Novel Processing, Preservation and Decontamination of Foods
- Advances in Green and Sustainable Nanomaterials: Applications in Energy, Biomedicine, Agriculture, and Environmental Science
- Biological and Chemical Hazards in Food and Food Products: Prevention, Practices, and Management

- Bioremediation and Phytoremediation Technologies in Sustainable Soil Management
    - Volume 1: Fundamental Aspects and Contaminated Sites
    - Volume 2: Microbial Approaches and Recent Trends
    - Volume 3: Inventive Techniques, Research Methods, and Case Studies
    - Volume 4: Degradation of Pesticides and Polychlorinated Biphenyls
- Dairy Engineering: Advanced Technologies and Their Applications
- Developing Technologies in Food Science: Status, Applications, and Challenges
- Emerging Technologies in Agricultural Engineering
- Engineering Interventions in Agricultural Processing
- Engineering Interventions in Foods and Plants
- Engineering Practices for Agricultural Production and Water Conservation: An Interdisciplinary Approach
- Engineering Practices for Management of Soil Salinity: Agricultural, Physiological, and Adaptive Approaches
- Engineering Practices for Milk Products: Dairyceuticals, Novel Technologies, and Quality
- Enzyme Inactivation in Food Processing: Technologies, Materials, and Applications
- Field Practices for Wastewater Use in Agriculture: Future Trends and Use of Biological Systems
- Flood Assessment: Modeling and Parameterization
- Food Engineering: Emerging Issues, Modeling, and Applications
- Food Process Engineering: Emerging Trends in Research and Their Applications
- Food Processing and Preservation Technology: Advances, Methods, and Applications
- Food Technology: Applied Research and Production Techniques
- Functional Dairy Ingredients and Nutraceuticals: Physicochemical, Technological, and Therapeutic Aspects
- Handbook of Research on Food Processing and Preservation Technologies:
    - Volume 1: Nonthermal and Innovative Food Processing Methods
    - Volume 2: Nonthermal Food Preservation and Novel Processing Strategies

    - Volume 3: Computer-Aided Food Processing and Quality Evaluation Techniques
    - Volume 4: Design and Development of Specific Foods, Packaging Systems, and Food Safety
    - Volume 5: Emerging Techniques for Food Processing, Quality, and Safety Assurance
- Modeling Methods and Practices in Soil and Water Engineering
- Nanotechnology Horizons in Food Process Engineering, 3-volume set:
    - Volume 1: Food Preservation, Food Packaging and Sustainable Agriculture
    - Volume 2: Scope, Biomaterials, and Human Health
    - Volume 3: Trends, Nanomaterials, and Food Delivery
- Nanotechnology and Nanomaterial Applications in Food, Health, and Biomedical Sciences
- Nanotechnology Applications in Agricultural and Bioprocess Engineering: Farm to Table
- Nanotechnology Applications in Dairy Science: Packaging, Processing, and Preservation
- Novel and Alternate Methods in Food Processing: Biotechnological, Physicochemical, and Mathematical Approaches
- Novel Dairy Processing Technologies: Techniques, Management, and Energy Conservation
- Novel Processing Methods for Plant-Based Health Foods: Extraction, Encapsulation and Health Benefits of Bioactive Compounds
- Novel Strategies to Improve Shelf-Life and Quality of Foods: Quality, Safety, and Health Aspects
- Processing of Fruits and Vegetables: From Farm to Fork
- Processing Technologies for Milk and Milk Products: Methods, Applications, and Energy Usage
- Quality Control in Fruit and Vegetable Processing: Methods and Strategies
- Scientific and Technical Terms in Bioengineering and Biological Engineering
- Soil and Water Engineering: Principles and Applications of Modeling
- Soil Salinity Management in Agriculture: Technological Advances and Applications

- State-of-the-Art Technologies in Food Science: Human Health, Emerging Issues and Specialty Topics
- Sustainable Biological Systems for Agriculture: Emerging Issues in Nanotechnology, Biofertilizers, Wastewater, and Farm Machines
- Sustainable Nanomaterials for Biomedical Engineering: Impacts, Challenges, and Future Prospects
- Sustainable Nanomaterials for Biosystems Engineering: Trends in Renewable Energy, Environment, and Agriculture
- Technological Interventions in Dairy Science: Innovative Approaches in Processing, Preservation, and Analysis of Milk Products
- Technological Interventions in Management of Irrigated Agriculture
- Technological Interventions in the Processing of Fruits and Vegetables
- Technological Processes for Marine Foods, from Water to Fork: Bioactive Compounds, Industrial Applications, and Genomics

OTHER BOOKS ON AGRICULTURAL AND BIOLOGICAL ENGINEERING FROM APPLE ACADEMIC PRESS, INC.

Management of Drip/Trickle or Micro Irrigation
Megh R. Goyal, PhD, PE, Senior Editor-in-Chief

Evapotranspiration: Principles and Applications for Water Management
Megh R. Goyal, PhD, PE, and Eric W. Harmsen, Editors

Book Series: Research Advances in Sustainable Micro Irrigation
Senior Editor-in-Chief: Megh R. Goyal, PhD, PE

Volume 1: Sustainable Micro Irrigation: Principles and Practices
Volume 2: Sustainable Practices in Surface and Subsurface Micro Irrigation
Volume 3: Sustainable Micro Irrigation Management for Trees and Vines
Volume 4: Management, Performance, and Applications of Micro Irrigation Systems
Volume 5: Applications of Furrow and Micro Irrigation in Arid and Semi-Arid Regions
Volume 6: Best Management Practices for Drip Irrigated Crops
Volume 7: Closed Circuit Micro Irrigation Design: Theory and Applications
Volume 8: Wastewater Management for Irrigation: Principles and Practices
Volume 9: Water and Fertigation Management in Micro Irrigation
Volume 10: Innovation in Micro Irrigation Technology

Book Series: Innovations and Challenges in Micro Irrigation
Senior Editor-in-Chief: Megh R. Goyal, PhD, PE

Volume 1: Management of Drip/Trickle or Micro Irrigation
Volume 2: Sustainable Micro Irrigation Design Systems for Agricultural Crops
Volume 3: Principles and Management of Clogging in Micro Irrigation
Volume 4: Performance Evaluation of Micro Irrigation Management

Volume 5: Potential Use of Solar Energy and Emerging Technologies in Micro Irrigation
Volume 6: Micro Irrigation Management: Technological Advances and Their Applications
Volume 7: Micro Irrigation Engineering for Horticultural Crops
Volume 8: Micro Irrigation Scheduling and Practices
Volume 9: Engineering Interventions in Sustainable Trickle Irrigation
Volume 10: Management Strategies for Water Use Efficiency and Micro Irrigated Crops
Volume 11: Fertigation Technologies In Micro Irrigation

ABOUT THE EDITORS

Junaid Ahmad Malik, PhD
Lecturer, Department of Zoology, Government Degree College, Bijbehara, Kashmir (J&K), India

Junaid Ahmad Malik, PhD, is a Lecturer in the Department of Zoology at Government Degree College, Bijbehara, Kashmir (J&K), India, and is actively involved with teaching and research activities. He has more than eight years of research experience. His areas of interest are ecology, soil macrofauna, wildlife biology, conservation biology, etc. Dr. Malik has published 21 research articles and technical papers in international peer-reviewed journals from Springer, Elsevier, and T&F, etc. He has authored three books, edited two books, 11 book chapters, and more than 10 popular editorial articles. He is also serving as editor and reviewer of several journals. He has participated in several state, national, and international conferences, seminars, workshops, and symposia, with more than 20 conference papers to his credit. He is a life member of SBBS (Society for Bioinformatics and Biological Sciences).

Dr. Malik received a BSc (2008) from the University of Kashmir, Srinagar, J&K; MSc (2010) in Zoology from Barkatullah University, Bhopal, Madhya Pradesh; and PhD (2015) in Zoology from the same university. He completed his BEd program in 2017 from the University of Kashmir, Srinagar, J&K. He started his career as a Lecturer in the School Education Department, Government of J&K, for two years.

Readers may contact him at: *malik.junaidahmad@gmail.com*.

Megh R. Goyal, PhD, PE

Retired Professor in Agricultural and Biomedical Engineering, University of Puerto Rico, Mayaguez Campus; Senior Acquisitions Editor, Biomedical Engineering and Agricultural Science, Apple Academic Press, Inc..

Megh R. Goyal, PhD, PE, is, currently a retired professor of agricultural and biomedical engineering from the General Engineering Department at the College of Engineering at the University of Puerto Rico–Mayaguez Campus (UPRM); and Senior Acquisitions Editor and Senior Technical Editor-in-Chief for Agricultural and Biomedical Engineering for Apple Academic Press Inc.

During his long career, he has worked as a Soil Conservation Inspector; Research Assistant at Haryana Agricultural University and Ohio State University; Research Agricultural Engineer/Professor at the Department of Agricultural Engineering of UPRM; and Professor of Agricultural and Biomedical Engineering in the General Engineering Department of UPRM. He spent a one-year sabbatical leave in 2002–2003 at the Biomedical Engineering Department of Florida International University, Miami, USA.

Dr. Goyal was the first agricultural engineer to receive the professional license in agricultural engineering from the College of Engineers and Surveyors of Puerto Rico. In 2005, he was proclaimed the "Father of Irrigation Engineering in Puerto Rico for the Twentieth Century" by the American Society of Agricultural and Biological Engineers, Puerto Rico Section, for his pioneering work on micro irrigation, evapotranspiration, agroclimatology, and soil and water engineering.

During his professional career of 52 years, he has received many awards, including Scientist of the Year, Membership Grand Prize for the American Society of Agricultural Engineers Campaign, Felix Castro Rodriguez Academic Excellence Award, Man of Drip Irrigation by the Mayor of Municipalities of Mayaguez/Caguas/Ponce and Senate/Secretary of Agriculture of ELA, Puerto Rico, and many others. He has been recognized as one of the experts "who rendered meritorious service for the development of [the] irrigation sector in India" by the Water Technology Centre of Tamil Nadu Agricultural University in Coimbatore, India, and ASABE who bestowed on him the 2018 Netafim Microirrigation Award. VDGOOD Professional Association of India awarded Lifetime Achievement Award at 12th Annual

Meeting on Engineering, Science and Medicine that was held on 20-21 of November of 2020 in Visakhapatnam, India.

Dr. Goyal has authored more than 200 journal articles and edited more than 100 books. AAP has published many of his books, including *Management of Drip/Trickle or Micro Irrigation; Evapotranspiration: Principles and Applications for Water Management;* ten-volume set on *Research Advances in Sustainable Micro Irrigation*. Dr. Goyal has also developed several book series with AAP, including Innovations in Agricultural & Biological Engineering, Innovations and Challenges in Micro Irrigation; and Innovations in Plant Science for Better Health: From Soil to Fork.

Dr. Goyal received his BSc degree in Engineering from Punjab Agricultural University, Ludhiana, India, and his MSc and PhD degrees from the Ohio State University, Columbus, Ohio, USA. He also earned a Master of Divinity degree from the Puerto Rico Evangelical Seminary, Hato Rey, Puerto Rico, USA.

Readers may contact him at goyalmegh@gmail.com.

Khursheed Ahmad Wani, PhD
Lecturer, Department of Environmental Science, Government Degree College, Bijbehara, Jammu and Kashmir, India

Khursheed Ahmad Wani, PhD, is a Lecturer in the Department of Environmental Science, Government Degree College, Bijbehara, J & K, India, and is actively involved in teaching and research. He was formerly Assistant Professor in the Department of Environmental Science at ITM University, Gwalior, India, for more than five years. He has more than 13 years of research experience. He has worked in the area of environmental monitoring, occupational health, vermicomposting, and environmental impact assessment. He has published 40 research papers in peer-reviewed journals, six edited and two authored books, 12 book chapters, more than 50 popular articles, and more than 50 conference abstracts. He has successfully guided five PhD students. He has worked as principal investigator in research projects sanctioned by the Madhya Pradesh Council of Science and Technology. He is a member of several national and international organizations and is on the editorial boards of various journals. He has organized different programs on environmental awareness and field trips. He has also completed two projects of the Indonesian Institute of Sciences, Indonesia.

He received a BSc (2004) from Kashmir University, Srinagar; MSc (2007) in Environmental Science from Jiwaji University, Gwalior, Madhya Pradesh, India; and PhD (2011) in Environmental Science from Jiwaji University, Gwalior, Madhya Pradesh, India.

Readers may contact him at: *kuurshidevs@gmail.com*.

CONTENTS

CONTRIBUTORS

Muhammad Zubair Akram
PhD Research Scholar, Department of Agricultural Genetic Engineering, Faculty of Agricultural Sciences and Technologies, Nigde Omer Halisdemir University, Nigde – 51240, Turkey, Mobile: +90-535-7608970, E-mail: zakram97@yahoo.com

Chanda Parulekar Berde
Assistant Professor, Marine Microbiology, SEOAS, Goa University, Goa – 403206, India, Mobile: +91-9860360635, E-mail: cvberde27@gmail.com

Vikrant B. Berde
Assistant Professor, Department of Zoology, Arts, Commerce, and Science College, Lanja – 416701, Ratnagiri, Maharashtra, India, Mobile: +91-7387766390, E-mail: berdevikrant31@gmail.com

Brij Bhushan
Associate Professor, Department of Chemistry, Graphic Era University, Dehradun – 248002, Uttarakhand, India, Mobile: +91-9837579633, E-mail: brijbhushaniitr@gmail.com

Trinath Biswal
Professor Department of Chemistry, Veer Surendra Sai University of Technology, Burla – 768018, Odisha, India, Mobile: +91-9439363041, E-mail: biswal.trinath@gmail.com

Pallaval V. Bramhachari
Assistant Professor, Department of Biotechnology, Krishna University, Machlipatnam – 521001, Andhra Pradesh, India, Mobile: +91 8897019580, E-mail: veerabramha@gmail.com

M. Subhosh Chandra
Assistant Professor and Raman Fellow, Department of Microbiology, Yogi Vemana University, Kadapa – 516005, Andhra Pradesh, India, Mobile: +91-9493372743, E-mail: chandra298@gmail.com

Ashaq Ahmad Dar
PhD Research Scholar, Department of Ecology and Environmental Sciences, School of Life Sciences, Pondicherry University, Puducherry – 605014, Tamil Nadu, India, Mobile: +91-7889433899, E-mail: bscashaq@gmail.com

Kai Dong
Assistant Professor, Key Laboratory for Space Bioscience and Biotechnology, School of Life Sciences, Northwestern Polytechnical University, 127 West Youyi Road, Xi'an – 710072, Shaanxi, China, Mobile: +86-15209281970, E-mail: dongkai120@nwpu.edu.cn

Megh R. Goyal
Senior Editor-in-Chief (Agriculture and Biomedical Engineering) for AAP, Retired Professor in Agricultural and Biomedical Engineering, University of Puerto Rico-Mayaguez, Mayaguez, PR-USA, USA, Mobile: 001-787-536-0039, E-mail: goyalmegh@gmail.com

Mir Zahoor Gul
Senior Research Associate, Department of Biochemistry, University College of Science, Osmania University, Hyderabad – 500059, Telangana, India, Mobile: +91-9848008648, E-mail: ziahgul@gmail.com

Kumar Kritartha Kaushik
Assistant Professor, Department of Zoology, Sipajhar College, Darrang – 784145, Assam, India, Mobile: +91-8638589259, E-mail: kkritartha@gmail.com

Mohammed Latif Khan
Department of Botany, Dr. Harisingh Gour Vishwavidyalaya (A Central University), Sagar – 470003, Madhya Pradesh, India, Mobile: +91-9425613661, E-mail: khanml61@gmail.com

Shreya Kotnala
PhD Research Scholar, Department of Chemistry, Graphic Era University, Dehradun – 248002, Uttarakhand, India, Mobile: +91-8791311332, E-mail: shreya.kotnala001@gmail.com

Tingli Lu
Professor, Key Laboratory for Space Bioscience and Biotechnology, School of Life Sciences, Northwestern Polytechnical University, 127 West Youyi Road, Xi'an – 710072, Shaanxi, China, Mobile: +86-13659188506, E-mail: lutinglixinxin@nwpu.edu.cn

Naga Raju Maddela
Research Professor, Faculty of Health Sciences, Technical University of Manabí, Portoviejo – 130105, Manabí, Ecuador, Mobile: +593-0962921508, E-mail: rmaddela@utm.edu.ec

Arunima Nayak
Associate Professor, Department of Chemistry, Graphic Era University, Dehradun – 248002, Uttarakhand, India, Mobile: +91-9557108497, E-mail: arunima_nayak@yahoo.com

Samreen Nazeer
PhD Research Scholar, Department of Agricultural Genetic Engineering, Faculty of Agricultural Sciences and Technologies, Nigde Omer Halisdemir University, Nigde – 51240, Turkey, Mobile: +90-535-7608970, E-mail: samreennazeer1@gmail.com

Joan Nyika
Tutorial Fellow, Geoscience, and the Environment Department, The Technical University of Kenya, Haile Selassie Avenue, P.O. Box – 52428, Postal Code – 00200, Nairobi, Kenya, Mobile: +254720697511; +254704390690, E-mail: joashmada2011@gmail.com

Mohd. Yousuf Rather
Project Research Fellow, Department of Botany, Dr. Harisingh Gour Vishwavidyalaya (A Central University), Sagar – 470003, Madhya Pradesh, India, Mobile: +91-7889439626, E-mail: yousufbashir786@gmail.com

Anila Sadia
Assistant Professor, Department of Botany, Faculty of Sciences, University of Agriculture, Faisalabad – 38040, Pakistan, Mobile: +92-315-6777635, E-mail: anilasadia45@gmail.com

Pimpi Sahu
PhD Research Scholar, Key Laboratory for Space Bioscience and Biotechnology, School of Life Sciences, Northwestern Polytechnical University, 127 West Youyi Road, Xi'an – 710072, Shaanxi, China, Mobile: +86-15686002296, E-mail: pimpisahu@mail.nwpu.edu.cn

Muhammad Abdullah Saleem
Assistant Professor, Department of Agronomy, Faculty of Agricultural Sciences, University of Agriculture, Faisalabad – 38040, Pakistan, Mobile: +90-300-7906326, E-mail: abdullahsaleem65@gmail.com

Podduturi Sindhura
Lecturer, Telangana Social Welfare Residential Degree College for Women, Bhupalapally – 506169, Telangana, India, Mobile: +91-8341294858, E-mail: sindhura.p2212@gmail.com

M. Srinivasulu
Academic Consultant, Department of Biotechnology, Yogi Vemana University, Kadapa – 516005, Andhra Pradesh, India, Mobile: +91-8333011076, E-mail: seenmahi452@gmail.com

Irfan-Ur-Rauf Tak
Lecturer, Department of Zoology, Government Degree College, Bijbehara – 192124, Anantnag, Jammu and Kashmir, India, Mobile: +91-7006784769, E-mail: takirfanrouf@gmail.com

Podduturi Vanamala
Lecturer, Telangana Social Welfare Residential Degree College for Women, Kamareddy – 503111, Telangana, India, Mobile: +91-9290753243, E-mail: pvanamalareddy@gmail.com

Tanaji Vasavilatha
Lecturer, Telangana Social Welfare Residential Degree College for Women, Nizamabad – 503001, Telangana, India, Mobile: +91-7207776918, E-mail: tanaji.vasavilatha@gmail.com

Qingyan Zeng
PhD Research Scholar, Key Laboratory for Space Bioscience and Biotechnology, School of Life Sciences, Northwestern Polytechnical University, 127 West Youyi Road, Xi'an – 710072, Shaanxi, China, Mobile: +86-17802920662, E-mail: zengqy@mail.nwpu.edu.cn

Zhuangzhuang Zhao
Masters in Biological Engineering, Key Laboratory for Space Bioscience and Biotechnology, School of Life Sciences, Northwestern Polytechnical University, 127 West Youyi Road, Xi'an – 710072, Shaanxi, China, Mobile: +86-18119343028, E-mail: zhaozhuangzhuang@mail.nwpu.edu.cn

Caiyun Zheng
PhD Research Scholar, Key Laboratory for Space Bioscience and Biotechnology, School of Life Sciences, Northwestern Polytechnical University, 127 West Youyi Road, Xi'an – 710072, Shaanxi, China, Mobile: +86-18900628206, E-mail: caiyun.zheng140205@mail.nwpu.edu.cn

Sofi Mohammad Zuber
Assistant Professor, Department of Zoology, Government Degree College, Bijbehara – 192124, Anantnag, Jammu and Kashmir, India, Mobile: +91-7006822147, E-mail: zubairhuma@gmail.com

ABBREVIATIONS

°C	degree Celsius
3-PBA	3-phenoxy benzaldehyde
ABC	ATP-binding cassette
ACC	1-Aminocyclopropane-1-carboxylate
ACCD	1-aminocyclopropane-1-carboxylate deaminase
ADS	acid detergent soluble
AFB	air force base
AHP	analytic hierarchy process
AMPR	arbuscular mycorrhizal phytoremediation
ANN	artificial neural network
API	American Petroleum Institute
APS	APT sulfurylase
As	arsenic
ASCE	American Society of Civil Engineers
ATP	adenosine triphosphate
B	boron
BCF	bioconcentration factor
BE	biometric extraction
Benzoyl-CoA	benzoyl coenzyme A
BM	buffalo manure
BSAF	biota-sediment accumulation factor
BTEX	benzene, toluene, ethylbenzene, and xylene
C	carbon
CaMs	calmodulins
CAX	cation exchange transporters
CBLs	calcineurin B-like
Cd	cadmium
CDF	cation diffusion facilitator
CFU	colony-forming unit
CH_4	methane
Co	cobalt
CO_2	carbon dioxide
COPT	copper transporter
Cr	chromium

Cu	copper
Cys	cysteine
DBT	dibenzothiophene
DDT	dichloro diphenyl trichloroethane
DMSO	dimethyl sulfoxide
DNA	deoxyribonucleic acid
DoD	Department of Defense
EDDS	ethylene diamines-di-succinic acid
EDTA	ethylene diamine tetra acetic acid
EGTA	ethylene glycol tetra acetic acid
Fe-S	iron-sulfur
FKW	food and kitchen waste
FRTR	federal remediation technology roundtable
GA	gibberellic acid
GEM	genetically engineered microorganism
Glu	glutamate
Gly	glycine
GSH	glutathione
H_2O_2	hydrogen peroxide
Hg	mercury
HMA	heavy metal associated
HMA	heavy metal ATPase
HMs	heavy metals
HMW PC HM	high molecular weight phytochelatins heavy metal complex
HMW	high molecular weight
HTS	high throughput sequencing
IAA	indole-3-acetic acid
IPA	indole 3-propionic acid
IRT	iron-regulated transporter
ITOPF	international tanker owners pollution federation
JA	jasmonic acid
K_{OW}	octanol/water partition coefficient
LCP	latex-clearing protein
LFG	landfill gas
LiP	lignin peroxidases
LMW PC-HM	low molecular weight phytochelatins heavy metal complex
LMW	low molecular weight

MBT	2-methyl-benzothiazole
MCs	metal chelators
MES	microbial electrochemical systems
ML	machine learning
Mn	manganese
MnP	manganese peroxidases
MRP	multidrug resistance-associated proteins
MSW	municipal solid waste
MTBT	2-methlthio benzothiazole
MTP	metal tolerance protein
MTs	metallothioneins
N	nitrogen
NA	natural attenuation
NAD	nicotinamide adenosine triphosphate
Ni	nickel
NPK	nitrogen, phosphorus, and potassium
NRAMP	natural resistance-associated macrophage protein
nZVI	zero-valent iron nanoparticles
OCPs	organochlorine pesticides
OEMs	oil-eating microorganisms
OHCB	obligate hydrocarbon clastic bacteria
ONGC	oil and natural gas corporation
OPs	organophosphate pesticides
P	phosphorous
PAHs	polycyclic aromatic hydrocarbons
Pb	lead
PC	phytochelatins
PCB	polychlorinated biphenyls
PCE	perchloroethylene
PCR	polymerase chain reaction
PCS	phytochelatin synthase
PDMS	polydimethylsiloxane
PEGs	polyethylene glycols
PGPB	plant growth-promoting bacteria
PGPR	plant growth-promoting rhizobacteria
PHs	petroleum hydrocarbons
POPs	persistent organic pollutants
RACER	remedial action cost engineering and requirements
rDNA	recombinant deoxyribonucleic acid

ReTec	remediation technologies
RF	random forest
ROS	reactive oxygen species
RoxA	rubber oxygenase
SA	salicylic acid
SDS	sodium dodecyl sulfate
Se	selenium
SOP	standard operating procedure
Sp.	species
SS	sewage sludge
TCA	tricarboxylic acid
TCE	trichloroethylene
TERI	Energy and Resources Institute
TEV	total economic value
TF	transcription factors
TFNs	triangular fuzzy numbers
TOPSIS	technique of order of preference by similarity to ideal solutions
TPH	total petroleum hydrocarbon
US EPA	the United States Environmental Protection Agency
VAM	vascular Arbascular Mychorhizae
VFAs	volatile fatty acids
VFW	vegetable and fruits waste
VOAs	volatile organic acids
VOCs	volatile organic compounds
WHC	water holding capacity
WRF	white-rot fungi
YSL	yellow strip-like
ZAT	zinc transporter
ZIP	zinc/iron-regulated transporters
Zn	zinc
ZRT	zinc regulated transporters
γ-Glu-Cys	glutamate-cysteine complex

FOREWORD

Münir Öztürk, PhD, DSc, FIAS
Vice President, Islamic World Academy of Sciences; Professor (Emeritus) of Ecology and Environmental Sciences; Consultant Fellow, Faculty of Forestry, University Putra Malaysia, Malaysia; Distinguished Visiting Scientist, ICCBS, Karachi University, Pakistan

We are facing severe environmental problems due to the interactions with our ecosystems. This includes the mother "Earth," i.e., our soils. Mostly our interactions lead to serious hazards for our health. Pollution of our surroundings is a serious problem facing humans on a global scale. It requires full attention for our own benefit. Excess input of hazardous wastes is ending up with the pollution of our soils, air, and waters.

The topics related to bioremediation and phytoremediation presented in this book share the facts related to the biological agents used to clean polluted environment. It is discussed under the title of "eco-biotechnology" these days and can be accepted as an effective management tool for sustainable soil management. The new strategies are coming to the forefront as successful ways. Some of these ways involve standardization as a helpful tool for studying plants on heavy metal (HM) polluted areas. Similarly, the use of various Archae domain organisms in remediation of contaminated natural resources is proposed together with microalgae.

Our soils are a part of the terrestrial environment, and their health needs full conservation against degradation processes. They possess a unique balance of different components, including microbial ones which contribute towards maintaining soil health. The ultimate goal of soil remediation must be not only to clean but at the same time restore the soil capacity for its full potential use. Every day, new bioremediation methods are emerging; some of these are based on the advances in molecular biology and genetic engineering.

This edited volume covers all these aspects within 11 chapters grouped under three parts, mostly written by researchers from India.

Part 1 includes the titles related to the "Fundamental Aspects." There are two chapters in this section, namely: "Phytoremediation and Phytotechnologies: An Overview" and "Phytoremediation and Environmental Factors." Chapter 1 discusses various aspects of phytoremediation as well as phytotechnologies. The chapter envisages the phytoremediation methods and techniques as well as their general remediation capacities in addition to throwing light on their advantages and limitations. In Chapter 2, the authors from Turkey and Pakistan have deliberated the effective approaches of bioremediation that can be employed for cleaning up the contaminated soils. The chapter also discusses some environmental factors, mainly biotic, which directly or indirectly affect the process of phytoremediation.

Part 2 has been assembled under the title "Technological Assessment" and includes three chapters. This part starts with Chapter 3, titled "Role of Phytoremediation to CleanUp Heavy Metal Polluted Soils." The chapter deliberates on phytoremediation as an important biotechnological technique to clean heavy metal polluted environs, particularly soils. The various strategies of this process have been discussed, including traditional and advanced ones, together with their effectiveness and economics in the light of current literature. The author has outlined these as heavy metal phytoremediation strategies and techniques, such as conventional techniques like phytoextraction, phytodegradation, phytostabilization, phytovolatilization, rhizofiltration, rhizodegradation, phytohydraulic control, phytodesalination, vegetative cover systems, riparian corridors and buffer strips, modified phytoremediation techniques, microbial-assisted systems, chemical-assisted systems, biochar-assisted systems, genetically engineered systems, factors affecting phytoremediation of heavy metals, effectiveness of phytoremediation mechanisms, and economic evaluation of phytoremediation in the clean-up processes.

The data on "Phytoremediation Technique for PAH Polluted Soils: Applications, Challenges, and Opportunities" are given in Chapter 4. The authors discuss the characteristics, sources, fate in the environment, hazards associated with PAHs. Established and emerging technologies used to date for their removal from the soil have been discussed together with the role of phytoremediation as a green and low-cost technology for remediation of contaminated soils. An overview of bioremediation with technological assessment and emphasis on the petroleum hydrocarbon bioremediation has been prepared by the authors from India and Ecuador jointly. They discuss the eco-friendly, sustainable, and cost-efficient methods, ideally suitable in bioremediation where microbial inoculants prove the removal efficiency of pollutants from the polluted sites in Chapter 5.

Part 3, under the title "Contaminated Sites and Xenobiotics," includes six chapters. The title "Phytoremediation of Soils Contaminated by Heavy Metals (Mercury, Cadmium, and Lead)" is the topic of Chapter 6. Authors comprehensively discuss the background, concepts, and future trends in phytoremediation of heavy metals. It also focuses on currently available phytoremediation techniques and their use in the remediation of soils contaminated with mercury, cadmium, and lead. In Chapter 7, bioremediation of soil adulterated with aromatic compounds has been considered in order to unravel the extent and feasibility of biological remediation of the soil spiked with PAHs and co-contaminated with heavy metals. Microbial Bioremediation and Amelioration of Pesticide-Contaminated Soils too has been written by the authors from India and Ecuador in Chapter 8. This chapter deals with the soil-microbe interactions with the pesticides of soil environment and ameliorating capability of pesticides.

An overview of bioremediation techniques using various microorganisms in the treatment of petroleum contaminated soils has been discussed in Chapter 9. The chapter describes the major environmental issues created due to the petroleum industry and human use of petroleum in day-to-day life. It also highlights the role of microorganisms in the rejuvenation of these contaminated soils with suitable examples and theories. The chapter is based on the hazardous effects of petroleum contaminated soils and remedial measures taken to treat these soils with the help of microorganisms. The role of various organisms in this process has been evaluated at length.

Chapter 10 has been written jointly by Chinese and Indian authors. It presents data on the phytoremediation of biomedical-waste-contaminated sites. They have stressed the fact that amongst all the categories of this waste, the liquid form poses a serious threat for human health as well as our environment together with our watersheds if improperly handled and disposed. They further point out that illegal and unethical reuse of untreated waste can be extremely dangerous and even fatal and can result in infection, genotoxicity, and cytotoxicity, chemical toxicity, radioactivity hazards, physical injuries, and public sensitivity.

Finally, Chapter 11 deals with the "Bioremediation of Waste Dumping Sites." The strategies and technologies adopted globally for degradation of wastes such as bioaugmentation, biopiles, biostimulation, bioventing, and bioattenuation have been discussed.

This is unique and would be an ideal source of scientific information for research scholars, faculty members, and scientists involved in the studies on environmental aspects. It will prove highly beneficial for the environmental engineering students.

I greatly appreciate the efforts spent by the editors Junaid Ahmad Malik, Megh R. Goyal, and Khursheed Ahmad Wani for successfully bringing together this impressive volume.

PREFACE

The world is evolving towards modernization, and there is a price to pay for that in the form of ecological contaminants. In the course of the last few decades, the investigation of sustainable way outs to numerous fast-growing global ecological issues has been the major theme of environmental pollution research. The existence of terrestrial, aerial, and aquatic organisms, as well as human beings, has been imperiled by pollutants produced owing to modernization and development. Although all the components of the environment are immensely lodged by the aftermath of modernization, the soil is the eventual sink for the contaminants. Environmental and land debasement is the outcome of human activities like urbanization, industrialization, and agricultural activities. In this perspective, the planned use of natural and integral traits of florae and their concomitant microorganisms to eliminate, accrue, immobilize, metabolize, or degrade different environmental contaminants expressively subsidizes to the ultimate product of various environmental contaminants and sustainably purify the land from unsolicited harmful compounds.

The living soils are the cornerstone not only of our food chain, but of our society as a whole. It is imperative that we keep them clean, recognizing that the soils are a source of our food. If we do not look after our land, our existence will be bleak. At present, fertile soil equal to approximately 30 soccer fields is being polluted every second, and about 52 million ha are impacted by numerous soil pollutants. Unsustainable agricultural activities contribute to the depletion of water and wind, the degradation of biomass, soil compaction, soil salinity, and degradation and the loss of nutrients. If global population and food production demands increase, it is of utmost importance to keep our soil fertile and profitable.

The addition of fertilizers, pesticides, and herbicides has created challenges for us and all animals in the 20th century. By releasing xenobiotics, which infiltrate our habitats and destroy the soils, industrialization has been the main contributor to pollution. The toxic substances reduce fertility, rendering the soil unfit for natural plant cover or agricultural growth.

The book *Bioremediation and Phytoremediation Technologies in Sustainable Soil Management (Volume 1: Fundamental Aspects and Contaminated Sites)* deals with the special strains and a combination of various processes that would make bioremediation and phytoremediation processes more prominent. In the chapters of this book, the researchers have introduced the recent status of pollution; the characteristics of bioremediation and phytoremediation using virus microorganisms that will provide readers with new ideas on effective bioremediation.

The book explores the different methods of remediation designed to combat soil and agricultural degradation around the globe. Biopesticides are also explained in the book as an alternative to conventional pesticides. Comparably, the potential possibilities for improving modern biopesticides are discussed. Furthermore, the role and value of biofertilizers as a renewable instrument in the field of modern agricultural sciences has been discussed by stressing their role in agriculture through the conservation of soil biodiversity and long-term fertility. A large part of the book is devoted to phytoremediation, which is known to be a healthy and eco-friendly mechanism for preserving the damaged habitats during the current times.

In this book, an effort has been made to provide biologists, agricultural engineers, environmental, and soil scientists, and chemists with a shared forum to collaborate together to identify sustainable alternatives to different environmental problems. The volume discusses the topics necessary for understanding sustainable growth approaches to the environment. In relation to different environmental issues and remedies, the book offers an analysis of ecological methods, bioremediation, phytoremediation, phytotechnology, and their combined importance. For teachers, scholars, professionals, and decision-makers, the book would be a beneficial asset.

For their collaboration, we are grateful to the contributors who make the present volume possible. We further express our gratitude for their exemplary kind help to Apple Academic Press, which turned our endeavors productive.

—*Editors*

PART I
Fundamental Aspects

CHAPTER 1

PHYTOREMEDIATION AND PHYTOTECHNOLOGIES: AN OVERVIEW

IRFAN-UR-RAUF TAK and SOFI MOHAMMAD ZUBER

ABSTRACT

Heavy metal (HM) contamination is considered a serious ecological pollution as it can impact the development of industries in various countries. This pollution can also cause several diseases in humans. A number of techniques have been employed to remove HMs from the environment and soil, such as phytoremediation, which aims at removing the pollutants from adulterated soils and water by means of plants. In contrast to this, "phytotechnologies make use of vegetation to contain and remove organic and inorganic contaminants that are present in the soil, surface waters and in the groundwater." This chapter provides a brief perspective on recent advancements in the research and practical applications of phytoremediation and phytotechnologies.

1.1 INTRODUCTION

Land, soil, and water have been subjected to a number of exploitations because of various anthropogenic activities, which have resulted in their degradation. Efforts have been made to reduce the pollution sources. Phytoremediation has been observed to be an effective technique because it has lower side effects than the conventional methods. In this context, the soil contamination has been a significant ecological problem due to its impacts on human health.

Contaminated sites are present throughout the world, which is in need of immediate cleanup to safeguard our wellbeing. In comparison, phytotechnologies can help to protect the environment by using plants to extract and immobilize pollutants that are present in soil, surface water, and groundwater. Phytotechnologies have found its applications in solid wastes related to mining. These can also help in stabilizing the tailings and to act as hydraulic control for drainage purposes, thus minimizing the contaminant exposure to humans. Phytotechnologies have also found its applications in the elimination of metals from the adulterated media, and this process requires the selection of plant species in a careful manner. These systems, once fully established are able to maintain with minimum efforts.

The present chapter focuses on various aspects of phytoremediation and phytotechnologies. Different methods and techniques and their general remediation capacities have been discussed besides highlighting on their advantages and limitations.

1.2 ROLE OF PHYTOREMEDIATION AND PHYTOTECHNOLOGIES

Contamination of land with heavy metals (HMs) is one of the most hazardous issues as it has been known to cause a decrease in the function of the soil. It has been found to affect the health of the humans and ecosystems and economic welfare [69]. Human intervention is needed in such cases to make use of existing technologies for the purpose of removing the pollutants. Physiochemical techniques can overcome this pollution. Providing chemical substances are one of the ways to remove the contaminants from the land and besides this.

Another way is to replace the contaminated land with other lands [70]. This later method encompasses the elimination of adulterated soil with clean soil. This technique has however many limitations, which mainly include the removal of the original soil [47]. To overcome these limitations, an eco-friendly favorable phytoremediation technique is employed at an affordable cost. This technique makes use of specific plants to remove the soil pollutants. Besides this, it also helps in reducing the toxic effects of contaminants from the surrounding environment [25, 71]. This technique has great potential to implement on different types of pollutants and also helps in recovering soil functions [10, 37]. This technique also includes soil fertility to allow the growth of the plants, which have added their economic values [63]. It has, however been seen that this technology could not be fully implemented by farmers, who lack knowledge, thus causing failure in the implementation of these technologies in HM contaminated lands. Therefore, farmer's knowledge and their training are essential for the success of these techniques.

In contrast to this, phytotechnologies are plant-based methods for the detection and removal of the contaminants present in the soil and in the groundwater [23]. The strategies are aimed at preventing the diseases that are caused by environmental exposures [9]. While the society is mainly concerned with the fact that sanitation helps in preventing the exposure to various types of communicable diseases, we remain less concerned about the fact that the environment plays a significant role in the prevention of the diseases. The success of various remediation techniques is determined by various types of clean-up goals, which are necessary.

Remediation tools and phytotechnologies have been known to have both advantages as well as disadvantages. The most common feature of using phytotechnologies when compared to other methods is their low cost and high acceptance for clean-up purposes [15, 23]. When evaluating the phytotechnologies in terms of capital costs, it is mainly involved with cultivating the plants *in situ* for proper conservation of energy and water. As a clean-up tool based on phytotechnologies, there is proper reduction of secondary pollutants that are associated with the remediation. The positive impacts of phytotechnologies involve protecting the surrounding environment besides providing various kinds of ecosystem services [21]. While assessing the phytotechnologies in terms of esthetic values, it has a great community appeal and acceptance [3] due to covering the contaminated site with various types of vegetations for reducing stress in urban environment [19]. It is because of these facts that phytotechnologies have a widespread public acceptance because it primarily helps in minimizing the exposure to various contaminants, besides helping in improving the quality of food, air, and water resources.

1.3 POTENTIAL OF PHYTOREMEDIATION

Phytoremediation is an *in-situ* and solar-powered methodology that helps in removing and transforming the pollutants besides metabolizing them [18, 44]. The plants that are used can subsequently be harvested and disposed-off very safely. It has been observed that plants can take up the pollutants through their root system and this property helps in detoxification and cleaning-up the pollutants. It is this property of plants for removing or elimination of the wastes, most importantly metals and phenolic compounds. It has also been observed that the plant-microbe interaction has an important role to play in maintaining the physiology and health of the plant. Plants have been for very long known to provide shelter to diverse kinds of microbes, which then ultimately help in degrading the toxic pollutants. Plants make use of a number of mechanisms to uptake and then degrade the inorganic and organic pollutants [48]. Some of the mechanisms used by the plants for the decontamination of polluted soils are discussed in this section.

1.3.1 PHYTOEXTRACTION

It is a cost-effective method through which plants can take up the soil contaminants, especially the metal contaminants through roots and then can transport them to shoots and leaves [33, 41]. Number of plants especially hyperaccumulator plants can accumulate a number of metals in their shoots without showing any visible symptoms [17]. These plants have been found to employ a number of metabolic processes for the purpose of up-taking and mobilizing the metal ions that are present in the soil and then convert them into less toxic forms [28]. After the metals have been accumulated, the resulting plant biomass is finally harvested and is then composted for the purpose of recycling the metals. A plant will be considered as an ideal hyperaccumulator, if it has a high biomass and is rapidly growing besides having high capability to accumulate the metals in its shoots. Both BCF (bioconcentration factor) and TF (translocation factor) have been seen to be very important parameters in the studies of HM uptake [67]. Such plants are considered to be a potential candidate for the phytoextraction of metals from the contaminated sites [20, 22, 33]. Also, plants with BCF value >1 and TF value <1 can have significant potential for phytostabilization [16].

1.3.2 PHYTOSTABILIZATION

It makes use of different types of plant species for immobilizing the HMs that are present in the soil. This is possible through different processes, which mainly include accumulation and precipitation. It helps in reducing the movement of metals and thereby prevents them from leaching into the groundwater and thereby preventing the metal ions from entering into the food chain [67]. Besides, this technique also helps in reducing the pollutant movement/transport that prevents their accumulation in the groundwater. One of the most important features of this technique is that there is no need of disposing of the metal-containing plants and therefore it has been seen that it can restore the polluted sites; and under normal conditions there is no sustainment of vegetation because of the presence of heavy concentration of metals [52].

1.3.3 PHYTODEGRADATION

This technique is also called phytotransformation that involves the uptake and subsequent degradation of pollutants, which are mainly organic in nature and are assisted by the microbes associated with the plant [33]. The technology has found its applications particularly in degrading: trichloroethylene (TCE), naphthalene, aromatic hydrocarbons, and pesticides. This technique is most suitable for hydrophobic organic compounds. In this regard, the octanol/water partition coefficient (K_{OW}) is an important physicochemical characteristic that describes the hydrophobic and hydrophilic properties of chemical compounds. In plants, K_{OW} is an important factor for determining the root access and the translocation of organic molecule. In this context, the organic contaminants with log K_{OW} <1 are considered to be highly water-soluble and those contaminants with log K_{OW} >3.5 are considered to have high sorption, which is mainly achieved through plant roots. It has therefore been seen that plants more often take up the contaminants with log K_{OW} values that are ranging from 0.5 to 3.5 [13, 53].

1.3.4 PHYTOVOLATILIZATION

Phytovolatilization makes use of the metabolic capabilities and rhizospheric microbes of plants for altering the toxic contaminants into less toxic volatile forms, which are then ultimately discharged into the environment. In this technique, metals in the soil are taken up by the roots and then are released into the environment through transpiration. This technique is most suitable and efficient especially when the metals are volatile in nature. The most important advantage of this technique is that it is able to remove the contaminants from the site without the need for harvesting the plant for the purpose of disposal. However, because of the fact that metals particularly Hg and Se are toxic in nature and this technique involves the volatilization of these metals into the atmosphere that makes the nature and implementation of this technique are highly debatable among all phytoremediation methods.

1.3.5 RHIZODEGRADATION

This technique is also called phytostimulation or microbe associated phytoremediation that involves plant substrates as agents to increase and stimulate the growth of microbial communities for the breakdown of organic pollutants [58]. Plant rhizosphere is an important factor to provide a suitable environment for the microorganisms, which are then in a position to breakdown the hazardous pollutants into less harmful products. In this process, plants release the nutrients in the form of exudates that prove very helpful for the microbial communities resulting in increased degradation of organic pollutants. These rhizospheric microbes in turn help in the enhancement of the plant growth by nitrogen fixation and the mobilization of these nutrients. In a number of cases, rhizosphere microbes are most responsible for the degradation of the pollutants. The rhizoremediation process is mainly augmented by the microbes and it is comparatively slower when compared to the *ex-situ* actions due to inhibition by competition among weeds and other abiotic and biotic stressors. Therefore, rhizoremediation is only effective in the root zone and is considered to be inappropriate for the treatment of deep subsurface layers.

1.3.6 RHIZOFILTRATION

In rhizofiltration, plants absorb and accumulate pollutants mainly through their roots. The pollutants are mainly toxic metals, which are being absorbed, are then precipitated, or accumulated from the polluted water and waste streams [54, 55]. A plant with the root system, which grows at a very fast rate, is considered as ideal for rhizofiltration and it helps to eliminate the wastes for a longer duration of time. This technique has worked along with other physical and biochemical processes; and in the case of filtration of metals, it involves the absorption and then subsequent precipitation of metals onto the roots of the plants. Besides this, several other biological processes include the precipitation of the metals from wastewater that are also associated with the rhizofiltration. This technique takes place along with phytoextraction and phytostabilization depending upon the nature of the pollutants that are present in the wastewater.

1.4 PHYTOREMEDIATION AS A SUSTAINABLE APPROACH

Sustainable development is considered in terms of the growth that is based on economic aspects, and that helps in ensuring that there is the fulfillment of the necessities without the need for compromising the future necessities. This concept mainly consists of three aspects that are based on economic, social, and environmental aspects [49, 60]. These three aspects are interrelated with one another. When we talk of sustainable agriculture, it is thought of in terms balanced nature that is in harmony with the environmental governance. Sustainable agricultural theory is based on the fact that there is awareness among the masses, which help in maintaining the environmental health and which also ensures that the products are safe and eco-friendly besides being safe to consume and also having high nutritional value [29]. This practice helps farmers in establishing sustainable agricultural practices [39]. Besides this, there is also sustainable intensification, which is a more eco-friendly technique, which besides utilizing no additional lands also causes no yield losses [66].

Sustainable agriculture is mainly aimed at overcoming the global issues, especially the economic problems, besides helping in maximizing the instant necessities. Depending on how sustainable agriculture is maintained, determines how well it is going to be sustained and being profitable and beneficial for the present and future generations.

When we look at the sustainable agricultural theory, the technique of phytoremediation is thought of as one of the ways for recovery of contaminated lands. It helps in increasing the soil fertility, which is considered a very important factor for determining the food security [22]. On the other hand, when we look at the economic aspects, the technique of phytoremediation is a dynamic tool as it helps in the absorption of the HMs at low costs. It is thought of in terms of sustainable land management, which ultimately helps in increasing the soil fertility, thus adding to the economic values [31].

Out of physicochemical methods, phytoremediation is considered to be most suitable [31]. However, the area of the contaminated land determines the success of this technique and ultimately determines the operating costs. The conventional techniques are thought to be incompatible [64]. It has been seen that the utilization of the phytoremediation technique on a land that has been proposed to be having low economic value is not considered good for the farming community [68].

Plants as phytoremediator agents must be in a position to give additional income to the farmers, and based on this, the technique of phytoremediation is considered to be the most eco-friendly with relatively low cost. It has been found that the use of phytoremediation technology is more than 80 times cheaper than other physicochemical techniques. When we look at the social aspects, phytoremediation also cares about how the farmers may be able to use this technology on their agricultural lands. Therefore, these factors make the farmers capable of choosing the right type of plants for effectively reducing the HM contaminants. In spite of all this, the farmer's knowledge is still considered to be very limited, and as a result, it has not been implemented at a large scale.

Out of various methods for increasing the knowledge of farmers, conventional agricultural extension is a successful method of passing on the important information to the farmers keeping in mind that they become aware and be able to overcome the agricultural problems for their own welfare. When compared to other techniques, phytoremediation has been found to be largely dependent on the contaminated land location, and under these circumstances, it becomes difficult to carry out the field trials on every piece of contaminated land. Also, in comparison to the phytoremediation technologies, the recovery of land is easier in case of more conventional techniques [36]. It has also been observed that the conventional techniques are easier to be applied than the phytoremediation technology, and it is because of these facts that phytoremediation should be carried out keeping in mind the environmental and commercial and social aspects.

1.5 FACTORS AFFECTING PHYTOREMEDIATION

Bioremediation has been known to depend on a number of interactions mainly among pollutants, nutrients, and environmental factors. It has been seen that the interaction between these factors can affect the biodegradability, bioavailability, and physiological requirements, which are crucial in determining the feasibility of the process of bioremediation. These factors are briefly discussed in this section.

1.5.1 POLLUTANT BIOAVAILABILITY

The bioavailability of pollutants is the main factor. It has been seen that the pollutants have the ability to interact with the surrounding environment to alter their availability to the organisms that have the ability to degrade these. Also, the bioavailability is varied amongst the species and organisms. Consequently, it can therefore be well-defined as the ability of the pollutant to move freely in an organism.

1.5.2 NUTRIENT AVAILABILITY

The availability of the nutrients can affect the microbial transformation, and it has also been found to affect the process of detoxification of contaminants. Also, microbes are in need of nitrogen, phosphate, and potassium, which help in their effective proliferation at the polluted site that mainly contain industrial waste. In such sites, the level of organic carbon is exceedingly high due to the natural toxins and easy availability of nutrients, which have been found to rapidly get degraded during the process of microbial development [2].

1.5.3 ABIOTIC AND ENVIRONMENTAL FACTORS

The environmental conditions include proper temperature, pH, water activity and moisture availability, which can affect the process of bioremediation by inhibiting the growth of microbes that help in the degradation of microbes. It has also been seen that microbes are able to divide and proliferate in the contaminated environment under ideal conditions, and are also in a better position to metabolize different pollutants. Therefore, if the environmental conditions are not favorable, microbes will grow very slowly or die in the process, and as a result, the contaminants are not easily cleaned up.

It has been found that pH is necessary for the proper biodegradation of toxins in maximum terrestrial and aquatic ecosystems [27]. Besides this, oxygen plays a very important role in the process of biodegradation. The solubility of the organic molecules is affected mainly by the temperature, which in turn increase their bioavailability to different kinds of microbes. It has also been found that the solubility of O_2 decreases with the increase in temperature, which mainly reduces the development and the metabolic activities of the aerobic microbes. In addition to this, moisture plays a very crucial role for all processes that take place inside the cell, such as, the transportation of waste products and nutrients in the microbial cell [56].

1.5.4 MICROBIAL ACTIVITY

The soil microbes help in the degradation of contaminants, which mainly depend on their population and their catabolic potency. A number of algae, fungi, and protozoa have been associated with the soil microflora, which work in a number of ways for the purpose of assailing the hydrocarbons. For effective biodegradation process, the number of soil microbes should be between 104 and 107 CFU (colony-forming unit); and the number should be <103 CFY per gram of soil for the effective biodegradation. It has been found that with CFU <103, there is a presence the toxic contaminants [32, 38].

Besides this, the activity of the soil microbes can be regulated in a number of ways, mainly by nutritional and environmental factors. For the purpose of effective biodegradation, it is very important that the microbes should change their catabolic activities by way of genetic alteration and the introduction of specific enzymes so that they are in a better position to effectively transform the target molecules [38].

1.6 ADVANTAGES OF PHYTOREMEDIATION TECHNOLOGY

- It offers solutions to a number of problems of soil contamination [50, 53].
- It can effectively reduce the heavy metal contamination [2].
- It has also been found to be very much effective for treating a wide range of toxic metals besides treating inorganic and organic contaminants [18, 53].
- It has been found to be effective for those water volumes with lower concentration of contaminants and also for those lands, which have lower contaminated surface soils [62].
- It is also most probably the cleanest and cheaper technology and can be used for the remediation of particular contaminated sites [38].
- It is also a very inexpensive technique for the purpose of remediating different environmental media and is also been found to be effective for such sites that have low level of contaminants [18, 65].
- It is considered as the cleanest technology and because of this reason; it is also called green technology.
- It is considered to be a very esthetically pleasing technology that rarely disrupts the environment and also preserves the topsoil.
- It is effective in reducing the contamination at low cost and it is eco-friendly.
- It is likewise a very important tool for the treatment of soils that are polluted with heavy metals, particularly when the biomass that is

produced during the process of phytoremediation can be converted into other forms of bioenergy. At a larger scale, this energy can be used for the production of thermal energy.

- Its advantage in comparison to many of the other conventional techniques is that it does not need the services of expert personal or high-cost equipment.
- Phytoextraction has been used at a very large scale for various types of clean-up operations and has also been applied for various types of heavy metal contaminants [50].
- Phytoremediation has also been considered to be a very important alternative to most harsh remediation techniques, which in comparison to phytoremediation completely destroy the biological components of the soil and alter their physical and chemical characteristics.
- Phytoremediation has been also found to be absolutely useful for treating soils that have high aluminum and salt levels [65].
- Phytoremediation has been found to be benefitting the soil leading to the formation of improved soil ecosystem [62].
- The most important feature of phytoremediation technology is that it is more easily acceptable as well as pleasing when compared to other techniques [40].
- The success of this technique is mainly dependent on the proper selection of plant species that have the ability to hyperaccumulate the heavy metals and during this process produce large amounts of biomass that is mainly used for crop production and other management practices [57].

1.7 LIMITATIONS OF PHYTOREMEDIATION

- It has been seen that for those sites with great risk to the humans, phytoremediation is not considered as a technique of choice. It has been found to be suitable for those areas where human contact is very least.
- The ground surface at the site of the remediation may have to be modified in order to prevent flooding or erosion of the site [51].
- The phytoremediation is time consuming in nature, and the large amount of biomass is produced.

- The phytoremediation technology can take up several seasons for cleaning up a particular site. The intermediates that are produced through the process of phytoremediation technology have been found to be highly toxic to some plants [62].
- The process of phytoremediation can be highly affected by the growth rate of the plants, and as a result, more time may be needed for cleaning up a particular site than other conventional techniques. Disposing off the contaminants may take years to accomplish, while as the process of degrading the contaminants may take a number of years [26, 32].
- The success of the phytoremediation technology is limited by the climate of the area, depth of the rooting system and also the level of contamination [66]. The contact of the root is the main limitation of the success of the phytoremediation technology. It has been seen that the contaminants should be in close contact with the root zone of that plant.
- It has been seen that either the roots of the plant should be in a position to reach out to the contaminants or the contaminants must be moved in the close range or vicinity of the plant [26, 32].

1.8 RESEARCH ADVANCES IN PHYTOTECHNOLOGY

1.8.1 PHYTOTECHNOLOGY AND CLEAN WATER

The main aim of phytotechnology is to improve the quality of water by carefully addressing the contaminants that are present in the water. It is also concerned with different processes that are used to treat the contaminated water before they are discharged into the environment [23]. It has recently been reported that some of the phytotechnologies have the unique ability to completely remove the trace contaminants that would otherwise require a lot of effort to remove [43].

A research work has found the unique ability of plants to detoxify the pharmaceuticals due to their diverse metabolic processes: a concept that was commonly referred to as green liver concept [42]. This approach has been very useful when it comes to addressing the diverse issues related to pharmaceuticals in the wastewater. Besides this, wetland approach with many ecological benefits has also been found to reduce the need of secondary oxidation. Therefore, plants have an unlimited potential to help in improving the quality of the wastewaters.

1.8.2 PHYTOTECHNOLOGY IN REDUCING EXPOSURE TO AIRBORNE CONTAMINANTS

Phytotechnologies are being constantly developed for the purpose of removing airborne pollutants, yet plants are known for their ability to help in improving the air quality [6]. Research studies during the recent past have mainly focused on understanding the role of plants as a means of transporting the pollutants and also for sequestering the pollutants, particularly airborne pollutants. In this regard, a research study has been carried out to stabilize the mine tailing sites in particularly semi-arid environments [34]. The greenhouse efficacy of plant species is native to establish a phytostabilization trial. It was seen that both arsenic and chromium concentrations were exceedingly high.

Also, the dust particles were found in numerous size ranges and these findings indicated the importance of the aerosols as a potential route of exposure for the communities that were living near the treatment site. The initial results showed a reduction of more than 60% in the particulate matter. The results were used to validate a potential dust model to predict the location of the windblown dust that was downwind from the treatment site [6, 12, 45]. This study has proved to be very helpful as it represented an overall integrated approach to quantify potential for remediation strategies that are mainly based on the plants for the purpose of stabilizing the mine tailings. It also served as a potential model of how communities that are affected by the mining practices may use an effective and sustainable technology using the native plants for the purpose of reducing the exposure to a number of harmful substances in the mine tailings.

1.8.3 USE OF PHYTOTECHNOLOGY IN EXPOSURE ASSESSMENT

The use of phytotechnology has also been found in assessing and mapping the exposure pathways. A research study investigated the use of plants as biosensors for the detection of contaminants in groundwater [8]. Research has also been carried out, in which plants were used to delineate the subsurface contaminants that are present both in saturated and subsurface horizons [46]. The approaches, which were initially used for analyzing the volatile compounds, have now been evaluated to *in planta* sampling, which is mainly used for rapid detection of pollutants besides acting as indicators of subsurface degradation [30, 59].

A number of methods have been used to identify vapor intrusion pathways, which have been found to link groundwater to that of the indoor air exposures. It has been seen that there is a correlation between VI and the uptake of the plant because roots of the plant occupy the same space that is occupied by the basements of the buildings. Both have been found to be related to drawing the pollutants into the above ground tissues of the plants.

Research has shown that the pollutant fraction that is inhaled from an indoor source is 1,000 times greater than that of the pollutants from an outdoor source [7]. It has been seen that in homes subject to the vapor intrusion, a considerably higher concentration of benzene, TCE, and PCE can considerably increase the risk of cancer [14]. The phytoforensic detection has been a useful tool to determine the fact that how the technique of phytotechnology can be used in a proper manner to evaluate the risk of hazardous substances. This is all possible without the need for entering the homes and without the use of complex procedures. This method has also been found to be most suitable for TCE and perchloroethylene (PCE), which have been found to be the most common groundwater contaminants [51].

Phytoforensic technology has been used in the recent past not only for detecting the contaminants that are present in the subsurface at the time of sampling but has also been very useful for detecting the previous history of contamination. A unique technology has been employed, which has been found to be very effective in telling the history of contamination at a particular site by employing a unique method of trapping the contaminants in the form of tree core rings. This is considered as a unique application of dendrochronology and dendrochemistry that have been successfully employed for reconstructing the potential contaminant exposure [4]. These methods have been tested and then subsequently validated as potential forensic tools [5] and have been found to play an important role in determining the public safety in terms of disease incidences, which are most commonly associated with previous exposures. This technology has been found to be a very important tool because it allows the detection and subsequent quantification of the previous exposures and also helps in establishing a link between the exposure and the diseases.

Phytoforensic technology, besides having the ability to accumulate the data regarding the present and past contaminations, has also been very effective in reducing the environmental impact as compared to other conventional technologies. Another advantage of phytoforensic technology is the sample size, which could be of the size of a pencil. In spite of all these facts, phytoforensic technology has certain limitations, such as the requirement of appropriate vegetations that are needed for sampling. Besides this, it has been found to be very effective in shallow soil profiles, where there is a greater risk of environmental contaminants getting in contact with the humans [46].

1.9 USE OF PHYTOTECHNOLOGIES IN SUSTAINABLE DEVELOPMENT

Phytotechnologies have been used in a number of environmental assessment procedures and can offer effective remediation tools to prevent the public health through the process of mitigating the exposure pathways. With the result of growing population and urbanization, more and more people have become exposed to different kinds of pollution, particularly air and water pollution. Therefore, the present challenge is to take steps to improve the contaminated sites and to reduce the exposure to a variety of contaminants [16]. This technique has also proven to be very useful for providing solutions to a number of exposures and are also affordable and acceptable on a global basis. These techniques are very important in remote areas and in many developing countries that have low infrastructure [21]. Further, because of the fact that many of the phytotechnologies require low cost for the establishment purpose, it is beneficial to make use of the phytobarriers to minimize the risks to different types of exposures, which may cause different types of diseases.

It has been seen that connecting the phytotechnologies with the research related to public health will help in ensuring the fact that developmental efforts are mainly focused to prevent the environmental exposures. A lot of research has been carried out with regards to exposure of humans to diverse kinds of toxicants, but there is still a need of preventive measures, which are more sustainable. It is where phytotechnologies play a very important role as they are highly sustainable both in terms of economic aspects and consumption of energy [25].

The need of preventive measures, which are phytotechnology-based, will increase in the near future because the climate is constantly changing and the need for the conservation of the ecosystem is ever increasing. If the studies on phytotechnology are coordinated with that of the epidemiological studies, they could provide very useful information of how important and effective these technologies are in preventing exposure to different types of diseases. In a nutshell, it can be said that phytotechnologies are essential for contributing to the fact that there is less exposure facing the world population [22]. Besides this, it has also been found to be very helpful to people from hazardous effects of various pollutants. It is in this regard that collaboration is very much needed with public health researchers that will effectively help in reducing the exposure to a number of contaminants. It will also help to establish an interdisciplinary model for helping in the development of other remediation technologies (ReTec). In this regard, this is a need of further research for solving the complex environmental contamination scenarios. It will enlighten the fact that phytotechnologies are not always the best solution, but it is very much important that all the remediation methods need to be considered for the prevention of different types of exposures in relation to public health.

1.10 PHYTOTECHNOLOGIES AND THEIR APPLICATIONS

Phytoremediation has been lately replaced by the term phytotechnologies, which mainly included the applications in which plants are used for the purpose of managing and controlling the pollutants for removing or destroying them [24]. This technique is mainly based on the different types of physiological mechanisms that are taking place in the plants and the microorganisms that are associated with these plants. These physiological mechanisms mainly include transpiration, metabolism, and photosynthesis [18].

It has been seen that plants make use of their roots for the purpose of digging their way into the soil and water in these roots can take up the organic compounds and inorganic substances. Roots have also been seen to stabilize and bind the substances onto their external surfaces. Once the roots take up different types of substances, they can then be transported, stored, and accumulated in the different cells and tissues of the plant [29]. Finally, it has been observed that the aerial parts of the plant may perform the subsequent process of exchange of gases with the environment, which will ultimately help in the release of different types of molecules. A number of phytotechnologies have been identified to help in eliminating different types of contaminants and have also been found to depend on a number of following facts [24]:

- Phytotransformation is suitable for organic contaminants;
- Rhizosphere remediation has been found to be most suitable for organic contaminants that are present in the soil;
- Phytostabilization is most suitable for inorganic and organic contaminants that are present in the soil;
- Phytoextraction is suitable for inorganic contaminants that are present in all substrates;
- Phytovolatilization is concerned with volatile substances;

- Evapotranspiration is suitable for controlling the hydraulic flow that is present in the contaminated environment.

From the knowledge on phytoremediation, a number of phytotechnologies have been designed to help in choosing the right type of plant and the right type of technology to address a particular contaminated site [2]. In this regard number of following methodologies have been identified:

- The vegetative cover is suitable for the control of infiltration;
- The vegetative cover is most suitable for soil remediation;
- The hydraulic barriers;
- The tree plantations for the remediation of soil and groundwater;
- The treatment of wetlands;
- The riparian buffers;
- The hydroponic systems.

All these approaches have been found to differ in both the purpose and goals, which can mainly be remediation, decontamination, and stabilization. It has been seen that once the goal is identified, the proper technique is employed by choosing the right type of the plant and the right type of phytotechnology.

1.11 LIMITATIONS OF PHYTOTECHNOLOGIES

Phytotechnologies are suitable under specific conditions. The major limitation is the area, depth, and time. The depth and area of the site depend on the plant species that are suitable for that particular site and also depend upon the soil characteristics. It has been seen that for the success of the phytotechnologies, large pieces of land are required than any other alternative [64]. Also, time plays a very important role for the success of phytotechnologies as they generally take a long time and are also affected by the seasonal and diurnal changes. All these limitations are taken into consideration while evaluating a particular phytotechnology.

A number of other limitations have also been identified, such as tolerance of the plant to different types of contaminants and the conditions of the site [59]. It is also affected by the water availability that is at the source of the irrigation and is also affected by the climate of that particular area. It has also been seen that it is affected by different types of pests and varied types of infestations.

All these limitations can be overcome by the careful selection of the plant species [51]. It is therefore necessary that the selection of the plant must be made based on the conditions of the specific site. It has been reported that for the success of phytotechnologies, some important elements play very essential role, which mainly include soil salinity, climate of the area, fertility, and location of the soil and commercial availability and plant hardiness. For this purpose, a number of techniques can be employed, which mainly include the site specific vegetations surveys and tests that are specially designed to evaluate the concerned species [23]. Besides the fact that there is a need for carefully selecting the plant species that is needed for remediation purposes, end-use considerations can also be taken into account in the initial phase of the selection of the plant species. For initial design, approximately 10% to 15% climax species must be selected.

1.12 NEW TRENDS IN PHYTOTECHNOLOGIES

1.12.1 NATURAL REMEDIATION

The latest trend in the use of the phytotechnologies has been called "Assisted Natural Remediation," which is basically a non-conventional application [1]. In this technique, soil amendments are used for the purpose of accelerating natural processes of remediation. In the case of metals, amendments are mainly aimed at immobilization, which mainly includes complexation, precipitation, and different types of chemical reactions. The main aim of this is to lower the bioavailability of the metals and not aimed at lowering their concentration. The authors of this chapter have presented successful examples to which the process of assisted natural remediation was applied [1].

1.12.1.1 BIOFORTIFICATION

Recent studies are based on the fact that interaction between different plant species and trace metals leads to biofortification. It is a process in which plants that are enriched with different types of micronutrients are seen as potential agents against malnutrition. As compared to the process of phyto-accumulation (which leads to the accumulation of HMs, which proves very dangerous to the food chain), biofortification has been seen as a technique, where crops containing special elements may become highly advantageous [61].

Fortified crops have been absolutely essential for the growth of soil that is low in micronutrients because it has been seen that their bioconcentration capacity leads to a higher concentration of micronutrients that are present in the edible tissues. Therefore, there should be a proper knowledge of mechanisms that are controlling metal contamination for determining the biochemical basis of these phenomena. In this regard, information is essential for those antinutrients that have been found to lead to a decrease in the availability of metals, and the most common examples include phytic acid and polyphenols.

1.12.1.2 BIOFUMIGATION AND GLUCOSINOLATES

A recent technique which makes use of the properties of the plant chemicals is known as biofumigation. It has been seen that *Brassicaceae* are used in fighting against pests and pathogens that are present in the agriculture because they are known to produce specific secondary metabolites, which are called glucosinolates [35]. These compounds have been found to have anticarcinogenic properties in humans, and they have also been found to contribute to the characteristic flavor that is peculiar of cruciferous plants. Their degradation products help in deterring the herbivores and also inhibit microorganisms [14]. In addition to this, there is a need to explore the connection between glucosinolates and HMs that are present in the environment, which may subsequently act as suppressors and repressors.

Glucosinolates contain sulfur so that they could have an impact in the phenomenon of sulfur metabolism of these HM sequestering peptides [4]. This has in turn led to the determination of the fact that has an interlock occurring during the process of pest and HM resistance.

1.13 FUTURE PERSPECTIVES

Bioremediation has been seen as a highly adaptable approach and an important field of research that is growing at a very fast rate with regards to the environmental restoration. This technique makes use of the capabilities of the microbes for the purpose of converting highly toxic environmental pollutants into less toxic forms. It has been seen that the presence of contaminants for a particular process of bioremediation requires the presence of nutrients that are in conjunction with appropriate growth conditions. This information has been found to be absolutely essential for manipulating the environmental conditions, which have been seen to constrain the biodegradation of organic contaminants for the purpose of *in situ* treatments.

While looking at the degradation capabilities of the microbes along with their metabolic processes, their resistance has revealed the fact that there is a suitable microbial formula, which can be appropriately amended keeping in mind the specific polluted site. Here it should also be kept in mind that the various field experiments have not been properly regulated and designed, which has led to the ambiguity. Thus, more research studies are needed in this field, which are mainly based on adapting the scientifically approved approaches and also have the ability to attain the high-quality data. Besides this, it is also necessary that there should be a proper understanding of the diversity of the microbes that are present in the contaminated site for procuring an improved insight about the potential degraders and makes us understand their biochemistry and genetics. This has all led to the development of suitable and more appropriate bioremediation approaches. It has been seen that the inadequate information is more harmful and affects the stability of the various bioremediation approaches. It is, therefore, necessary that proper research should be carried out focusing on the specific mechanisms through which the organic pollutants are metabolized by the bacterial communities.

Different mechanisms should be studies, which are pertaining to movement of the organic pollutants in the bacterial cells. It should also focus on their disintegration pathways and their regulation of the degradation enzymes. The future research should focus on the fact that there is a proper study of the enzyme systems and the genes should be carefully studied that are involved in the disintegration of the organic pollutants.

There should be development of a bimolecular engineering approach, which should further increase the degradation abilities of the microbes and also the enzymes. Besides, the bacteria have improved abilities of degrading the different types of pesticides that are present under field conditions. It has also been seen that the advances in the field of bioremediation have mainly occurred because of the support it has received from biochemistry, molecular biology, and environmental engineering.

Also, phytotechnology is an emerging tool that has been very useful in the process of remediating the contaminated soils. Research in this field has led to important findings that are related to the plant and soil sciences. However, in spite of all this, there has been little commercial success as far as this technology is concerned. Therefore, there is need for looking into the fact whether it is really a fertile area of research. The terminologies that have been used in phytotechnologies have been found to be the most important factor in limiting the commercial success of this technology. It has been propagated that there should be the formation of valuable biomass on the piece of the contaminated land. This valuable biomass mainly includes timber and biofortified products and ecologically important species.

1.14 SUMMARY

The phytoremediation technology is eco-friendly and cost-effective compared to other technologies. Also, the selection of appropriate plant not only helps in removing the soil pollutants but also can provide extra income to farmers. Research in the field of phytoremediation and phytotechnologies has been thriving at a very fast rate despite the fact that there is a lack of appropriate funding. Several factors should be kept in mind for the purpose of accomplishing a good remediation, such as careful selection of the plant species.

There is a need of techniques, which are commercially feasible as well as effective at the same time. There should be a clear distinction among different processes and techniques to improve the communication as well as cooperation among different commercial sectors. These should also be exploitation of bioenergy production and production of biofortified crops. It can be concluded that phytoremediation and phytotechnologies will always remain very promising tools for cleaning up the soil, but they should not be linked with the production of valuable biomass products.

KEYWORDS

- **contaminants**
- **groundwater**
- **heavy metals**
- **phytoremediation**
- **phytotechnologies**
- **soil sciences**

REFERENCES

1. Adriano, D. C., Wenzel, W. W., Vangronsveld, J., & Bolan, N. S., (2004). Role of assisted natural remediation in environmental cleanup. *Geoderma, 122*, 121–142.
2. Atagana, H. I., Haynes, R. J., & Wallis, F. M., (2003). Optimization of soil physical and chemical conditions for the bioremediation of creosote-contaminated soil. *Biodegradation, 14*, 297–307.
3. ATSDR, (2010). *Leading Change for Healthy Communities and Successful Land Reuse* (p. 96). Washington-DC: U.S. Environmental Protection Agency (EPA), Agency for Toxic Substances and Disease Registry (ATSDR). https://www.atsdr.cdc.gov/sites/brownfields/docs/ATSDR_LandReuse.pdf (accessed on 1 August 2021).
4. Balouet, J. C., Burken, J. G., & Karg, F., (2012). Dendro-chemistry of multiple releases of chlorinated solvents at a former industrial site. *Environmental Science and Technology, 46*(17), 9541–9547.
5. Balouet, J. C., Smith, K. T., Vroblesky, D., & Oudijk, G., (2009). Use of dendrochronology and dendrochemistry in environmental forensics: Does it meet the Daubert criteria? *Environ. Forensics, 10*(4), 268–276.
6. Baumgardner, D., Varela, S., Escobedo, F. J., Chacalo, A., & Ochoa, C., (2012). The role of a peri-urban forest on air quality improvement in the Mexico city megalopolis. *Environ. Poll., 163*, 174–183.
7. Bennett, D. H., McKone, T. E., & Evans, J. S., (2002). Defining intake fraction. *Environmental Science & Technology, 36*(9), 206A–211A.
8. Burken, J. G., Vroblesky, D. A., & Balouet, J. C., (2011). Phytoforensics, dendrochemistry, and phytoscreening: New green tools for delineating contaminants from past and present. *Environmental Science and Technology, 45*(15), 6218–6226.
9. Cengage, G., & Breslow, L., (2002). Primary prevention. In: Breslow, L., (ed.), *Encyclopedia of Public Health* (Vol. 3, p. 1480). Alpha Books, Penguin Putnam, London, UK.
10. Chen, C., Wang, X., & Wang, J., (2019). Phytoremediation of cadmium-contaminated soil by *Sorghum bicolor* and the variation of microbial community. *Chemosphere, 235*, 985–994.
11. Conesa, H. M., Evangelou, M. W. H., Robinson, B. H., & Schulin, R., (2012). A critical view of current state of phytotechnologies to remediate soils: Still a promising tool? *Scientific World Journal, 2012*, 110–115.
12. Csavina, J., Landazuri, A., & Wonaschutz, A., (2011). Metal and metalloid contaminants in atmospheric aerosols from mining operations. *Water Air Soil Pollution, 221*, 145–157.
13. Doucette, W. J., & Bugbee, B., (1998). Phytoremediation of dissolved phase trichloroethylene using mature vegetation. In: Wickramanayake, G. B., & Hinchee, H. E., (eds.), *Bioremediation and Phytoremediation: Chlorinated and Recalcitrant Compounds* (pp. 251–256). Battelle Press, Columbus – Ohio.
14. Environmental Protection Agency, (2008). *EPA's Vapor Intrusion Database: Preliminary Evaluation of Attenuation Factors* (p. 118). Washington (DC): EPA.
15. Environmental Protection Agency, (2010). *Phytotechnologies for Site Cleanup* (p. 89). EPA Report 542-F-10-009.
16. Fitz, W. J., & Wenzel, W. W., (2002). Arsenic transformation in the soil-rhizosphere-plant system, fundamentals and potential application of phytoremediation. *J. Biotechnol., 99*, 259–278.

17. Garbiscu, C., & Alkorta, I., (2001). Phytoextraction: Cost-effective plant based technology for the removal of metals from the environment. *Bioresour. Technol., 77*, 229–236.
18. Germaine, K. J., McGuinness, M., & Dowling, D. N., (2012). Improving phytoremediation through plant associated bacteria. In: De Bruijn, F. J., (ed.), *Molecular Ecology of the Rhizosphere* (Vo. 2, p. 1328). Hoboken, Wiley-Blackwell.
19. Grahn, P., & Stigsdotter, U., (2003). Landscape planning and stress. *Urban Forestry & Urban Greening, 1*, 1–18.
20. Gupta, A. K., & Sinha, S., (2007). Phytoextraction capacity of the *Chenopodium album* L. grown on soil amended with tannery sludge. *Bioresour. Technol., 98*, 442–446.
21. Holzman, D. C., (2012). Accounting for nature's benefits: The dollar value of ecosystem services. *Environ. Health Perspect., 120*(4), 152–157.
22. Hosman, M. E., El-Feky, S. S., Elshahawy, M. I., & Shaker, E. M., (2017). Mechanism of phytoremediation potential of flax (*Linum usitatissimum* L.) to Pb, Cd and Zn. *Pelagia Res. Libr. Asian J. Plant Sci. Res., 7*, 30–40.
23. Interstate Technology & Regulatory Council (ITRC), (2009). *Phytotechnology Technical and Regulatory Guidance and Decision Trees* (p. 110). Revised. PHYTO-3. Washington, D.C.: Interstate Technology & Regulatory Council, Phytotechnologies Team, Tech Reg Update.
24. Interstate Technology and Regulatory Cooperation (ITRC), (2001). *Phytotechnology Technical and Regulatory Guidance Document* (p. 89). Washington, D.C.: Interstate Technology & Regulatory Council, Phytotechnologies Team, Tech Reg Update.
25. Iqbal, M. P., (2012). Lead pollution: Risk factor for cardiovascular disease in Asian developing countries. *Pak. J. Pharm. Sci., 25*, 289–294.
26. Jeddi, K., & Chaieb, M., (2018). Evaluation of the potential of *Erodium glaucophyllum* L. for phytoremediation of metal-polluted arid soils. *Environ. Sci. Pollut. Res., 25*(36), 36636–36644.
27. Jourtey, N. T., Bahafid, W., Sayel, H. E. I., & Chachtouli, N., (2013). Biodegradation: Involved microorganisms and genetically engineered microorganisms. In: Chamy, R., & Rosenkranz, F., (eds.), *Biodegradation-Life Science* (pp. 289–319). Intech Publisher, China.
28. Kumar, P. B. A. N., Dushenkov, V., Motto, H., & Raskin, I., (1995). Phytoextraction: Use of plants to remove heavy metals from soils. *Environ. Sci. Technol., 29*(5), 1232–1238.
29. Lestari, Y. K., & Suryana, A. T., (2013). Sustainability of organic rice farming in Indonesia. *Sustain. Org. Rice Farming Indones., 10*, 55–60.
30. Limmer, M. A., Balouet, J. C., & Karg, F., (2011). Phytoscreening for chlorinated solvents using rapid *in-Vitro* SPME sampling: Application to urban plume in Verl, Germany. *Environmental Science and Technology, 45*(19), 827–882.
31. Mahar, A., Wang, P., & Ali, A., (2016). Challenges and Opportunities in the phytoremediation of heavy metals contaminated soils: A review. *Ecotoxicol. Environ. Saf., 126*, 111–121.
32. Margesin, R., Labbe, D., & Schninner, F., (2003). Characterization of hydrocarbon-degrading microbial populations in contaminated and pristine contaminated soils. *Appl. Environ. Microbiol., 69*(6), 3085–3092.
33. McGrath, S. P., & Zhao, F. J., (2003). Phytoextraction of metals and metalloids from contaminated soils. *Curr. Opin. Biotechnol., 14*, 277–282.

34. Mendez, M. O., & Maier, R. M., (2008). Phytostabilization of mine tailings in arid and semiarid environments: Emerging remediation technology. *Environ. Health Perspec., 116*, 278–283.
35. Mithen, R., (2001). Glucosinolates: Biochemistry, genetics and biological activity. *Plant Growth Regul., 34*, 91–103.
36. Mwegoha, W. J. S., (2008). The use of phytoremediation technology for abatement soil and groundwater pollution in Tanzania: Opportunities and challenges. *J. Sustain. Dev. Africa, 10*, 140–156.
37. Ou, L., Wang, H., Chen, C., Chen, L., Zhang, W., & Wang, X., (2018). Physiologically based pharmacokinetic (PBPK) modeling of human lactational transfer of methylmercury in China. *Environ. Int., 115*, 180–187.
38. Pawar, R. M., (2012). *The. Effect of Soil pH on Degradation of Polycyclic Aromatic Hydrocarbons (PAHs)* (pp. 124–138). PhD Dissertation; University of Hertfordshire, Hatfield, United Kingdom.
39. Pretty, J., & Bharucha, Z. P., (2014). Sustainable intensification in agricultural systems. *Ann. Bot., 114*, 1571–96.
40. Salido, A. L., Hasty, K. L., Lim, J. M., & Butcher, D. J., (2003). Phytoremediation of arsenic and lead in contaminated soil using Chinese brake ferns (*Pteris vittata*) and Indian mustard (*Brassica juncea*). *International Journal of Phytoremediation, 5*, 89–103.
41. Salt, D. E., Smith, R. D., & Raskin, L., (1998). Phytoremediation. *Ann. Rev. Plant Physiol. Plant Mol. Biol., 49*(1), 643–668.
42. Sandermann, H., (1994). Higher plant metabolism of xenobiotics: The 'green liver' concept. *Pharmacogenetics, 4*(5), 225–241.
43. Schroder, P., Navarro-Avino, J., & Azaizeh, H., (2007). Using phytoremediation technologies to upgrade wastewater treatment in Europe. *Environmental Science and Pollution Research, 14*(7), 490–497.
44. Segura, A., & Ramos, J. L., (2013). Plant-bacteria interactions in the removal of pollutants. *Curr. Opin. Biotechnol., 24*, 467–473.
45. Solís-Dominguez, F., & White, S. A., (2012). Response of key soil parameters during phytostabilization in extremely acidic tailings: Effect of plant species. *Environ. Sci. Technol., 46*, 1019–1027.
46. Struckhoff, G., Burken, J. G., & Schumacher, J. G., (2005). Phytoremediation of vadose zone VOCs. *Environmental Science and Technology, 39*(6), 1563–1568.
47. Sukalic, A., Mackic, S., Leto, A., & Dzubur, A., (2018). Human health risk assessment of heavy metals from the agricultural soil in south Herzegovina. *Agric. Conspec. Sci., 83*, 45–50.
48. Tangahu, B. V., Abdullah, S. R. S., & Basri, H., (2011). Review on heavy metals (As, Pb, and Hg) uptake by plants through phytoremediation. *Int. J. Chemi. Eng., 2011*, 1–31.
49. Tanguay, G. A., Rajaonson, J., Lefebvre, J. F., & Lanoie, P., (2010). Measuring the sustainability of cities: An analysis of the use of local indicators. *Ecol. Indic., 10*, 407–418.
50. Teschner, N., Orenstein, D. E., Shapira, I., & Keasar, T., (2017). Socio-ecological research and the transition toward sustainable agriculture. *Int. J. Agric. Sustain., 15*, 99–101.
51. Toccalino, P. L., Norman, J. E., & Hitt, K. J., (2010). Quality of source water from public-supply wells in the United States: 1993–2007. In: *U.S. Geological Survey Scientific Investigations Report 2010–5024* (p. 206).

52. Tordoff, G. M., Baker, A. J. M., & Willis, A. J., (2004). Current approaches to the revegetation and reclamation of metalliferous mine wastes. *Chemosphere, 41*, 219–228.
53. Trapp, S., Zambrano, K. C., Kusk, K. C., & Karlson, U., (2000). Phytotoxicity test using transpiration of willows. *Arch. Environ. Contam. Toxicol., 39*, 154–160.
54. United States Environmental Protection Agency (USEPA), (2000). *Engineered Approaches to In Situ Bioremediation of Chlorinated Solvents: Fundamentals and Field Applications* (p. 119). EPA-542-R-00-008.
55. United States Environmental Protection Agency (USEPA), (2000). *Ground Water Issue* (p. 67). EPA 540/S-01/500.
56. Vidali, M., (2001). Bioremediation. An overview. *Pure Appl. Chem., 73*(7), 1163–1172.
57. Vijayan, V. D., & Sushama, P. K., (2018). Phytoremediation as an effective technology for the removal of heavy metals from dump yard soils *Nat. Environ. Pollut. Technol., 17*, 1353–1358.
58. Vishnoi, S. R., & Shrivastava, P. N., (2008). Phytoremediation-green for environmental cleanup. In: *The 12th World Lake Conference*, (pp. 1016–1021).
59. Vroblesky, D. A., Nietch, C. T., & Morris, J. T., (1999). Chlorinated ethenes from groundwater in tree trunks. *Environ. Sci. Technol., 33*(3), 510–515.
60. Wang, B., Xie, H. L., Ren, H. Y., Li, X., Chen, L., & Wu, B. C., (2019). Application of AHP, TOPSIS, and TFNs to plant selection for phytoremediation of petroleum-contaminated soils in shale gas and oil fields. *J. Clean. Prod., 233*, 13–22.
61. Welch, R. M., & Graham, R. D., (2004). Breeding for micronutrients in staple food crops from a human nutrition perspective. *J. Exp. Bot., 55*, 353–364.
62. Witters, N., Van, S. S., & Ruttens, A., (2009). Short-rotation coppice of willow for phytoremediation of a metal-contaminated agricultural area: A sustainability assessment. *Bioenergy Res., 2*, 144–152.
63. Wuana, R. A., & Okieimen, F. E., (2009). Heavy metals in contaminated soils: A review of sources, chemistry, risks, and best available strategies for remediation. *Heavy Met. Contam. Water Soil Anal. Assessment, Remediat. Strateg., 2011*, 1–50.
64. Wuana, R. A., Okieimen, F. E., & Imborvungu, J. A., (2010). *Removal of Heavy Metals from a Contaminated Soil using Organic Chelating Acids, 7*, 485–496.
65. Yang, B., Xu, T., & Shi, L., (2017). Analysis on sustainable urban development levels and trends in China's cities. *J. Clean. Prod., 141*, 868–880.
66. Yao, Z., Li, J., Xie, H., & Yu, C., (2012). *Review on Remediation Technologies of Soil Contaminated by Heavy Metals, 16*, 722–729.
67. Yoon, J., Cao, X., Zhou, Q., & Ma, L. Q., (2006). Accumulation of Pb, Cu, and Zn in native plants growing on a contaminated Florida site. *Sci. Tota. Environ., 368*(2), 456–464.
68. Zehra, A., & Sahito, Z. A., (2020). Assessment of sunflower germplasm for phytoremediation of lead-polluted soil and production of seed oil and seed meal for human and animal consumption. *J. Environ. Sci. (China), 87*, 24–38.
69. Zhang, W., Chen, L., Zhang, R., & Lin, K., (2015). Effects of decabromodiphenyl ether on lead mobility and microbial toxicity in soil. *Chemosphere, 122*, 99–104.
70. Zhang, X., Zhong, T., Liu, L., & Ouyang, X., (2015). Impact of soil heavy metal pollution on food safety in China. *PLoS One, 10*, 1–14.
71. Zhou, J., Du, B., Hu, Y., & Liang, J., (2019). New criterion for the health risk assessment of se and pb exposure to residents near a smelter. *Environ. Pollut., 244,* 218–227.

CHAPTER 2

PHYTOREMEDIATION AND ENVIRONMENTAL FACTORS

MUHAMMAD ZUBAIR AKRAM, SAMREEN NAZEER, ANILA SADIA, and MUHAMMAD ABDULLAH SALEEM

ABSTRACT

Agricultural productivity is greatly affected by toxic pollutants accumulation in the environment. Green plants are used in phytoremediation for removing toxic pollutants with phytoextraction, phytostabilization, phytofiltration, etc. Among all these techniques, phytoextraction is the best strategy for removing heavy metals (HMs) because of its eco-friendly behavior, low cost, and easy operation. The hyperaccumulator plants can accumulate toxic materials in higher concentrations in their organs and have excellent adaptability to such contaminated conditions. These plants are specific to the area. Plant roots play an essential role in the absorption of these contaminants. The plant growth-promoting rhizobacteria (PGPR) can increase tolerance in such contaminated conditions.

2.1 INTRODUCTION

In this modern age and current world, different forms of pollution are continuously destroying our surroundings and causing many environmental problems including global warming [141], due to industrialization [145]. As a consequence, the environment has out-turned with harmful hazardous chemicals that can affect our environment [136]. Low production of crops and soil pollution has occurred due to these toxic agents resulting in the imbalance of proper lifestyle [57].

The toxic contaminants in nature have been classified into: metals, radioactive substances, organic, and inorganic chemicals. Metallic toxins include mercury, cadmium, selenium, copper, and cadmium, causing various diseases. Among radioactive elements, cesium, uranium, and strontium has the affinity to disturb nature. Inorganic substances are mainly sodium, nitrate ions, arsenic, and ammonium ions [167]. Nowadays, pesticide compounds (atrazine, toluene, and benzene) and chlorinated solvent (trichloroethylene (TCE)) have caused pollution in the environment [82, 95]. Besides these, many other organic pollutants persist in the environment, as they are directly released and become part of the biological system (like animal tissue) making themselves resistant to degradation [77]. In this way, these lipophilic chemicals present in the environment remain in the state for a long time, leaving a question mark for its remediation [88].

The deposition of contaminants in the land has been increasing since the last decade that has impacted adversely globally [16]. The removal of pollutants has also become a critical problem for soil and other living entities on the earth as these can move to plant and animal bodies through soil interaction and may impact the food chain [173]. Heavy metals (HMs) are non-biodegradable, which cause a significant threat to humans and our environment [153].

This chapter discusses some environmental factors (mainly biotic), which directly or indirectly affect the process of phytoremediation.

2.2 ENVIRONMENTAL POLLUTION AND PHYTOREMEDIATION

To overcome environmental issues, the reclamation of accumulated contaminants from the soil and aquatic environment is necessary [51]. The prevention of HMs in the terrestrial ecosystem is another efficient way in their mitigation [60]. There are several techniques for the removal of such pollutants, However, they have their own limitations. Besides this, these technologies and advancements need remodeling of sites with labor that is a costly operation. Instead of using these techniques, one can take a revolutionary step to get rid of the toxic pollutants with the use of microorganisms, plants, or their bilateral interaction [95].

The phytoremediation includes phytovolatilization, rhizofiltration, and phytoextraction. These plants can grow in the soil, which is rich in metals. The most dominant side of this technology is that it is a useful, novel, and environmentally acceptable with low cost and eco-friendly [166, 173]. These plants do not disturb the top layers of soil, grow well by absorbing contaminants, and thus, can conserve the soil fertility and structure of the land along with the addition of organic contents [106].

The phytoremediation by using plant extracts was introduced by Chaney [23]. The cost-effective nature of this technique made it widely acceptable by the industry and consumers. Furthermore, it is suitably practicable in larger fields. In case of cost, it can be priced as 5% lower than that of other techniques. The plantation of different trees can prevent leaching of metals and abrasion of the topsoil layer. There are energy productive plants like Jatropha, Poplar, and Willow with efficient ability of fast-growing and high biomass production [2]. This technology is also known as "green clean," as the use of chemicals is restricted. The collaborative work of plants and microbes plays a major role in agriculture with minimum usage of pesticides and chemical fertilizers [111].

2.3 SOURCES OF HEAVY METALS (ZN, CD, CU, CR, AND PB) IN THE ENVIRONMENT

An environment is a combination of numerous factors, such as land, water, microorganisms, flora, fauna, air, and many others that interact with our daily life directly or indirectly. Our environment has been contaminated with different HMs that exert dangerous effects on living organisms. Although the real source of these HMs is natural, yet anthropogenic activities have been the most significant source to introduce these contaminants into the environment [118]. Generally, metal with atomic weight five times higher than that of water is considered as HM. Almost all HMs are available in low concentrations in the environment, but some of these can cause toxicity at very low concentrations.

HMs can also be evolved from natural sources and anthropogenic activities. Considering natural events, volcanic eruptions, weathering of rocks, and forest fires are the most significant sources of these metal emissions [81]. And the anthropogenic cause of HMs includes big industries, wastewater, metallurgic processes, runoff; and the agricultural sector is recognized to release HMs in different sections of the environment [143]. Some most essential sources releasing HMs are:

- Automobile exhaust, which is a key contributor of lead in the environment;
- Smelting: which is a releaser of arsenic and insecticides [28] that generate nickel, selenium, mercury vanadium and tin;
- Most of the cadmium is deposited by the burning of fossil fuels [53];

- Fertilizer usage in agricultural land is also a contribution to the addition of heavy metals in the environment;
- Concentration of heavy metals increases due to urbanization and industrialization [105].

The emission of HMs during mining, smelting, and industrial sources can restore the HM through wet and dry deposition [127].

2.4 EFFECTS OF HEAVY METALS (HMS)

HMs are elements with an atomic mass of >20 and a density >5 g/cm^3 [8]. HMs exhibit metallic properties, such as stability of cation, ductility, ligand specificity and conductivity [25]. Cd, As, Cr, Zn, Pb, Sn, and Hg are some examples of HMs [16]. Some essential and non-essential metals can affect vegetation, plants, and environment. Non-essential metals include: Pb, As, Zr, Sb, Cd, and Hg. And Zn, Se, Mn, and Cu are essential metals [156]. These HMs are among the top dangerous elements classified by the USPA and US Agency for Toxic and Disease Registry [74, 124, 172].

The Cr, Ni, Zn, and Cu have severe effects on the growth of plant, such as limited growth, chlorosis of leaves, necrosis, and finally death of plants [47]. The maximum accumulation of contaminants in plants may convert the plant from normal to stress conditions, such as growth and developmental stages of plant are profoundly affected by these contaminants. Agricultural land productivity significantly reduces the action of these pollutants [18] that can affect the plants by altering their cellular functions (e.g., thylakoid membrane functions, etc.). Indirectly, they affect the molecular and biochemical aspects of crops. Directly, they reduce the biomass of the plants, the seed germination, the photosynthesis, respiration; and alter plant water status and synthesis of proteins [171]. ROS (reactive oxygen species) and free radicals can increase under metal stress. Many of the enzymes inactivated and lipid peroxidation become maximized by HMs. While at the molecular level, these contaminants may cause DNA (deoxyribonucleic acid) damage, abnormality in meiosis and mitosis phases and finally can change the chromosomal sequences [131].

Metals badly affect the biosphere, which consists of living and non-living things. Plants and animals are living while water, soil, and air are non-living [104]. HMs can change the physical, biological, and chemical features of the environment, land, air, and water [170]. Organic and inorganic are two types of environmental pollutants: (i) The inorganic contaminants may enter into the environment through various anthropogenic activities [86]; (ii) Organic contaminants in a different biosphere can also severely affect the environment [164].

2.5 PHYTOREMEDIATION AND ITS STRATEGIES

Phytoremediation is used by green plants for clean-up of contaminates and organic pollutants. Phytoremediation is simple, eco-friendly, well-organized method for fixing the contaminants. It also has economic value and inexpensive [68]. Phytoremediation is an emerging and easy technique to remediate excessive metals and other pollutants [14]. Different approaches exist for phytoremediation. Based on bioavailability, soil properties and contaminants, five classes are [114]:

- Phytoextraction;
- Phytostabilization;
- Phytovolatilization;
- Phytofilteration; and
- Phytodegradation.

The selection of plants should rely on the requirements of pollutants. For example, under the low-polluted area, select that type of plant that can uptake metal from the rhizosphere and convert it into utilized form [134]. If plant can store the impurities into their tissues, then the phytoaccumulation approach will be used. Besides this, if HM is removed from the membrane, then phytovolatilization and phytoextraction approach may be used. In pollutant-rich areas, the plant cannot transport the contaminants from roots to aerial parts and fixes them around the rhizosphere [178].

2.5.1 PHYTOEXTRACTION

Plant roots can uptake HMs and can transfer them to the aerial parts by xylem loading. The plants use transporters and other agents for movement and absorption of pollutants from soil [108]. Two methods of phytoextraction are described in this section.

2.5.1.1 PHYTOEXTRACTION METHOD

It is a natural process in which the accumulation of contaminants is done naturally or without any additional agents. The unassisted process is also known as continuous phytoextraction method. It is based on the natural hyperaccumulator plants that have a typical accumulating potential [91]. By high trichrome density, efficient regulation of stomatal conductance, proper plant water status, and osmotic adjustment, these plants can uptake maximum amounts of contaminants and can translocate them to aerial parts of the plant [130]. Brassicaceae, Asteraceae, Fabaceae, and Scrophulariaceae are several examples of hyperaccumulator plants. They can accumulate 100 to 1,000 times higher manganese, nickel, lead, copper, zinc, and cadmium than the non-accumulating plants [97].

2.5.1.2 INDUCED METHOD

It is an accommodation process in which plants uptake and accumulate HMs by using helping agents, such as transporters, microbes, hormones, chelators, and other transporters media [52]. It is also called induced method. For example, ZIP transporter helps plants to uptake cadmium and nickel. Various hormones (e.g., SA (salicylic acid), GA (gibberellic acid), brassinosteroid, and JA (jasmonic acid)) are involved in the transportation of contaminants [149]. Some Phytochelators and glutathione (GSH) synthases may help in uptake HMs to different parts of plants [15, 52].

2.5.2 PHYTOSTABILIZATION

This process includes uptake of HMs by the root tissues and accumulation near the rhizospheric zone [134]. Plants change their chemistry for immobilization and reduction in the availability of HMs [12]. The phytostabilization immobilizes contaminants either by accumulation, valence reduction or by absorption in plant roots [125]. By stabilizing these contaminants into a substrate, reduction in the bioavailability and immobilization is carried out [84]. Immobilization rate in plants is increased by utilizing some organic and inorganic modifications. Organic modifications (such as biosolid) can maximize pH rate, enhance soil physical structure; while the addition of NPK fertilizer and lime comes under the inorganic amendment.

Mechanism of phytostabilization depends on root exudates, metal-binding proteins on the cell wall and the presence of microorganisms in the rhizosphere zone. Chelator transporter and vacuole sequestration of HMs by binding protein also help in phytostabilization [178].

The efficiency of phytostabilization can be increased by the best selection of an area and plant species. Phytostabilization can be characterized by two crucial elements: (i) an organic and inorganic modification that is attached to the system; and (ii) the plant itself. The principal objectives of phytostabilization are [1, 134]:

- Reduce the mobility of contaminants to maximize biodiversity;
- Avoid direct uncovering of a living organism (that exists in soils) to the pollutants;
- To stable the vegetation cover that helps to restrict the uptake of trace elements.

2.5.3 PHYTOVOLATILIZATION

It is an intake of contaminants by roots that convert them into a vapor form. These vapors are then transpired to the atmosphere through stomata [6]. Some beneficial bacteria are available in the soil to aid in the biotransformation of HMs. The rate of phytovolatilization is also elevated by biotransformation [26]. Volatilization of HMs can be done directly and indirectly [92]: (i) under direct mechanism, contaminants volatile from leaves and stem; and (ii) volatilization through roots is an indirect method.

In plants, HMs split into air space of plant cells because of the volatilization process. Then, these metals are disseminated into the medium air of the cell. It is considered that contaminants are less toxic in medium air of plant cells than in the air spaces of the cells.

2.5.4 PHYTOFILTRATION

Removal of pollutants by using different plant organs (such as roots, stems, and leaves) is known as phytofiltration (Figure 2.1):

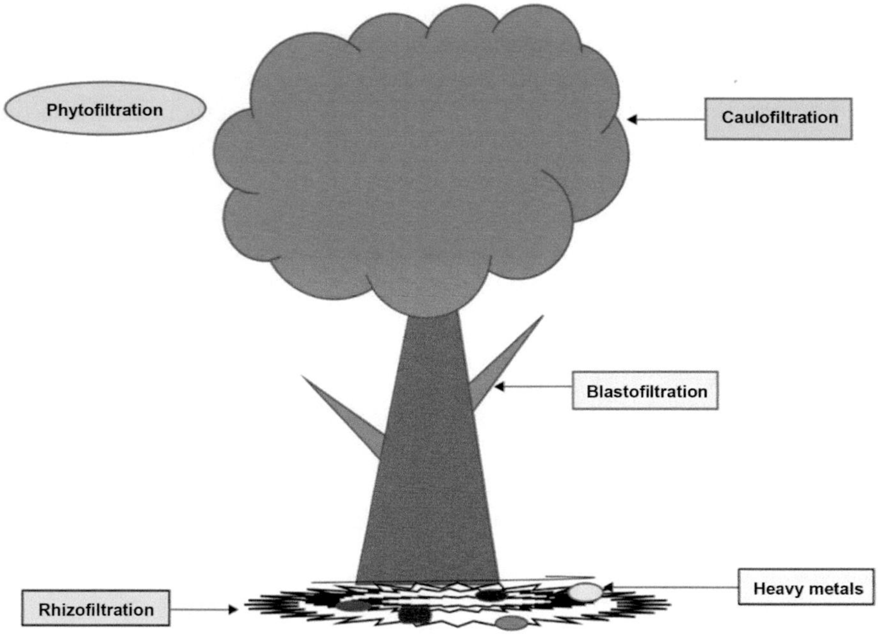

FIGURE 2.1 Phytofiltration of contaminant.

- If the plants use their shoots for removal of contaminants, then it is called caulofiltration [46, 126];
- If plants remove these contaminants from the soil by using its roots, it is known as rhizofiltration;
- If the plant uses its seedlings for removing contaminants, then called blastofiltration.

It can be classified by three different methods that depend on kinetics which are discussed in subsections [29].

2.5.4.1 SPONGING ON THE TOP OF THE ROOT

This method involves the quick deletion of HMs from the surface by positive response of chemical and physical mechanisms: particular uptake, ion exchange capacity and also because of some chelation transporter. However, there is not any involvement of biological activity in the whole process.

2.5.4.2 DEPENDENCE ON PLANT METABOLISM

This process manages the uptake of metals from the solution. This slows the speed mainly due to intracellular uptake, storage in plant vacuole and finally translocated to the shoots. Then from the shoots, these HMs are filtered.

2.5.4.3 RELEASE OF ROOT EXUDATES

In metal removal, the slowest component is the release of root exudates. It moderates the precipitation of metals from the solution. In solution, it presents in the form of insoluble compounds. By using intra- and extra-cellular mechanisms, phytofiltration can remove the accumulation of the metal.

2.5.5 PHYTODEGRADATION

In the phytodegradation process, plants absorb the contaminants by metabolic process or by some enzymes from the polluted soil, they digest them and finally are pulled down into their tissues [19]. Degradation breaks down the heavy complex organic compounds and converts them into light and simple molecules. Degradation also involves the incorporation of metals into the tissues by turning it into a less toxic compounds [142]. Phytodegradation is a speedy and quick process than the biodegradation. The focus of phytodegradation is to convert the contaminants into NH_3, CO_2, Cl^- and NO^{3-} [150]. Mostly breakdown of contaminants occurs in two ways [112]:

- Firstly, by metabolic processes present inside the plants; and
- Secondly, some enzymes that are produced by the plants.

These enzymes act on the polluted organic compounds and convert them into the inorganic compounds. Some compounds are converted in a stable and simple compound by degradation [122]. The enzymes that degrade and transmute the nitroaromatic compounds are known as nitroreductases. Another catalyst group is dehalogenases, which degrade and breakdown the pesticides and other chlorinated solvents. Under degradation, the enzyme Laccases is also produced to deteriorate the anilines [46]. It is concluded that C4 plants have more potential to phytodegradation as compared to C3 plants [75]. Phytodegradation is affected by the efficiency of contaminants uptake, the concentration of toxin and the presence of water [112]. Table 2.1 indicates the advantages and disadvantages of strategies for phytoremediation.

TABLE 2.1 Advantages and Disadvantages of Different Phytoremediation Strategies

Strategy	Advantages	Disadvantages
Phytoremediation	• Eco-friendly; • Not any costly equipment needed; • No any most expert's persons required; • It upgrades the landscape; • Very less disturbance to environments; • Improve the water quality; • No need for any specific life design [78].	• Take too many years to rectify the pollutant areas; • It is not a fast method than the other conventional; • Risk of mobilization of the contaminants into groundwater; • The response of this technique is not active under winter seasons [45].
Phytoextraction	• Phytoextraction does not create any change in the topography; • It provides the benefit for the conservation of the environments; • It is considered to be the best mechanism for removing the contaminants from soil, water, and sediments from all resting techniques [27].	• In the rhizosphere for metals removal, it has very less bioavailability; • From the root, it has a low rate of absorption of heavy metals; • Here is some risk that some metals under the phytoextraction process may retain within the plant root system [137].
Phytostabilization	• The allocation of the phytobiomass is considered that it is not more critical than other techniques [100]; • Prevent the contaminants in the deeper layers of the soil to get leach down [134]; • It has more economic value, has owned more sustainability and also reduces the exposure of metals; • It enhances the organic content in the soil, add some growth enhancer nutrients, water holding capacity even higher [117].	• The major drawback is that the metals do not move from the other side; they remain on the soil site [135]; • It requires long time to monitor the mobilization of metals and other impacts like toxic and ecological ways [117]; • It needs a high amount of fertilizer application that is more expensive and costly.
Phytovolatilization	• It converts the contaminants into a deficient, toxic form; • The pollutants released into the atmosphere rapidly degrade by a natural process like phytodegradation [70, 146].	• Risk of contaminants that they may accumulate in some plant's parts and vegetation like lumber and fruits; • Here is a very less amount of the metabolite in the plant tissue [70, 146].

TABLE 2.1 *(Continued)*

Strategy	Advantages	Disadvantages
Phytofiltration	• The best mechanism is that contaminants that are present in the plants do not move to the aerial part of the plants [70, 146].	• It continually needs the maintenance of the plant pH; • For phytofiltration plant must be needed to grow firstly in the greenhouse or other nursery [70, 146].
Phytodegradation	• It is economically feasible; • Environmentally friendly; • It is a quick process; • Use some enzymes that react fast as compared to other mechanisms [150].	• For efficient performance, it needs about the growing season that is more than one; • There is also a risk that these pollutants may enter in foods through insects and plants [70].

2.6 ENVIRONMENTAL FACTORS AFFECTING PHYTOREMEDIATION

2.6.1 CHARACTERISTICS OF PLANT SPECIES FOR PHYTOREMEDIATION

Phytoremediation and translocation efficiency differs from plant to plant [109]. Cu mainly accumulates in the roots and Pb accumulates in the roots of monocotyledon plants. Therefore, *Atriplex verucifera* and *Chenopodium album* were suitable for the phytoremediation of Pb. Besides, this availability of Pb and its interaction also depends on plant species [76].

Another example can be *Triglochin striata*, which is a phytostabilizer storage of metals in its below-ground parts (roots and rhizomes). In contrast to this, *Phragmites australis* is a pytoextractor, which accumulates metals in its above-ground tissues (stems and leaves). In addition to this, *Juncus maritimus* can store metals in both its above-ground and below-ground parts [7].

2.6.2 CHARACTERISTICS OF SOIL MEDIUM

Mobilization, uptake, and bioavailability of metals are much influenced by soil properties [95]. The soil pH can affect the availability and uptake of HMs. The pH can increase and decrease the concentration of metals in soil solution. As soil becomes more clayey, the availability of HM ions decreases, and cation exchange capacity is reduced [139]. It has been revealed that the increase in soil organic matter contents decreases the availability of trace elements [49].

2.6.3 THE ROOT ZONE

The root zone significantly affects the mobilization of HM ions. The root morphology of *Suaeda salsa* affects the absorption of Cd from the rhizosphere. It also regulates Ca^{++} channels in the root cell, e.g., Cd^{2+} inflow is maximum near the tip of the root [94, 168]. The phytoremediation efficiency of metals depends significantly on root exudates, which affect the root zone in various ways by providing nutrients to the microorganisms present in the root zone. The roots can enhance the solubility of HMs [80]. Plant growth-promoting bacteria (PGPB) can play an important role in producing siderophores and phytohormones [155].

2.6.4 CHEMICAL CHARACTERISTICS OF CONTAMINANTS

Microbes are able to enhance the phytoremediation process. HMs above a certain limit can affect the specific bacterial strains. HMs toxicity can affect microbes in the root zone. Changes in the community of bacteria lead to disturbance in phytoremediation [110]. Marquez-Garcia et al. [103] suggested that the concentration of HMs may limit colonization, seedling development that results in disordered phytoremediation.

2.6.5 ENVIRONMENTAL FACTORS

HMs accumulation percentage is in the order of Zn < Cd < Pb < Cu in *Spartina alterniflora*. Besides, *Spartina alterniflora* can also change metals from their immobilized state to mobilized state. These can also increase the accumulation of HMs [22]. $Mn^{3+/4+}$ oxides and Fe^{+++} oxides could supply vital binding sites for HMs under metal toxicity conditions. Under anoxic conditions, sulfide provides essential binding sites for attachment of Cu and Pb. The mobility of HMs highly depends on environmental conditions that is affected by biogeochemical processes [169].

2.6.6 ADDITION OF CHELATING AGENTS

Most of the metals have higher binding capacities. Once they are absorbed in the surface, it is hard to remove them. For this purpose, chelating agents play a vital role. Examples of chelating agents in use are: EDTA (ethylene diamine tetraacetic acid) and EDDS (ethylene diamines-di-succinic acid) [162]. These compounds can extract metals from the soil in the form of organic solvents. First used chelating agent was *Vetiveria zizanioides* for Pb phytostabilization. These agents can also enhance antioxidant activities in the system [10].

2.7 PHYTOREMEDIATION AND PLANT GROWTH PROMOTING RHIZOBACTERIA (PGPR)

Contaminations in the biosphere are one of the significant environmental risks, which is continuously increasing due to urbanization [11, 34, 179]. A variety of organic pollutants, such as solvents, hydrocarbons, and volatile composites, contaminates the soil. Some inorganic contaminants (such as HMs) are constantly releasing toxic substances into the environment [11, 33].

This technique has many advantages, because they require only a small quantity of nutrients [36]. It is manageable and acceptable by society due to environmental sustainability and its esthetic appearance [38, 56, 73]. Many plant species that are grown either in a contaminated or natural environment have the potential to extract toxic metals and store them into their vegetative and reproductive parts [54]. Phytoremediation plants have high tolerance, fast-growing, and adaptability to gather toxic materials [34, 37, 54].

Plant growth-promoting rhizobacteria (PGPR) are vital microbes that can remove contaminants, such as HMs [24, 62]. They reduce the HM toxicity by immobilizing, mobilization, absorption, and transformation. Many bacteria are naturally present in free-living or in symbiotic states that help roots in the uptake of nutrients and improve the growth of the plant [113]. Various studies have shown tolerance of these microbes against metal stress [4, 159]. These bacteria limit the availability of these HMs by attaching with them in the form of siderophores. Some rhizobacteria have also been involved in metal accumulation. Together with hyperaccumulating plants, they aid in the uptake of metals [157].

The PGPR are bacteria that can enhance root development of plants due to which growth increases [37, 56, 101]. Some plant species show maximum phytoremediation in slightly metal contaminated areas [11, 33, 161]. However, whenever a plant is exposed to high HMs concentration, the plant growth rate may slow down and produce less biomass [101]. The biological resource may help in the reduction of these drawbacks [34, 39]. One of the primary natural sources is rhizobacteria, as it allows the plant to accumulate more HMs. These also help in maintaining soil structure, recycling nutrients, and reducing toxicity levels of HMs [55]. They reduce toxicity through various biodegradation, biosorption, biotransformation, biomineralization, bioleaching, and biovolatization [39]. By promoting growth and metal availability [179], some rhizobacteria may facilitate the plant and enhance its phytoremediation [36, 96]. PGPR promotes the process of phytoremediation and improves the power of HM tolerance of plants. Based on functions and properties, PGPR are classified into four different types (Figure 2.2) [54, 179, 181].

However, most of the known phytoremediators are gylcophytes (plants tolerate relatively low concentration of salt) [101]. They are not suitable for remediation of toxic metals from saline soil [101]. Halophytes have a better ability to phytoremediation for saline soil instead of gylcophytes. Halophytes can tolerate more HMs uptake. HMs mobility increases as the soil become saline. The translocation of HMs also improves by these microbes [54].

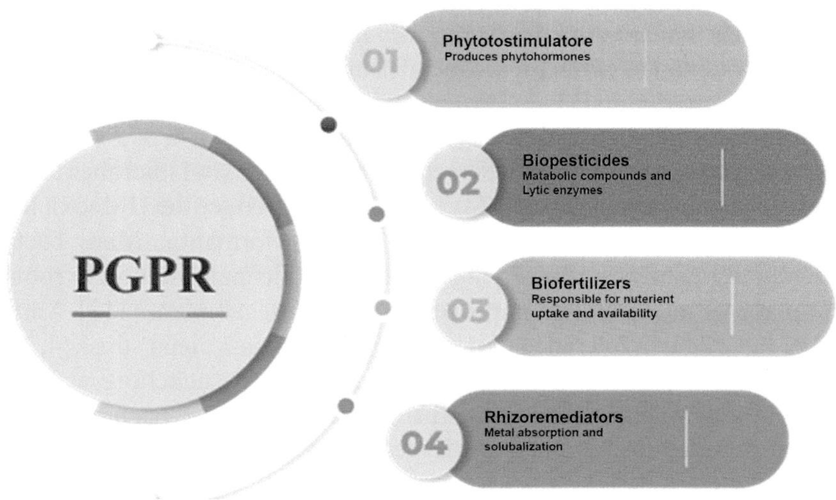

FIGURE 2.2 PGPR classification based on their properties.

2.7.1 PHYTOHORMONE PRODUCTION

Phytohormones are plant growth regulators to enhance growth. GA, auxins, cytokinin, abscisic acid, jasmonates, ACCD (1-aminocyclopropane-1-carboxylate deaminase) and SA are examples of phytohormones. These microscopic substances directly impact the growth through various metabolic and physiological plant functions under stress conditions [123].

2.7.1.1 AUXINS

Proliferation and induction of roots is highly dependent on auxins. Some of the auxins, phenylacetic acid, IPA (indole 3-propionic acid), etc., are synthesized by rhizobacteria. They help in the promotion of tissue expansion, vascular bundles, and root nodule development. IAA (indole-3-acetic acid) is a notable phytochrome. Karthik and Arulselvi [73] showed that rhizobacterial strain with IAA was active and could promote root growth and cell elongation in many crops. An interesting fact about phytohormones is that PGPR synthesized them and they inhibit the toxic effect of excessive metals [38, 43, 69].

2.7.1.2 OTHER PHYTOHORMONES

The other phytohormones (such as gibberellins) can stimulate elongation of shoots through cell division and differentiation. They stop growth by actions of the signaling system of gibberellins [101]. PGPR strains could produce other various phytohormones (jasmonates and SA, etc.), which helps in the metabolic process and tolerance to HMs [35].

2.7.2 PRODUCTION OF ACC (1-AMINOCYCLOPROPANE-1-CARBOXYLASE)

ACC deaminase are enzymes used by PGPR to convert ACC (1-aminocyclopropane-1-carboxylase) into α-ketobutyrate and also in ammonia. This ACC is an ethylene precursor in all developed plants. Under stress conditions, both ethylene and ACC level could increase [55]. Therefore, the activity of ACC is a crucial trait to decrease the level of ethylene by PGPR under stress conditions, including toxic metals and salinity [123].

ACC synthase plays a significant role during the synthesis of ethylene. However, during stress, the level of ethylene in plants increases. At high concentrations, ethylene may disturb plant metabolism and growth of roots [34, 55]. It may harm crop productivity due to which ethylene regulation is significant. For this purpose, ACC deaminase makes plant growth-promoting strain that regulates the ethylene level through ACC enzyme degradation [69]. ACCD enzyme is a member of the enzyme group that alters ACC into ammonia and α-ketobutyrate. Plants can use ammonia and α-ketobutyrate as a source of nitrogen and carbon, respectively. Furthermore, through ACC reduction, PGPR decreases the toxic effects of ethylene and promote the growth of plant [55].

A recent study shows that the addition of ACC rhizobacterial strains significantly contributes to the phytoremediation. When the strains of ACC are used as bioinoculant in the form of Serratia K120, Enterobacter N9 and Escherichia N16, it showed significantly effective growth of roots and shoot length of *H. annuus* plant. As a result, phytoremediation of Ni, Pb, As, and Cu is increased [54].

High levels of ethylene are released under stress conditions. This stress ethylene is the result of plant response under HM stress conditions [123]. Besides, ethylene synthesis inhibitors can reduce the severity of environmental stress [43]. The ACC deaminase containing bacteria are useful against growth inhibition due to inundation and HM accumulation [123].

2.7.3 NITROGEN FIXATION

One of the significant mechanisms of PGPR is to assist the plant growth through the activity of biological nitrogen [55, 69]. The proficient use of rhizobacteria in the symbiosis of legume-*Rhizobium* under environmental or HM stress not only helps in soil fertility but also in the growth of plants [54].

Under stressful conditions, rhizobial bacteria in the form of symbiosis could improve plant growth and soil fertility. The process involves the changing of inert nitrogen into valuable ammonia. This process is facilitated by certain species of bacteria and actinomycetes [101]. The *Rhizobium leguminosarum biovartrifolii* was active for nitrogen fixation even in the existence of inorganic substance (such as HMs) that continuously enhances elongation and biomass of shoots and roots of the desired crop.

Glick et al. [55] revealed that the inoculation of plant promoting-rhizobacteria stimulates the absorption of nitrogen. It also enhances the ROS scavenging by antioxidant activity in nickel-contaminated soil [34, 38]. The PGPR nitrogen-fixing ability is increased by the synthesis of GSH reductase enzyme, which is responsible for reducing the toxic effects of peroxide radicles. Nitrogen-fixing inoculations not only help in growth promotion but also improves phytoremediation efficiency [56, 123].

2.7.4 SIDEROPHORE PRODUCTION

Siderophores are the iron-containing proteins with low molecular weight (LMW) having a powerful attraction to Fe^{3+}. It is involved in biochemical events of enzymatic and transpiration reactions. Iron is essential for electron transport chain, photosynthesis, and oxidative phosphorylation [56]. The insoluble form of Fe^{3+} is converted into a soluble form of Fe^{2+} by siderophore protein. The sidosphoric microbial compounds are playing a significant role in the rhizosphere [101].

2.7.4.1 MECHANISM OF SIDEROPHORES IN ASSISTING PHYTOREMEDIATION

Siderophores increase the availability of iron and plant growth. Furthermore, siderophores formed by microbes enhance the mobilization of trace metals. The mobility of trace elements increases due to binding, which increases their availability and movement in the rhizosphere [56]. Overall, the plant-specific growth-promoting substance shows positive results on the growth through influencing metabolic and physiological functioning of plants under stress and normal conditions. These effects enhance the tolerance level of plant stress and assist in reducing the barriers, such as a decrease in biomass and growth during the process of phytoremediation [54, 56, 101]. Iron-siderophores complex provides iron to the cell under stress conditions.

Outer membrane receptor proteins sense the iron-siderophores complex. When these complex binds, it causes conformational changes. It starts signaling to initiate TonB interaction. By utilizing the energy of the TonB complex (i.e., proton motive force), this iron-siderophores complex transfer to the periplasm. Here it binds with periplasmic binding protein. This transfers the iron-siderophore complex to the cytoplasm through ATP-binding cassette (ABC) transporter. In the cytoplasm, iron is released from this complex in two ways [71]: (i) Ferric reductase can be used for reduction; and (ii) some chemical modifications like acetylation and esterases are used to extract Fe. Table 2.2 presents plant growth-promoting rhizobacteria that aid in phytoremediation.

TABLE 2.2 Plant Growth Promoting Rhizobacteria Aiding in Phytoremediation

Bacterial Strain	PGP Activity of PGPR	Accumulator Plant	Heavy Metal	References
Acinetobacter howfii	Siderophores, ACCD, N fixation and P solubilization	*V. radiata*	As	[101, 123]
Aeroginosa CPSB1	HCN, P solubilization and IAA	*T. aestivum*	Cu, Cd, and Cu	[76]
Arthrobacter sp.	Siderophores, ACCD, and IAA	*B. napus, S. alfredii*	Cd	[101]
Azotobactor sp.	Siderophores, IAA, and P solubilization	*Eichhornia crassipes*	As	[123]
B. magaterium	Siderophores, and IAA	*B. rapa*	Pb and Cd	[54, 56]
C. funkri	IAA, ammonia, hydrolytic enzyme, catalase	*Phaseolus vulgaris*	Cr	[56, 123]
Enterobacter N9	Siderophores, ACCD, and IAA	*H. annus*	Cu, Pb, Ni, Zn	[54]
P. stewartii ASI11	Siderophores, ACCD, and IAA	*Leptochloa fusca*	Cr	[54, 56]
Pseudomonas sp.	Siderophores, ACCD, N fixation and P solubilization	*H. tuberostes*	Zn, Cd	[123]
Rhodobacter sphaeroides	Production of IAA	*T. aestivum*	Zn and Cd	[37, 39, 101]

2.8 MECHANISMS OF TOLERANCE OF HEAVY METALS (HMS) IN PLANTS

2.8.1 TRANSGENIC PLANTS

The concept of transgenic plant was introduced for phytoremediation by Van Aken [165]. HMs are uptaken by plants through different active and passive pathways. Metal movement from soil to the epidermis of the cells is through a passive way, while the concentration gradient, that is built according to require demand, is the active pathway.

HM, such as, Lead has no biological role in the plant growth, so that it may be toxic for both plants and microbes, causing health concerns [50]. Nickle comes from different sources, such as ultramafic bedrocks. Under high concentration, it can be dangerous for the land because it affects soil fertility and plant health [132]. A study conducted on *brassica* shows a higher accumulation of selenium (almost three times more) than the wild variety in the APT sulfurylase gene (APS) overexpressed line [32]. Another significant advantage of these transgenic plants is green chemistry. They are used to replace the utilization of toxic chemicals and helps in the formation of phytosiderophores that aid in metal absorption [40, 133].

Many genes have been identified in different organisms involved in the absorption, translocation, and sequestration of chemicals that help in the detoxification of metals [99]. These genes can be transformed into contender's fast-growing plants to assess HM remediation. By action of these genes, overexpression of recombinant protein (involved in metal chelation, such as phytochelatins (PC), metallothioneins (MTs)) formation and the transportation will occur (Figure 2.3).

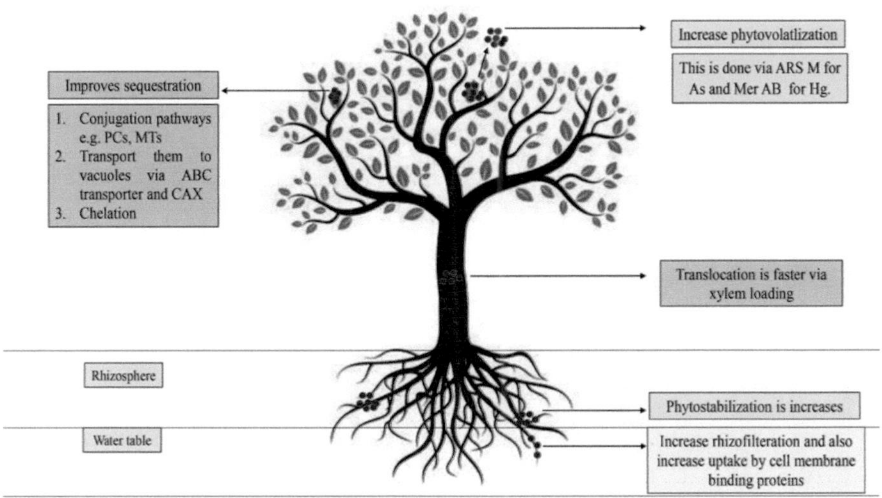

FIGURE 2.3 Properties of transgenic plants involve in phytoremediation.

Homeostasis of trace elements mainly involves membrane transporters. Genes that encode these transporters (proteins) play a crucial role in this process [115]. Many transporter proteins have been identified that encode for homeostasis of these metals, such as, ZIP regulated transporter, HMA (heavy metal ATPase) and MTP (metal tolerance protein), etc. [41]. Manipulation of these transporters at the genetic level may enhance the root potential to uptake metals, translocate these metals and may sequester these metals into vacuoles (Figure 2.3).

These plants must-have properties of high phytovolatilization and phyto-stabilization efficiencies. They have an improved sequestration process in it. These plants have the rizofilteration ability that helps them to uptake more metals (e.g., Pb via CBP) to sequester them.

2.8.2 ROLE OF GENES/TRANSPORTERS IN ACCUMULATION OF HEAVY METALS (HMS) IN VACUOLES

There are many genes/transporters in a cell that helps the plant to manage against HMs. Once HMs enter into the cell, the first strategy is to remove it from the cell by moving them out. Another approach is to sequester them into the cell vacuole by various transporters [175]. In terms of phytoremediation efficiency, these metal transporters mainly act at three stages (Figure 2.4): (i) they absorb these metals from the soil by their roots; (ii) they translocate them from roots to shoots through xylem; and (iii) they store them into the cell vacuoles.

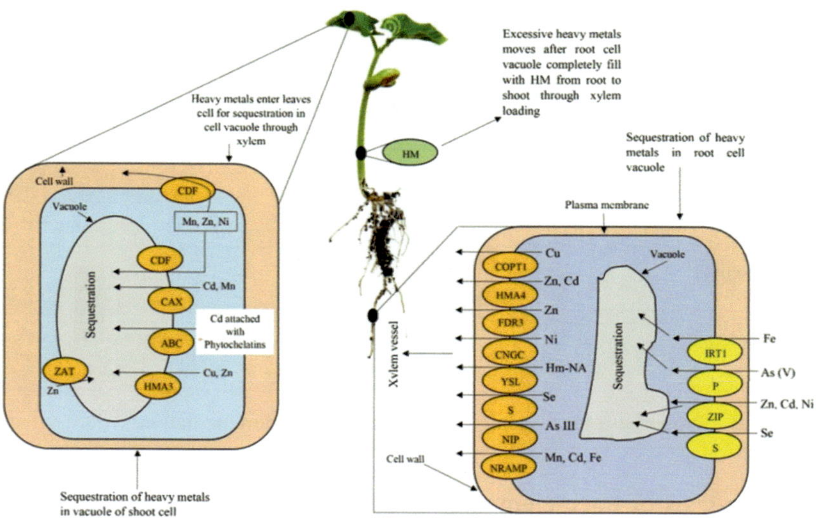

FIGURE 2.4 Schematic diagram of heavy metal sequestration into cell vacuoles by different genes/transporters.

Source: Modified from Ref. [66].

HMs enter into the cell by various carriers and sequester them into the vacuole and excessive HM is then transferred from roots to shoots through xylem and stored in the vacuole in shoot cells as shown in Figure 2.4, where: HM: heavy metal, ZIP: zinc regulated transporter, HMA: heavy metal associated transporter, YSL: Yellow stripe transporter, NRAMP: Natural resistance-associated macrophage protein, CDF: cation diffusion facilitator, ABC: ATP (adenosine triphosphate) binding cassette transporter.

Some ATP pumps are present on the vacuolar membrane that can provide energy in the sequestration process [3]. Many transporters/carriers exist on the cell membrane and tonoplast, such as:

- ABC transporter (ATP binding cassette) on tonoplast [98];
- CAX (cation antiporter) [107];
- CDF proteins (cation diffusion facilitator);
- COPT proteins (copper transporter);
- HMA (heavy metal associated) [87];
- IRT (iron regulated transporter) on the plasma membrane [67];
- MRP (multidrug resistance-associated proteins);
- NRAMP family (natural resistance-associated macrophage protein) on the plasma membrane and tonoplast [65, 140];
- YSL transporter (yellow strip-like) [175];
- ZRT (zinc regulated transporters) [65].

Zn enters the cell by ZIP transporter. Fe and Mn uptake is by IRT transporter (Figure 2.4). Two *IRT* genes have been extracted from *Populus tremula* (*PtIRT1* and *PtIRT3*). *PtIRT1* gene only expresses in roots while *PtIRT3* expresses in the whole plant. These genes are expressing under Fe and Zn deficiencies that show that they contribute to the homeostasis of these metals [67]. Through NRAMP transporter, HMs move from the cytoplasm to the storage compartments (vacuole or extracellular spaces) (Figure 2.4).

A study showed that two *NRAMP* genes (*NRAMO2* and *NRAMP6*) were overexpressed in the roots of *S. integra* that were treated with Cd [107]. Through these transporters, HM was translocated to the shoots; and these HMs go into the vacuole with the activity of some transporters (e.g., ABC transporter, CAX, CDF, HMA3 and ZAT (zinc transporter)) and are stored there. PC also act to combine with HMs to sequester them into the vacuoles (Figure 2.4).

2.8.3 SEQUESTRATION OF HEAVY METALS (HMS) IN CELL VACUOLE

Chelation helps in the sequestration of HMs in the cells and aids in tolerance against HM stress. Different compounds (such as MTs, organic compounds (organic acids), amino acids (glutamine, cysteine, and glycine (Gly)) and PC) can enhance tolerance against HM stress (Figure 2.5). The leading role is played by PC that are proteins with higher cysteine contents, LMW, and unique metal binding capacity [174]. MTs are also proteins. PC have already been identified in many plant species, including monocots and dicots. These are synthesized from the signaling of GSH [64].

GSH is a non-protein antioxidant that helps plants against many abiotic stresses like HMs stress [72]. This GSH activates the PC. Two steps involved in the formation of GSH (Figure 2.5) [177]:

- The γ-Glu-Cys complex structures formed by the combination of cysteine and glutamate (Glu) (Figure 2.5). This step is catalyzed by the GSH1 enzyme (i.e., glutamate-cysteine ligase) (Figure 2.5); and
- This γ-Glu-Cys complex combines with Gly and result in the formation of GSH (Figure 2.5).

GSH is present in almost all organelles of the cell that includes mitochondria, chloroplast, and nucleus. It decreases the signaling of ROS by increasing its concentration in the cytosol [31]. It triggers the activity of PC by phytochelatins synthase (PCS).

PC-HM complex (phytochelatins heavy metal complex) is formed by attachment of HMs with PC. This complex is also known as a LMW complex [64]. This complex is transferred to the cell vacuole by ABC transporter [116]. Apart from PC activity, HMs can also enter into vacuole by direct exchange through CAX transporter. CAX is a proton exchanger that exchanges a proton with HM. In vacuole, HMs are already present that goes to the cell by the action of CAX and MTPs. These HMs are attached to the LMW complex to form a high molecular weight (HMW) complex [151] as shown in Figure 2.5.

In Figure 2.5, ZIP: zinc/iron-regulated transporters, Cys: cysteine, Glu: glutamate, γ-Glu-Cys: glutamate cysteine complex, GSH1: glutamate-cysteine ligase, GSH2: glutathione synthetase, Gly: glycine, PCS: phytochelatin synthase, PC: phytochelatins, GSH: glutathione, HM: heavy metal, MTPs: metal tolerance proteins, LMW PC-HM: low molecular weight phytochelatins heavy metal complex, HMW PC HM: high molecular weight phytochelatins heavy metal complex, ABC: ATP binding cassette transporter, CAX: cation exchange transporters, and NRAMP: Natural resistance-associated macrophage protein.

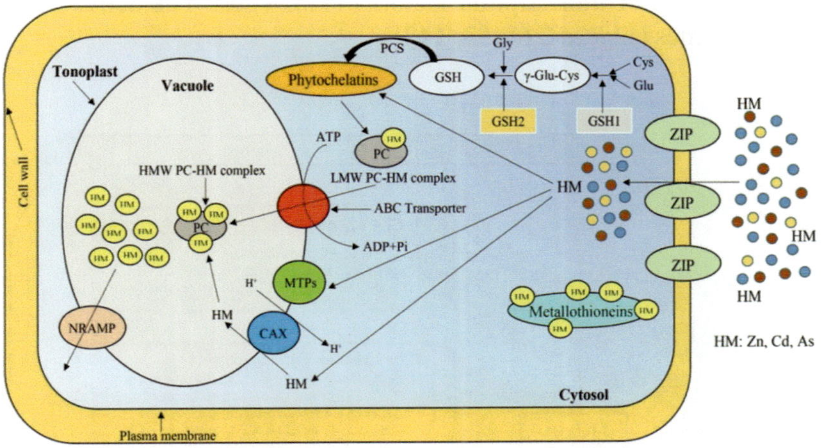

FIGURE 2.5 Schematic representation of the sequestration of heavy metals in cell vacuole by phytochelatins and some other transporters.

MTs also bind HM with them to make them in the binding state as much as they can [176]. First, MTs were identified from horse kidney in 1957 that act a cadmium binding protein [102]. NRAMP transporters are present on the vacuolar membrane. It not only helps in HM compartmentalization but also helps in the remobilization of HMs from the vacuole to extracellular spaces [15]. This way, these HMs are sequestered by PC activity.

When HMs enter the cell via various carriers, a signal has been generated to produce GSH that activates PC to bind with HM and form a LMW complex (Figure 2.5). This complex goes into vacuole via ABC transporter. HMs already go into vacuole via cation exchange carriers. NRAMP proteins present on the tonoplast help in compartmentalization and remobilization of metals.

2.8.4 HEAVY METALS (HMS) SIGNALING CASCADE

Plants sense HMs and generate some responses that result in modulation of molecular and biochemical reactions. Finally, plants make metal transporter proteins (MTPs) and metal-binding proteins that help to counter this metal stress [148]. During the early stages of metal stress, plant response is similar to other abiotic stresses (e.g., osmotic stress and oxidative stress) [128]. This similarity gives an idea of the interconnection between these pathways. This interconnection and convergence of one channel into another finally results in the generation of transcription factors (TFs). These TFs activate stress-responsive genes (Figure 2.6) [83, 148].

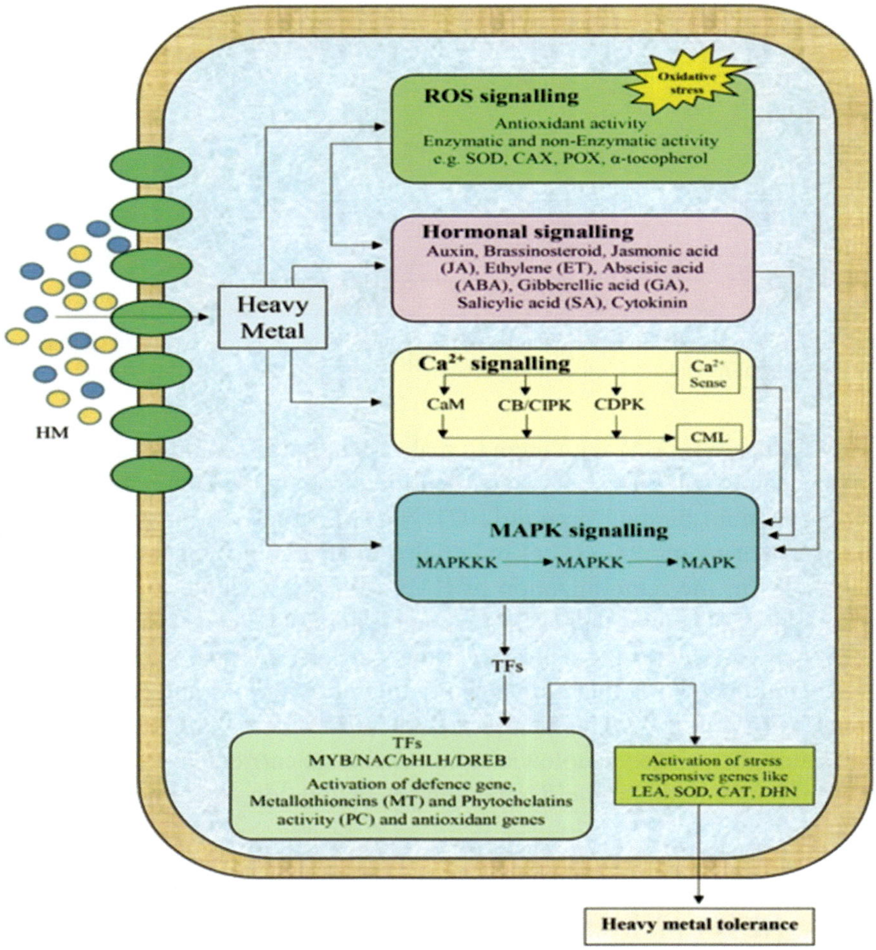

FIGURE 2.6 Schematic representation of signaling cascade in heavy metal stress.

Source: Modified from Ref. [182].

Different signal transduction pathways regulated under HM stress include Ca^{2+} signaling, MAPK signaling, ROS signaling and hormonal signaling. Change in cytosolic Ca^{2+} concentration due to metal stress is sensed by different calcium sensors present in plants (such as CaMs (calmodulins), CBLs (calcineurin B-like protein), CDPK (Ca^{2+} dependent protein kinases and CaM-like proteins) (Figure 2.6) [152]. Ca^{2+} attach with them and trigger different downstream signaling pathways. Ca^{2+} activates antioxidant enzymes that in return, give tolerance against HMs [42]. Foliar application of Ca^{2+} enhances the activity of APX and SOD enzymes [5].

In the same way, during MAPKs signaling, various TFs (MYB, MYC, NAC, WRKY, etc.), phosphorylate, and aids in stress response (Figure 2.6) [159]. Multiple studies have also shown that this MAPKs signaling can be activated by ROS signaling, specific metal-ligand and even by a change in the antioxidant system (Figure 2.6) [83, 89].

ROS also disturbs phytohormone signaling (auxin, ethylene, and JA signaling). By altering root architecture, plants can survive in HM stress conditions. Phytohormones (such as auxin, ethylene, and cytokines) can control root architecture and lateral roots [30]. Photosynthetic pigments and chloroplast membranes that inhibit as a result of HM stress can be recovered by Cytokinin [119]. A study revealed that rice plants exposed to JA had an improved antioxidant system against Cd stress [147]. Some other studies on phytohormone under metal stress indicate that changes in phytohormones lead to changes in plant response.

In Figure 2.6, whenever HM enters the cell, it regulates many signaling pathways that end up with some TFs to regulate stress-responsive genes, which ultimately lead to tolerance against HM stress.

2.8.5 ROLE OF PHYTOCHELATINS (PC), GLUTATHIONE (GSH) AND ANTIOXIDANTS IN HEAVY METAL (HM) ACCUMULATION

Oxidative stress arises under HM stress that produces an excess of ROS [89]. These ROS damages DNA, affects the DNA recovery mechanisms, nutrient balance, and membrane stability [154]. To avoid HMs toxic effects, the plant develops several strategies to combat the metal's excessiveness. Firstly, it prevents its uptake by restricting them to the apoplast. They can limit them by combining these metals with cell walls or with cellular exudates. Restrictions can also be possible by inhibition of long-distance transport [61]. These HMs go to the cell in higher concentrations, plant activates some tolerance mechanisms like chelation by PC (activated by GSH), MTs, and sequestration them to the vacuole of the cell [180].

At the early stages of HM stress, plant synthesize some cysteine-rich compounds named PC [120]. They formed by signaling of GSH. So, GSH plays a vital role in defense against HMs stress. Some studies reveal that overexpression of the PC gene does not enhance tolerance. A study in Arabidopsis revealed that PC increase up to 2.1 folds, but tolerance of plant against metal stress does not improve [85]. Generally, PC in excess accumulate more HMs instead to enhance plant tolerance [48]. HMs detoxification is not just related to chelation. It involves the production of HMW compounds in the vacuole [48] (Figure 2.7). These metals ions that combine with PC are then transferred to the vacuole from cytosol by transports (ABC transporter) on tonoplast [138].

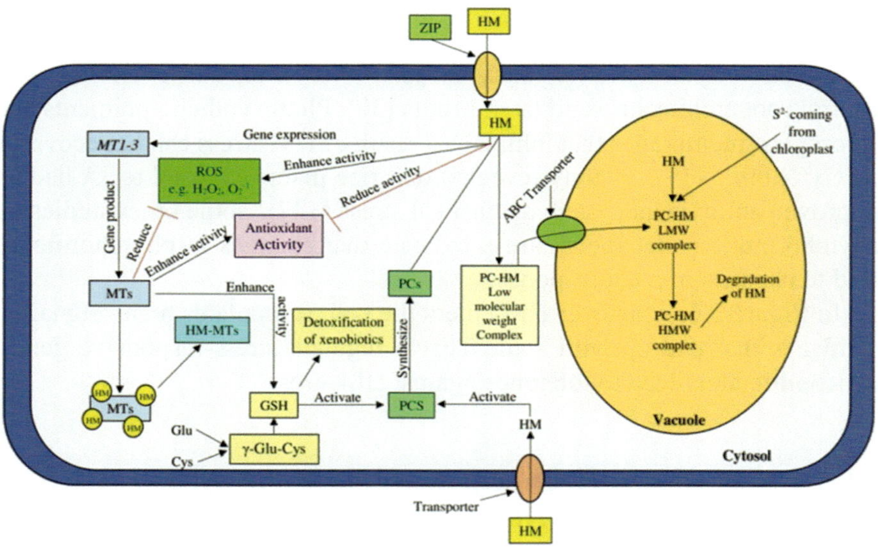

FIGURE 2.7 Schematic representation of metallothioneins (MTs), phytochelatins (PC), and glutathione (GSH) activity under heavy metal stress.

Besides PC, MTs are also crucial in HM detoxification. These are also cysteine rice compounds that are metal-binding proteins [20]. MTs play a different role in the cell includes maintaining homeostasis of essential metal ions, sequestration of HMs and protection from oxidative stress (ROS) [66]. HMs bind with these MTs for sequestration of HM (Figure 2.7) [129].

Overexpressed MTs gene plant shows a higher tolerance against HMs and better compartmentalization strategy [58, 93, 160].

MTs proposed to help maintain metal homeostasis. It is well-documented that ROS produced during excessive accumulation of HMs affect macro-molecules of the cell (proteins and DNA) and leads to metabolic and physiological disorders [59]. In such conditions, MTs protect the plant by enhancing antioxidant activities that scavenge excess ROS (Figure 2.7) [13]. Different classes of *MTs* genes exist (*MT1a*, *MT2a*, *MT2b* and *MT3*). From this, *MT1a* and *MT2b* expressed in phloem while *MT2a* and *MT3* are expressed in mesophyll cells [63, 96]. An MT-biosensor was developed that helps to produce signals associated with the metal of interest. Such type of tools is very much beneficial in metal contaminated areas for examining the environment.

According to Figure 2.7, HM enters the cell through transporters. It activates PCS that synthesizes PC. They bind with HM to form the LMW complex. This composite transfer to vacuole through a transporter (ABC transporter) present on tonoplast. MTs encodes by *MT1–3* gene expressed under HM stress. They also enhance the activity of GSH produced by a combination of Glu, cysteine, and Gly. Under HM stress, antioxidant system activity is reduced, and this results in the accumulation of ROS species. MTs lowered the generation of ROS and enhanced the activity of antioxidant enzymes. In the vacuole, LMW complex converts into the HMW complex by combining S^{2-} (coming from chloroplast) and more HMs.

2.9 ADVANTAGES AND LIMITATIONS OF PHYTOREMEDIATION

The technique of phytoremediation is considered easy and fastest way to remediate soils [17, 45]. Phytoremediation seems to be very fruitful as compared to other customary methods of soil remediation. A realistic look at this technique has some constraints and limitations. Table 2.3 indicates the advantages and limitations of phytoremediation.

TABLE 2.3 Advantages and Limitations of Phytoremediation

Advantages	Limitations
It is economically feasible and requires less energy to operate [158]	It is difficult to work for hyperaccumulators at those sites, which are mixed up with multiple pollutants because those plants have a limited ability to accumulate a specific element only [9].

TABLE 2.3 *(Continued)*

Advantages	Limitations
It is eco-friendly as it reduces the exposure of pollutants without harming the environment [91].	Besides this, non-native, and wild plants often show typical problems of risk analysis as not enough information is available for the plant species used in phytoremediation [90].
It can be applied to a vast area and is also easily manageable.	It needs field experimentations in polluted areas along with the formation of gene banks and herbariums [21].
It is also helpful in reducing nutrients losses by minimizing erosion [90].	One of the main problems is that some unusual plants that use in the phytoremediation process are scarce [163].
It removes secondary pollutants of water and air that can accumulate in the ecosystems [91].	The distribution of phytoremediation plants is confined to limited places.
Phytoremediation is also considered a cost-effective source of cleaning hazardous pollutants [95].	Bud or spores collected from hyperaccumulators may have reproductive issues as some of the plants are sporadic [9].
Plants cover the whole ground, protect the soil by reducing the potential spread of waste metals and sensibly fix the polluted soil [90].	Mining projects often create problems for the plants grown in metal-rich soil [44].
The best advantage of this technique is that it is easy to manage, which is premium to assemble the future public support and interest [144].	Another disadvantage is of slow-growth of hyperaccumulators and so as less production of biomass [144].

The contaminants with high solubility may leach from the root zone (outside), exhibiting low proficiency of plant uptake. Microclimates may be harsh at the sites following the low germination and establishment of plant species. Biological interactions of species may have stimulated or limiting effect on clean-up. Time and effort are required to have a complete concept of biotic interactions among plant species and other inhabitants in the system.

2.10 FUTURE PROSPECTS OF PHYTOREMEDIATION

The remediation of toxic compounds by an eco-friendly combination of plants and microbes has received much attention in the present era.

Previously, *in situ* and *ex situ* evaluations have been carried out to examine the importance of remediation [54]. For toxic metals contaminated sites, we need more research and methods to quantify, identify, and clean-up standards for contaminations, significant health risks and technologies for remediation of sites [73].

The ultimate goal of soil remediation is not only to remove toxic materials but to improve soil structure and rhizobacteria growth [43, 79]. Up-to-date techniques of biotechnology (such as polymerase chain reactions (PCR), phospholipid fatty acid analysis, and fluorescence *in situ* hybridization) will be helpful to determine the genetic evolution, structures, and to find more effective methods of bioremediation [95].

Future studies of phytoremediation must concern with exploring genetic and molecular mechanisms related to phytoremediation. More research work is needed to understand the effect of different HMs on plant growth and productivity. There is a need to explore the role of phytohormones and PGPR in phytoremediation [43, 56, 123]. The most critical challenges of phytoremediation are public awareness and implementation of this green technology. We need to promote phytoremediation by different platforms, such as media and NGO's. We need to show that phytoremediation programs are trustable and can be applied without any harm.

2.11 SUMMARY

HMs must be removed to avoid serious health issues to humans, plants, and animals. Phytoremediation has become the best choice and it includes phytoextraction, phytostabilization, phytovolatilization, phytodegradation, and phytofiltration. Plant species selection, medium, root zone, and chelating agents, etc., affect these processes. In general, hyperaccumulator plants are used in phytoremediation through PC, signaling mechanisms, and antioxidant systems. They can sequester relatively high concentrations of HMs in their vacuoles and detoxify them. Phytoremediation has an advantage that it reduces the HMs to go into the food chain. However, climatic changes and hydrologic conditions (such as waterlogging and water deficit) pose great challenges for this technique.

KEYWORDS

- **glutathione**
- **phytochelatins**
- **phytohormones**
- **phytoremediation**
- **plant growth-promoting bacteria**
- **sequestration**
- **transgenic plants**

REFERENCES

1. Abdel-Shafy, H. I., & Mansour, M. S., (2018). Phytoremediation for the elimination of metals, pesticides, PAHs and other pollutants from wastewater and soil. In: *Phytobiont and Ecosystem Restitution* (pp. 101–136). Singapore; Springer.
2. Abhilash, P. C., Powell, J. R., Singh, H. B., & Singh, B. K., (2012). Plant-microbe interactions: Novel applications for exploitation in multipurpose remediation technologies. *Trends in Biotechnology, 30*, 416–420.
3. Adams, J. P., Adeli, A., Hsu, C. Y., & Harkess, R. L., (2011). Poplar maintains zinc homeostasis with heavy metal genes *HMA4* and *PCS1*. *Journal of Experimental Botany, 62*, 3737–3752.
4. Ahemad, M., (2012). Implication of bacterial resistance against heavy metals in bioremediation: A review. *Journal of Institute of IntegrativeOmics and Applied Biotechnology, 3*, 39–46.
5. Ahmad, A., Hadi, F., & Ali, N., (2015). Effective phytoextraction of cadmium (Cd) with increasing concentration of total phenolics and free proline in *Cannabis sativa* (L) plant under various treatments of fertilizers, plant growth regulators, and sodium salt. *International Journal of Phytoremediation, 17*, 56–65.
6. Ali, S., Abbas, Z., Rizwan, M., & Zaheer, I. E., (2020). Application of floating aquatic plants in phytoremediation of heavy metals polluted water: A review. *Sustainability, 12*(5), 1927.
7. Almeida, C. M. R., Mucha, A. P., & Teresa, V. M., (2011). Role of different salt marsh plants on metal retention in an urban estuary (Lima estuary, NW Portugal). *Estuarine, Coastal and Shelf Science, 91*(2), 243–249.
8. Alvarez-Vázquez, L. J., & Martínez, A., (2019). Mathematical analysis and optimal control of heavy metals phytoremediation techniques. *Applied Mathematical Modeling, 73*, 387–400.
9. Alyazouri, A. H., Jewsbury, R. A., & Tayim, H. A., (2013). Phytoextraction of Cr(VI) from soil using *Portulaca oleracea*. *Toxicological and Environmental Chemistry, 95*(8), 1338–1347.
10. Andra, S. S., Datta, R., Reddy, R., Saminathan, S. K. M., & Sarkar, D., (2011). Antioxidant enzymes response in vetiver grass: A greenhouse study for chelant-assisted

phytoremediation of lead-contaminated residential soils. *Clean - Soil, Air, Water, 39*(5), 428–436.

11. Andrea, M. M. E., & Carolina, T. E., (2019). A evaluation of contaminants in agricultural soils in an irrigation district in Colombia. *Heliyon, 5*(8). Article ID: e02217.
12. Anjum, N. A., Ahmad, I., & Válega, M., (2014). Saltmarsh halophyte services to metal-metalloid remediation: Assessment of the processes and underlying mechanisms. *Critical Reviews in Environmental Science and Technology, 44*(18), 2038–2106.
13. Ansarypour, Z., & Shahpiri, A., (2017). Heterologous expression of a rice metallothionein isoform (*OsMTI-1b*) in *Saccharomyces cerevisiae* enhances cadmium, hydrogen peroxide and ethanol tolerance. *Brazilian Journal of Microbiology, 48*, 537–543.
14. Arora, P. K., (2020). *Microbial Technology for Health and Environment*. Springer.
15. Asgher, M., Per, T. S., & Anjum, S., (2017). Contribution of glutathione in heavy metal stress tolerance in plants. In: *Reactive Oxygen Species and Antioxidant Systems in Plants: Role and Regulation under Abiotic Stress* (pp. 297–313). Springer, Singapore.
16. Ashraf, S., Ali, Q., Zahir, Z. A., Ashraf, S., & Asghar, H. N., (2019). Phytoremediation: Environmentally sustainable way for reclamation of heavy metal polluted soils. *Ecotoxicology and Environmental Safety, 174*(15), 714–727.
17. Bell, T. H., Joly, S., Pitre, F. E., & Yergeau, E., (2014). Increasing phytoremediation efficiency and reliability using novel omics approaches. *Trends in Biotechnology, 32*, 271–280.
18. Bhat, J. A., Shivaraj, S. M., & Singh, P., (2019). Role of silicon in mitigation of heavy metal stresses in crop plants. *Plants, 8*(3), 71–77.
19. Bisht, R., Chanyal, S., & Srivastava, R. K., (2020). Systematic review on phytoremediation technology: Removal of pollutants from wastewater and soil. *International Journal of Research in Engineering, Science and Management, 3*, 54–59.
20. Capdevila, M., & Atrian, S., (2017). Metallothionein protein evolution: A miniassay. *JBIC Journal of Biological Inorganic Chemistry, 16*(7), 977–989.
21. Chai, M., Shi, F., Li, R., Liu, L., Liu, Y., & Liu, F., (2013). Interactive effects of cadmium and carbon nanotubes on the growth and metal accumulation in a halophyte *Spartina alterniflora* (Poaceae). *Plant Growth Regulation, 71*(2), 171–179.
22. Chai, M., Shi, F., Li, R., Qiu, G., Liu, F., & Liu, L., (2014). Growth and physiological responses to copper stress in a halophyte *Spartina alterniflora* (Poaceae). *Acta Physiologiae Plantarum, 36*(3), 745–754.
23. Chaney, R. L., (1983). Plant uptake of inorganic waste constituents. In: Parr, J. F., & Marsh, P. B., & Kla, J. M., (eds.), *Land Treatment of Hazardous Wastes* (pp. 50–76). Park Ridge - NJ: Noyes Data Corp.
24. Chen, M., Xu, P., Zeng, G., Yang, C., Huang, D., & Zhang, J., (2015). Bioremediation of soils contaminated with polycyclic aromatic hydrocarbons, petroleum, pesticides, chlorophenols and heavy metals by composting: Applications, microbes and future research needs. *Biotechnology Advances, 33*, 745–755.
25. Chibuike, G. U., & Obiora, S. C., (2014). Heavy metal polluted soils: Effect on plants and bioremediation methods. *Applied and Environmental Soil Science, 2014*, 34–40.
26. Copat, C., Vinceti, M., D'Agati, M. G., & Arena, G., (2013). Ferrante mercury and selenium intake by seafood from the Ionian sea - risk evaluation. *Ecotoxicol. Environ. Saf., 100*, 87–92. doi: 10.1016/j.ecoenv.2013.11.009.

27. Cristaldi, A., Conti, G. O., & Jho, E. H., (2017). Phytoremediation of contaminated soils by heavy metals and PAHs. Brief review. *Environmental Technology and Innovation, 8*, 309–326.
28. Cui, J., Zhao, Y., Lu, Y., Chan, T., Zhang, L., Tsang, D. C. W., & Li, X., (2019). Distribution and speciation of copper in rice (*Oryza sativa* L.) from mining-impacted paddy soil: Implications for copper uptake mechanisms. *Environmental International, 126*, 717–726.
29. DalCorso, G., Fasani, E., Manara, A., Visioli, G., & Furini, A., (2019). Heavy metal pollutions: State of the art and innovation in phytoremediation. *International Journal of Molecular Sciences, 20*(14), 3412–3418.
30. De Smet, S., Cuypers, A., Vangronsveld, J., & Remans, T., (2015). Gene networks involved in hormonal control of root development in *Arabidopsis thaliana*: A framework for studying its disturbance by metal stress. *International Journal of Molecular Sciences, 16*, 19195–19224. doi: 10.3390/ijms160819195.
31. Diaz-Vivancos, P., Simone, A. D., & Kiddle, G., (2015). Glutathione - linking cell proliferation to oxidative stress. *Free Radical Biology and Medicine, 89*, 1154–1164.
32. Dimkpa, C., Bindraban, P., & McLean, J. E., (2017). Methods for rapid testing of plant and soil nutrients. In: Lichtfouse, E., (ed.), *Sustainable Agriculture Reviews* (Vol. 25, pp. 45–55) Cham. Springer.
33. Ding, W., Zhang, J., Wu, S., Zhang, S., Christie, P., & Liang, P., (2019). Responses of the grass *Paspalum distichum* L. to Hg stress: A Proteomic study. *Ecotoxicology and Environmental Safety, 183,* 109–549.
34. Dixit, P., Mukherjee, P. K., Sherkhane, P. D., Kale, S. P., & Eapen, S., (2011). Enhanced tolerance and remediation of anthracene by transgenic tobacco plants expressing a fungal glutathione transferase gene. *Journal of Hazardous Material, 192*, 2704–2706.
35. Doty, S. L., Freeman, J. L., & Cohu, C. M., (2017). Enhanced degradation of TCE on a superfund site using endophyte-assisted poplar tree phytoremediation. *Environmental Science and Technologies, 51*(17), 10050–10058.
36. Doty, S. L., (2008). Enhancing phytoremediation through the use of transgenics and endophytes. *New Phytologist, 179*(2), 318–333.
37. Du, Z., Arias, T., Meng, W., & Chye, M., (2016). Plant acyl-CoA-binding proteins: An emerging family involved in plant development and stress responses. *Progress in Lipid Research, 63*, 165–181.
38. Duan, G., Kamiya, T., Ishikawa, S., Arao, T., & Fujiwara, T., (2012). Expressing *ScACR3* in rice enhanced arsenite efflux and reduced arsenic accumulation in rice grains. *Plant and Cell Physiology, 53,* 154–163.
39. Duhoux, A., Carrère, S., Gouzy, J., Bonin, L., & Délye, C., (2015). RNA-Seq. Analysis of rye-grass transcriptomic response to an herbicide inhibiting acetolactate-synthase identifies transcripts linked to non-target-site-based resistance. *Plant Molecular Biology, 87*, 4–5.
40. Escande, V., Garoux, L., & Grison, C., (2014). Ecological catalysis and phytoextraction: Symbiosis for future. *Applied Catalysis B: Environmental, 146*, 279–288.
41. Fan, H. L., Li, L., Zhou, S. F., & Liu, Y. Z., (2016). Continuous preparation of Fe_3O_4 nanoparticles combined with surface modification by L-cysteine and their application in heavy metal adsorption. *Ceramics International, 42*(3), 4228–4237.

42. Fang, H., Jing, T., Liu, Z., Zhang, L., Jin, Z., & Pei, Y., (2014). Hydrogen sulfide interacts with calcium signaling to enhance the chromium tolerance in *Setariaitalica. Cell Calcium, 56*, 472–481.
43. Farh, M. E., Kim, Y. J., Sukweenadhi, J., Singh, P., & Yang, D. C., (2017). Aluminum. Resistant, plant growth-promoting bacteria induce overexpression of aluminum stress-related genes in arabidopsis thaliana and increase the ginseng tolerance against aluminum stress. *Microbiological Research, 200*, 45–52.
44. Farraji, H., (2014). Wastewater treatment by phytoremediation methods. In: Aziz, H. A., (ed.), *Wastewater Engineering: Types, Characteristics and Treatment Technologies* (pp. 205–218). Malaysia: IJSRPUB.
45. Farraji, H., Zaman, N. Q., Tajuddin, R., & Faraji, H., (2016). Advantages and disadvantages of phytoremediation: Review. *International Journal of Environmental Technology and Sciences, 2*, 69–75.
46. Favas, P. J., Pratas, J., Paul, M. S., Sarkar, S. K., & Prasad, M. N. V., (2016). Phytofiltration of metal (Loid)-contaminated water: The potential of native aquatic plants. In: *Phytoremediation* (pp. 305–343). Springer, Cham.
47. França, F. C., & Albuuerque, A. M., (2017). Heavy metals deposited in the culture of lettuce (*Lactuca Sativa L.)* by the influence of vehicular traffic in Pernambuco, Brazil. *Food Chemistry, 215*, 171–176.
48. Furini, A., (2012). *Plants and Heavy Metals* (p. 119). Netherlands, Springer.
49. Gabrijel, O., Davor, R., Zed, R., Marija, R., & Monika, Z., (2009). Cadmium accumulation by muskmelon under salt stress in contaminated organic soil. *Science of the Total Environment, 407*(7), 2175–2182.
50. Gall, J. E., & Rajakaruna, N., (2013). The physiology, functional genomics, and applied ecology of heavy metal-tolerant *Brassicaceae. Brassicaceae: Characterization, Functional Genomics and Health Benefits, 2013*, 121–148.
51. Gerhardt, K. E., Gerwing, P. D., & Greenberg, B. M., (2017). Opinion: Taking phytoremediation from proven technology to accepted practice. *Plant Science, 256*, 170–185.
52. Ghori, Z., Iftikhar, H., Bhatti, M. F., & Sharma, I., (2016). Phytoextraction: The use of plants to remove heavy metals from soil. In: *Plant Metal Interaction* (pp. 385–409). Elsevier.
53. Gill, S. S., & Tuteja, N., (2011). Cadmium stress tolerance in crop plants: Probing the role of sulfur. *Plant Signaling and Behavior, 6*, 215–222.
54. Glick, B. R., (2012). Plant growth-promoting bacteria: Mechanisms and applications. *Scientifica, 2012,* 1–15.
55. Glick, B. R., Cheng, Z., Czarny, J., & Duan, J., (2007). Promotion of plant growth by ACC deaminase-producing soil bacteria. In: *New Prospectives and Approaches in Plant Growth-Promoting Rhizobacteria Research* (pp. 329–339). Dordrecht: Springer.
56. Gopinath, A., Krishna, K., & Karthik, C., (2020). Adsorptive removal and recovery of heavy metal ions from aqueous solution/effluents using conventional and non-conventional materials. In: *Modern Age Waste Water Problems* (pp. 309–328). Springer.
57. Goutam, S. P., Saxena, G., & Singh, V., (2018). Green synthesis of TiO_2 nanoparticles using leaf extract of *Jnfropho career* L. for photocatalytic degradation of tannery wastewater. *Chemical Engineering Journal, 336*, 586–396.

58. Gu, C. S., Liu, L. Q., Deng, Y. M., Zhu, X. D., Huang, S. Z., & Lu, X. Q., (2015). The heterologous expression of the *Iris lactea* var. Chinensis type-2 metallothionein *IlMT2b*-gene enhances copper tolerance in *Arabidopsis thaliana. Bulletin of Environmental Contamination and Toxicology, 94*, 247–253.
59. Hasan, M. K., Liu, C., Wang, F., Ahammed, G. J., Zhou, J., & Xu, M. X., (2016). Glutathione-mediated regulation of nitric oxide, S-nitrosothiol, and redox homeostasis confers cadmium tolerance by inducing transcription factors and stress response genes in tomato. *Chemosphere, 161*, 536–545.
60. Hasan, M. M., & Uddin, M. N., (2019). Assisting phytoremediation of heavy metals using chemical amendments. *Plants, 8*, 295–299.
61. Hasan, M., Ahammed, G. J., Yin, L., Shi, K., Xia, X., & Zhou, Y., (2015). Melatonin mitigates cadmium phytotoxicity through modulation of phytochelatins biosynthesis, vacuolar sequestration, and antioxidant potential in *Solanum Lycopersicum* L. *Frontiers in Plant Science, 6*, 601–604.
62. Hassan, T. U., Bano, A., & Naz, I., (2017). Alleviation of heavy metals toxicity by the application of plant growth-promoting rhizobacteria and effects on wheat grown in saline-sodic field. *International Journal of Phytoremediation, 19*, 522–529.
63. Hassinen, V. H., & Tervahauta, A. I., (2011). Plant metallothioneins - metal chelators with ROS scavenging activity. *Plant Biology, 13*, 225–232.
64. Hayashi, S., Kuramata, M., Abe, T., Takagi, H., & Ozawa, K., & Ishikawa, S., (2017). Phytochelatin synthase *OsPCS1* plays a crucial role in reducing arsenic levels in rice grains. *The Plant Journal, 91*, 840–848.
65. He, J. L., Li, H., Ma, C. F., & Zhang, Y. L., (2015). Overexpression of bacterial γ-glutamyl cysteine synthetase mediates changes in cadmium influx, allocation and detoxification in poplar. *New Phytologist, 205*, 240–254.
66. Hossain, M. A., Piyatida, P., Da Silva, J. A. T., & Fujita, M., (2012). Molecular mechanism of heavy metal toxicity and tolerance in plants: central role of glutathione in detoxification of reactive oxygen species and methylglyoxal and in heavy metal chelation. *Journal of Botany, 2012*. Article ID: 872875.
67. Huang, D., & Dai, W., (2015). Two iron-regulated transporter (*IRT*) Genes showed differential expression in poplar trees under iron or zinc deficiency. *Journal of Plant Physiology, 186*, 59–67.
68. Iori, V.,Pietrini, F., Massacci, A., & Zacchini, M., (2015). Morphophysiological responses, heavy metal accumulation and phytoremoval ability in four willow clones exposed to cadmium under hydroponics. In: Ansari, A. A., Gill, S. S., Gill, R., & Lanza, G. R., & Newman, L., (eds.), *Phytoremediation: Management of Environmental Contaminant* (pp. 87–98). Switzerland; Springer International Publishing.
69. James, C. A., & Strand, S. E., (2009). Phytoremediation of small organic contaminants using transgenic plants. *Current Opinions in Biotechnology, 20,* 237–241.
70. Jee, C., (2016). Advances in phytoremediation and rhizoremediation. *Octa Journal of Environmental Research, 4*, 18–32.
71. Jha, C. K., & Saraf, M., (2015). Plant growth-promoting rhizobacteria (PGPR): A review. *Journal of Agricultural Research and Development, 5*, 108–119.
72. Kao, C. H., (2015). Role of glutathione in abiotic stress tolerance of rice plants. *Journal of Taiwan Agricultural Research, 164*, 167–176.

73. Karthik, C., & Arulselvi, P. I., (2017). Biotoxic effect of chromium (VI) on plant growth-promoting traits of novel *Cellulosimicrobium funkei* strain AR8 Isolated from *Phaseolus vulgaris* rhizosphere. *Geomicrobiology Journal, 34*(5), 434–442.
74. Khalid, S., Shahid, M., & Niazi, N. K., (2017). Comparison of technologies for remediation of heavy metal contaminated soils. *Journal of Geochemical Exploration, 182*, 247–268.
75. Khandare, R. V., & Govindwar, S. P., (2015). Phytoremediation of textile dyes and effluents: Current scenario and future prospects. *Biotechnology Advances, 33*(8), 1697- 1714.
76. Khodaverdiloo, H., & Taghlidabad, R. H., (2013). Phytoavailability and potential transfer of Pb from a salt-affected soil to *Atriplex verucifera*, *Salicornia europaea* and *Chenopodium album*. *Chemistry and Ecology, 30*(3), 216–226.
77. Kim, Y. H., Khan, A. L., & Kim, D. H., (2014). Silicon mitigates heavy metal stress by regulating p-type heavy metal ATPases, *Oryza sativa* low silicon genes and endogenous phytohormones. *BMC Plant Biology, 149*(1), 1–13.
78. Kinidi, L., Salleh, S. F., Wahab, N. A., & Wei, T., (2017). Phytoremediation, a ammoniacal nitrogen uptake by macrophytes with phytoremediation. *Research and Reviews: Journal of Engineering and Technology, 6*, 28–35.
79. Kong, Z., & Glick, B. R., (2017). The role of bacteria in phytoremediation: Innovations and future directions. *Applied Bioengineering, 327*–353.
80. Koo, B. J., Chen, W., Chang, A. C., & Page, A. L., (2010). Root exudates based approach to assess the long-term phytoavailability of metals in biosolids-amended soils. *Environmental Pollution, 158*(8), 2582–2588.
81. Kumar, S., Dubey, R. S., Tripathi, R. D., Chakrabarty, D., & Trivedi, P. K., (2015). Omics and biotechnology of arsenic stress and detoxification in plants: Current updates and prospective. *Environmental International, 74*, 221–230.
82. Kumar, S., Singh, R., Behera, M., & Kumar, V., (2019). Restoration of pesticide-contaminated sites through plants. In: *Phytomanagement of Polluted Sites* (pp. 313–327). New York: Elsevier.
83. Kumar, S., & Trivedi, P. K., (2016). Heavy metal stress signaling in plants. In: Ahmad, P. (ed.), *Plant Metal Interaction-Emerging Remediation Techniques* (pp. 585–603). Amsterdam: Elsevier, (2016.
84. Laghlimi, M., & Baghdad, B., El Hadi, H., & Bouabdli, A., (2015). Phytoremediation mechanisms of heavy metal contaminated soils: A review. *Open Journal of Ecology, 5*(8), 375.
85. Lee, S., Moon, J. S., Ko, T. S., & Petros, D., (2003). Overexpression of *Arabidopsis* phytochelatin synthase paradoxically leads to hypersensitivity to cadmium stress. *Plant Physiology, 131*, 656–663.
86. Lepp, N. W., (2012). *Effect of Heavy Metal Pollution on Plants: Metals in the Environment* (Vol. 2, pp. 217). Springer Science and Business.
87. Li, D. D., Xu, X. M., & Hu, X. Q., (2015). Genome-wide analysis and heavy metal-induced expression profiling of HMA gene family in *Populus trichocarpa*. *Frontiers in Plant Science, 6*, 1149–1153.
88. Li, J., Pu, L., Han, M., Zhu, M., Zhang, R., & Xiang, Y., (2014). Soil salinization research in China: Advances and prospects. *Journal of Geographical Sciences, 24*(5), 943–960.

89. Li, M., & Ahammed, G. J., (2016). Brassinosteroid ameliorates zinc oxide nanoparticles -induced oxidative stress by improving antioxidant potential and redox homeostasis in tomato seedling. *Frontiers in Plant Science, 7*, 615–620.
90. Li, M., Lou, Z., Wang, Y., Liu, Q., Zhang, Y., Zhou, J., & Qian, G., (2015). Alkali and alkaline earth metallic (AAEM) species leaching and Cu(II) sorption by biochar. *Chemosphere, 119*, 778–785.
91. Liang, L., Liu, W., Sun, Y., Huo, X., Li, S., & Zhou, Q., (2017). Phytoremediation of heavy metal contaminated saline soils using halophytes: Current progress and future perspectives. *Environment Reviews, 25*, 269–281.
92. Limmer, M., (2016). Phytovolatilization of organic contaminants. *Environmental Science and Technology, 50*(13), 6632–6643.
93. Liu, J., Shi, X., Qian, M., Zheng, L., Lian, C., & Xia, Y., (2015). Copper induced hydrogen peroxide upregulation of a metallothionein gene, *OsMT2c*, from *Oryza sativa* L. confers copper tolerance in *Arabidopsis thaliana*. *Journal of Hazardous Material, 294*, 99–108.
94. Liu, S., Yang, C., Xie, W., Xia, C., & Fan, P., (2012). The effects of cadmium on germination and seedling growth of *Suaeda* salsa. *Procedia Environmental Sciences, 16*, 293–298.
95. Liu, W. T., Ni, J. C., & Zhou, Q. X., (2013). Uptake of heavy metals by trees: Prospects for phytoremediation. In: *Material Science Forum* (Vol. 743, pp. 768–781). Trans Tech Publications.
96. Liu, W., Wang, Q., Wang, B., Hou, J., Luo, Y., Tang, C., & Franks, A. E., (2015). Plant growth-promoting rhizobacteria enhance the growth and Cd uptake of *Sedum plumbizincicola* in a Cd-contaminated soil. *Journal of Soils and Sediments, 15*, 1191–1199.
97. Lutts, S., & Lefèvre, I., (2015). How can we take advantage of halophyte properties to cope with heavy metal toxicity in salt-affected areas? *Annals of Botany, 115*(3), 509–528.
98. Ma, Y., He, J., Ma, C., & Luo, J., (2014). Ectomycorrhizas with *Paxillus involutus* enhance cadmium uptake and tolerance in *Populus× canescens*, *Plant Cell and Environment, 37*, 627–642.
99. Maestri, E., & Marmiroli, N., (2011). Transgenic plants for phytoremediation. *International Journal of Phytoremediation, 13*, 264–279.
100. Mahar, A., Wang, P., Ali, A., & Awasthi, M. K., (2016). Challenges and opportunities in the phytoremediation of heavy metals contaminated soils: A review. *Ecotoxicology and Environmental Safety, 126*, 111–121.
101. Manoj, R. S., & Chinnannan, K., (2020). Understanding the molecular mechanism for the enhanced photoremediation of heavy metal through plant growth-promoting rhizobacteria. *Journal of Environmental Management, 254*, 109–779.
102. Margoshes, M., & Vallee, B. L., (1957) A Cadmium protein from equine kidney cortex. *Journal of American Chemical Society, 79*, 4813–4814.
103. Marquez-Garcia, B., & Marquez, C., (2013). The effects of heavy metals on germination and seedling characteristics in two halophyte species in Mediterranean marshes. *Marine Pollution Bulletin, 70*(1, 2), 119–124.
104. Masindi, V., & Muedi, K. L., (2018). Environmental contamination by heavy metals. *Heavy Metals, 10*, 115–132.

105. McLaughlin, M., (2018). Drivers of soil pollution in agricultural fields. In: *Proceedings of the Global Symposium on Soil Pollution*, (pp. 68–162). Italy food and agriculture organization (FAO), Rome.
106. Mench, M., Schwitzguebel, J. P., Schroeder, P., & Bert, V., (2009). Assessment of successful experiments and limitations of phytotechnologies: Contaminant uptake, detoxification and sequestration and consequences for food safety. *Environmental Science and Pollution Research, 16*, 876–900.
107. Migeon, A., & Blaudez, D., (2010). Genome-wide analysis of plant metal transporters, with an emphasis on poplar. *Cellular and Molecular Life Sciences, 67*, 3763–3784.
108. Mishra, A., Mishra, S. P., Arshi, A., Agarwal, A., & Dwivedi, S. K., (2020). Plant-microbe interactions for bioremediation and phytoremediation of environmental pollutants and agro-ecosystem development. In: *Bioremediation of Industrial Waste for Environmental Safety* (pp. 415–436). Singapore; Springer.
109. Mosekiemang, T., & Dikinya, O., (2012). Efficiency of chelating agents in retaining sludge-borne heavy metals in intensively applied agricultural soils. *International Journal of Environmental Science and Technology, 9*(1), 129–134.
110. Mucha, A. P., Teixeira, C., & Reis, I., (2013). Response of a salt marsh microbial community to metal contamination. *Estuarine, Coastal and Shelf Science, 130*, 81–88.
111. Mustapha, M. U., & Halimoon, N., (2015). Screening and isolation of heavy metal tolerant bacteria in industrial effluent. *Procedia Environmental Science, 30*, 33–37.
112. Muthusaravanan, S., & Sivarajasekar, N., (2018). Phytoremediation of heavy metals: Mechanisms, methods and enhancements. *Environmental Chemistry Letters, 16*(4), 1339–1359.
113. Nadeem, S. M., Ahmad, M., Zahir, Z. A., & Javaid, A., (2014). Role of mycorrhizae and plant growth-promoting rhizobacteria (PGPR) in improving crop productivity under stressful environments. *Biotechnology Advances, 32*, 429–448.
114. Ojuederie, O. B., & Babalola, O. O., (2017). Microbial and plant-assisted bioremediation of heavy metal polluted environments: Review. *International Journal of Environmental Research and Public Health, 14*(12), 1504–1509.
115. Paape, T., Hatakeyama, M., & Shimizu-Inatsugi, R., (2016). Conserved but attenuated parental gene expression in allopolyploids: Constitutive zinc hyperaccumulation in the allotetraploid *Arabidopsis* kamchatica. *Molecular Biology and Evolution, 33*(11), 2781–2800.
116. Pang, S., Duan, L., & Liu, Z., (2012). Co-induction of a glutathione-S-transferase, a glutathione transporter and an ABC transporter in maize by xenobiotics. *PLoS One, 7*. Article ID: e40712.
117. Parmar, S., & Singh, V., (2015). Phytoremediation approaches for heavy metal pollution: A review. *Journal of Plant Sciences and Research, 2*(2), 135–140.
118. Pierart, A., Shahid, M., Séjalon-Delmas, N., & Dumat, C., (2015). Antimony bioavailability: Knowledge and research perspectives for sustainable agriculture. *Journal of Hazardous Material, 289*, 219–234.
119. Piotrowska-Niczyporuk, A., Bajguz, A., Zambrzycka, E., & Godlewska-Żyłkiewicz, B., (2012). Phytohormones as regulators of heavy metal biosorption and toxicity in green alga *Chlorella vulgaris* (Chlorophyceae). *Plant Physiology and Biochemistry, 52*, 52–65.

120. Pochodylo, A. L., & Aristilde, L., (2017). Molecular dynamics of stability and structures in phytochelatin complexes with Zn, Cu, Fe, Mg and Ca: Implications for metal detoxification. *Environmental Chemistry Letters, 15,* 495–500.
121. Pollard, A. J., Reeves, R. D., & Baker, A. J. M., (2014). Facultative hyperaccumulation of heavy metals and metalloids. *Plant Science, 217*, 8–17.
122. Qixing, Z., Zhang, C., Zhineng, Z., & Weitao, L., (2011). Ecological remediation of hydrocarbon contaminated soils with weed plant. *Journal of Resources and Ecology, 2*(2), 97–105.
123. Rai, K. P., Kim, H. K., Sang, S. L., & Jin, H. L., (2020). Molecular mechanism in phytoremediation of environmental contaminants and prospects of engineered transgenic plant/microbes. *Science of the Total Environment, 705*. Article ID: 135858.
124. Rai, P. K., Lee, S. S., Zhang, M., Tsang, Y. F., & Kim, K. H., (2019). Heavy metals in food crops: Health risks, fate, mechanisms and management. *Environment International, 125*, 365–385.
125. Ranieri, E., & Moustakas, K., (2020). Phytoextraction technologies for mercury-and chromium-contaminated soil: A review. *Journal of Chemical Technology and Biotechnology, 95*(2), 317–327.
126. Rezania, S., Taib, S. M., & Din, M. F. M., (2016). Comprehensive review on phytotechnology: Heavy metals removal by diverse aquatic plants species from wastewater. *Journal of Hazardous Materials, 318*, 587–599.
127. Rodríguez-Eugenio, N., McLaughlin, M., & Pennock, D., (2018). Soil pollution: Hidden reality. In: *Proceedings of the Global Symposium on Soil Pollution* (pp. 304–382). Italy Food and Agriculture Organization (FAO), Rome.
128. Rucinska-Sobkowiak, R., (2016). Water relations in plants subjected to heavy metal stresses. *Acta Physiologiae Plantarum, 38*, 257–260; doi: 10.1007/s11738-016-2277-5.
129. Sácký, J., Leonhardt, T., & Boroviˇcka, J., (2014). Intracellular sequestration of zinc, cadmium and silver in *Hebeloma mesophaeum* and characterization of its metallothionein genes. *Fungal Genetics and Biology, 67*, 3–14. doi: 10.1016/j.fgb.2014.03.003.
130. Safari, A. A., Tahmasbian, I., & Safari, M., (2015). Chelating agents and heavy metal phytoextraction. In: *Heavy Metal Contamination of Soils* (pp. 367–393). Springer.
131. Sahebi, M., Hanafi, M. M., SitiNorAkmar, A., Rafii, M. Y., Azizi, P., Tengoua, F., NurulMayzaitulAzwa, J., & Shabanimofrad, M., (2015). Importance of silicon and mechanisms of biosilica formation in plants. *BioMed Research International, 2015*.
132. Seneviratne, M., Gunaratne, S., Bandara, T., Weerasundara, L., Rajakaruna, N., Seneviratne, G., & Vithanage, M., (2016). Plant growth promotion by *Bradyrhizobium japonicum* under heavy metal stress. *South African Journal of Botany, 105*, 19–24.
133. Seneviratne, M., & Vithanage, M., (2015). The role of siderophores on plants under heavy metal stress: A view from the rhizosphere. *Journal of Botanical Sciences, 4*(3), 23–29.
134. Shackira, A. M., & Puthur, J. T., (2019). Phytostabilization of heavy metals: Understanding of principles and practices. In: *Plant-Metal Interactions* (pp. 263–282). Springer Cham.
135. Shah, V., & Daverey, A., (2020). Phytoremediation: Multidisciplinary approach to clean up heavy metal contaminated soil. *Environmental Technology and Innovation*. Article ID: 100774.

136. Sharma, B. M., Bharat, G. K., & Tayal, S., (2014). Environment and human exposure to persistent organic pollutants (POPs) in India: Systematic review of recent and historical data. *Environment International, 66*, 48–64.
137. Sharma, P., & Pandey, S., (2014). Status of phytoremediation in world scenario. *International Journal of Environmental Bioremediation and Biodegradation, 2*(4), 178–191.
138. Sharma, S. S., Dietz, K. J., & Mimura, T., (2016). Vacuolar compartmentalization as indispensable component of heavy metal detoxification in plants. *Plant Cell and Environment, 39*, 1112–1126.
139. Sheoran, V., Sheoran, A. S., & Poonia, P., (2016). Factors affecting phytoextraction: A review. *Pedosphere, 26*, 148–166.
140. Shi, X., Sun, H., Chen, Y., Pan, H., & Wang, S., (2016). Transcriptome sequencing and expression analysis of cadmium (Cd) transport and detoxification related genes in Cd-accumulating *Salix Integra*. *Frontiers in Plant Sciences, 7*, 1577–1582.
141. Shrivastava, P., & Kumar, R., (2015). Soil salinity: A serious environmental issue and plant growth-promoting bacteria as one of the tools for its alleviation. *Saudi Journal of Biological Sciences, 22*(2), 123–131.
142. Shukla, K. P., & Sharma, S., (2013). Rhizoremediation: A promising rhizosphere technology. In: *Applied Bioremediation: Active and Passive Approaches* (pp. 331–334).
143. Silva, V., Mol, H. G., Zomer, P., Tienstra, M., Ritsema, C. J., & Geissen, V., (2019). Pesticide residues in European agricultural soils-a hidden reality unfolded. *Science of the Total Environment, 653*, 1532–1545.
144. Sinegani, A. A. S., Tahmasbian, I., & Sinegani, M. S., (2015). Chelating agents and heavy metal phytoextraction. In: Sherameti, I., & Varma, A., (eds.), *Heavy Metal Contamination of Soils: Monitoring and Remediation* (pp. 367–393). Springer International Publishing; Cham.
145. Singh, A., (2015). Soil salinization and waterlogging: A threat to environment and agricultural sustainability. *Ecological Indicators, 57*, 128–130.
146. Singh, D., Tiwari, A., & Gupta, R., (2012). Phytoremediation of lead from wastewater using aquatic plants. *Journal of Agricultural and Technology, 8*(1), 1–11.
147. Singh, I., & Shah, K., (2014). Exogenous application of methyl jasmonate lowers the effect of cadmium-induced oxidative injury in rice seedlings. *Phytochemistry, 108*, 57–66.
148. Singh, S., Parihar, P., Singh, R., Singh, V. P., & Prasad, S. M., (2015). Heavy metal tolerance in plants: Role of transcriptomics, proteomics, metabolomics, and ionomics. *Frontiers in Plant Science, 6*, 1143–1148.
149. Singh, S., Parihar, P., Singh, R., Singh, V. P., & Prasad, S. M., (2016). Heavy metal tolerance in plants: Role of transcriptomics, proteomics, metabolomics and ionomics. *Frontiers in Plant Science, 6*, 1143–1149.
150. Singh, T., & Singh, D. K., (2017). Phytoremediation of organochlorine pesticides: Concept, method and recent developments. *International Journal of Phytoremediation, 19*(9), 834–843.
151. Song, W. Y., & Mendoza-Cózatl, D. G., (2014). Phytochelatin-metal(loid) transport into vacuoles shows different substrate preferences in barley and *Arabidopsis*. *Plant Cell and Environment, 37*, 1192–1201.
152. Steinhorst, L., & Kudla, J., (2014). Signaling in cells and organisms-calcium holds the line. *Current Opinion Plant Biology, 22*, 14–21.

153. Suman, J., Uhlik, O., Viktorova, J., & Macek, T., (2018). Phytoextraction of heavy metals: A promising tool for clean-up of polluted environment? *Frontiers in Plant Science, 9*, 1476.
154. Tamás, M. J., Sharma, S. K., Ibstedt, S., Jacobson, T., & Christen, P., (2014). Heavy metals and metalloids as a cause for protein misfolding and aggregation. *Biomolecules, 4*, 252–267.
155. Teixeira, C., & Almeida, C. M. R., (2014). Development of autochthonous microbial consortia for enhanced phytoremediation of salt-marsh sediments contaminated with cadmium. *Science of the Total Environment, 493*, 757–765.
156. Tepanosyan, G., Sahakyan, L., Belyaeva, O., Maghakyan, N., & Saghatelyan, A., (2017). Human health risk assessment and riskiest heavy metal origin identification in urban soils of Yerevan, Armenia. *Chemosphere, 184*, 1230–1240.
157. Thijs, S., Langill, T., & Vangronsveld, J., (2017). The bacterial and fungal microbiota of hyperaccumulator plants: Small organisms, large influence. In: *Advances in Botanical Research* (Vol. 83; pp. 43–86). New York: Academic Press.
158. Tian, S., Jia, Y., & Ding, Y. Z., (2014). Elevated atmospheric CO_2 enhances copper uptake in crops and pasture species grown in copper-contaminated soils in a micro- plot study. *Clean-Soil Air Water, 42*(3), 347–354.
159. Tiwari, S., Lata, C., Singh, C., P., & Prasad, V., (2017). Functional genomic perspective on drought signaling and its crosstalk with phytohormone-mediated signaling pathways in plants. *Current Genomics, 18*, 469–482.
160. Tomas, M., Pagani, M. A., & Andreo, C. S., (2015). Sunflower metallothionein family characterization. Study of the Zn (II) and Cd (II)-binding abilities of the *HaMT1* and *HaMT2* isoforms. *Journal of Inorganic Biochemistry, 148*, 35–48.
161. Ullah, A., Heng, S., Munis, M. F. H., Fahad, S., & Yang, X., (2015). Phytoremediation of heavy metals assisted by plant growth-promoting (PGP) bacteria: A Review. *Environmental and Experimental Botany, 117*, 28–40.
162. Ullmann, A., & Brauner, N., (2013). New biodegradable organic-soluble chelating agents for simultaneous removal of heavy metals and organic pollutants from contaminated media. *Journal of Hazardous Material, 260*, 676–688.
163. Vamerali, T., Bandiera, M., Lucchini, P., Dickinson, N. M., & Mosca, G., (2014). Long-term phytomanagement of metal-contaminated land with field crops: Integrated remediation and biofortification. *European Journal of Agronomy, 53*, 56–66.
164. Van, A. E., Covaci, A., Blust, R., & Bervoets, L., (2012). Persistent organic pollutants in the Scheldt estuary: Environmental distribution and bioaccumulation. *Environmental International, 48*, 17–27.
165. Van, A. B., (2008). Transgenic plants for phytoremediation: Helping nature to clean up environmental pollution. *Trends in Biotechnology, 26*(5), 225–227.
166. Van, D. E. A., Baker, A. M., & Reeves, R., (2013). Hyperaccumulators of metal and metalloid trace elements: facts and fiction. *Plant and Soil, 362*(1–2), 319–334.
167. Van, O. M. J., & Maggio, A., (2015). Functional biology of halophytes in the phytoremediation of heavy metal contaminated soils. *Environmental and Experimental Botany, 111*, 135–146.
168. Wang, B., Cheng, D., Chen, Z., & Zhang, M., (2019). Bioinformatic exploration of the targets of xylem sap miRNAs in maize under cadmium stress. *International Journal of Molecular Sciences, 20*, 1474–1480.

169. Wang, Y. J., Zhou, L. M., Zheng, X. M., Qian, P., & Wu, Y. H., (2013). Influence of *Spartina alterniflora* on the mobility of heavy metals in salt marsh sediments of the Yangtze River Estuary, China. *Environmental Science and Pollution Research, 20*(3), 1675–1685.
170. Wong, M. H., (2012). Environmental. contamination. Chapter 1. In: *Health Risks and Ecological Restoration* (pp. 5–18). Boca Raton: Taylor & Francis Group.
171. Wu, J., Shi, Y., Zhu, Y., Wang, Y., & Gong, H., (2013). Mechanisms of enhanced heavy metal tolerance in plants by silicon: A review. *Pedosphere, 23*(6), 815–825.
172. Xiong, T., Austruy, A., & Pierart, A., (2016). Kinetic study of phytotoxicity induced by foliar lead uptake for vegetables exposed to fine particles and implications for sustainable urban agriculture. *Journal of Environmental Sciences, 46*, 16–27.
173. Yan, A., Wang, Y., & Tan, S. N., (2020). Phytoremediation: A promising approach for revegetation of heavy metal-polluted land. *Frontiers in Plant Science, 11*, 1–15.
174. Yang, J., Gao, M. X., Hu, H., Ding, X. M., Lin, H. W., & Wang, L., (2016). *OsCLT1*, a CRT-like transporter 1, is required for glutathione homeostasis and arsenic tolerance in rice. *New Phytologist, 211*, 658–670. doi: 10.1111/nph.13908.
175. Yang, Y., Liang, Y., Ghosh, A., Song, Y., Chen, H., & Tang, M., (2015). Assessment of arbuscular mycorrhizal fungi status and heavy metal accumulation characteristics of tree species in a lead-zinc mine area: Potential applications for phytoremediation. *Environmental Science and Pollution Research, 22*, 13179–13193.
176. Yang, Z., & Chu, C., (2011). Towards understanding plant response to heavy metal stress. *Abiotic Stress in Plants-Mechanisms and Adaptations, 10*. Article ID: 24204.
177. Zechmann, B., (2014). Compartment-specific importance of glutathione during abiotic and biotic stress. *Frontiers in Plant Science, 5*, 566–569.
178. Zeng, P., Guo, Z., Cao, X., Xiao, X., Liu, Y., & Shi, L., (2018). Phytostabilization potential of ornamental plants grown in soil contaminated with cadmium. *International Journal of Phytoremediation, 20*(4), 311–320.
179. Zloch, M., Kowalkowski, T., Tyburski, J., & Hrynkiewicz, K., (2017). Modeling of phytoextraction efficiency of microbially stimulated *Salix dasyclados* L. *International Journal of Phytoremediation, 19*(12), 1150–1164.
180. Zhao, Y., & Chengcai, C., (2011). Towards understanding plant response to heavy metal stress: Chapter 04. In: Shanker, A., (ed.), *Abiotic Stress in Plants-Mechanisms and Adaptations* (p. 26). https://www.intechopen.com/books/abiotic-stress-in-plants-mechanisms-and-adaptations Online; Intechopen (accessed on 1 August 2021).
181. Zhou, C., Zhu, L., Ma, Z., & Wang, J., (2017). *Bacillus amyloliquefaciens* SAY09 increases cadmium resistance in plants by activation of auxin-mediated signaling pathways. *Genes, 8*(7), 173–178.
182. Tiwari, S., & Lata, C. (2018). Heavy metal stress, signaling, and tolerance due to plant-associated microbes: an overview. *Frontiers in plant science, 9,* 452.

PART II

Technological Assessment

CHAPTER 3

ROLE OF PHYTOREMEDIATION TO CLEANUP HEAVY METAL POLLUTED SOILS

JOAN NYIKA

ABSTRACT

Heavy metal (HM) contamination of soil and water is the greatest environmental concern in the 21st century due to health and environmental risks associated with pollutants infringement of food chains. Several biological and physicochemical techniques have been traditionally applied to cleanse HMs off the contaminated soils, though they are expensive and often lead to additional pollution. Phytoremediation involves the use of plants to take up, sequester, and/or decontaminate HMs; and it has proven to be a successful alternative to conventional methods due to its eco-friendly nature and low cost. This chapter focuses on phytoremediation to cleanse HM polluted soils.

3.1 INTRODUCTION

Environmental pollution by heavy metal (HM) is a growing environmental concern worldwide. The pollutants include metals and metalloids of density beyond 6 g cm^{-3}; and these metals include: arsenic (As), boron (B), cadmium (Cd), chromium (Cr), copper (Cu), cobalt (Co), manganese (Mn), mercury (Hg), selenium (Se) and zinc (Zn). These pollutants are either from geologic processing of rocks and soils or anthropogenic sources, such as industrial, and municipal activities. These anthropogenic sources are linked to the dominant trends of urbanization and industrialization [33]. HMs migrate to non-polluted areas as leachates or dusts through land and water resources to contaminate the ecosystems [55]. In the migrated areas, the pollutants can bioaccumulate, become non-biodegradable and are biomagnified to pose environmental and health risks through their capacity to interfere with natural biogeochemical cycles and food chains [3].

Number of clean-up methods (such as chemical oxidation, solvent extraction, thermal treatment, vitrification, stabilization/solidification, incineration, isolation, and removal) have been used to cleanse polluted environments (soils, water, and residues) off the HMs [14]. However, these methods are cost-ineffective and result in the transport of contaminated materials, which may result in secondary contamination. These challenges necessitate *in-situ* cleansing of these pollutants that is economical and less disruptive to the environment [61].

In this chapter, biotechnological advances focusing on phytoremediation are proposed as suitable alternatives. This bioremediation procedures use plants and their inherent microbes to isolate, degrade, purify, and remove toxic HMs. Phytoremediation is suitable compared to conventional methods for bioremediation of HMs, because plants occur as primary recipients of the pollutants from soils where they are released. Also, this chapter discusses and reviews the strategies and techniques of phytoremediation of HMs, factors influencing the process, its effects, and economic evaluation.

3.2 HEAVY METAL (HM) PHYTOREMEDIATION STRATEGIES AND TECHNIQUES

Phytoremediation is a result of two Greek words, 'phyto' and 'remedium,' which means plant and elimination of evil, respectively. Phytoremediation describes the use of plants and plant-based microorganisms to reduce the toxic effects and concentrations of environmental contaminants, such as radionuclides, HMs, and organic pollutants (such as pesticides, polychlorinated biphenyls (PCB) and polynuclear aromatic hydrocarbons) [59]. The preference to the method is due to its cost-effectiveness, novelty, *in-situ* applicability, eco-friendly nature, and its effectiveness with utilization of solar energy [64].

In this process, plants process pollutants without affecting top-soils and hence it is possible to retain soil fertility and its utility for agricultural purposes. Chaney [8] first suggested the concept of phytoremediation, whose acceptability has since been taken up since the 1990s. The "green clean" nature, low installation and management costs and applicability in large fields make the concept favorable in sustainable environmental management [1].

3.2.1 CONVENTIONAL PHYTOREMEDIATION TECHNIQUES

3.2.1.1 PHYTOEXTRACTION

Phytoextraction (also referred to us phytosequestration, phytoabsorption or phytoaccumulation) is a phytoremediation technique to cleanse sediments, biosolids, water, and soils off HMs [3]. It involves uptake of pollutants from the water and soil by plant roots and their subsequent transport and accumulation in the shoots [51, 61]. The technique is influenced by metal speciation, availability of the metal to the plant, soil characteristics and the specific plant characteristics. Plants useful for this process should have the following features according to Shabani and Sayadi [58]:

- Ability to accumulate high levels of metal contaminants from the soils without affecting plant physiology;
- Ability to adapt to the dominant climatic and environmental conditions;
- Ability to endure the potent effects of target heavy metals;
- Ability to translocate the pollutants to the shoots from roots;
- Capacity to produce more above-ground biomass or shoots;
- Easy growth and harvesting ability;
- Faster growth rate;
- Highly branched and widely spread root system;
- Not palatable to herbivores to prevent contamination of food chains;
- resistance to pests and pathogens.

The aim of this technique is to reduce HM concentrations in polluted soils to the required levels and within a reasonable timeframe (Figure 3.1). The extraction process depends on the above characteristics and can occur as two approaches [14, 42]: (i) the use of hyperaccumulators, which are plants with natural HM-amassing ability; and (ii) use of chemically enhanced crops (such as Indian mustard, rice, oats, peas, barley, and corn) with high biomass.

Plants, such as *Trifolium* spp. have several harvests in one growth period and are more preferred [3] due to their fast growth rate, short life-cycle, tolerance to abiotic stress, and high biomass production compared to trees and shrubs. Some of the HMs, their sources, harmful effects, and plants species that can hyper-accumulate them are presented in Table 3.1.

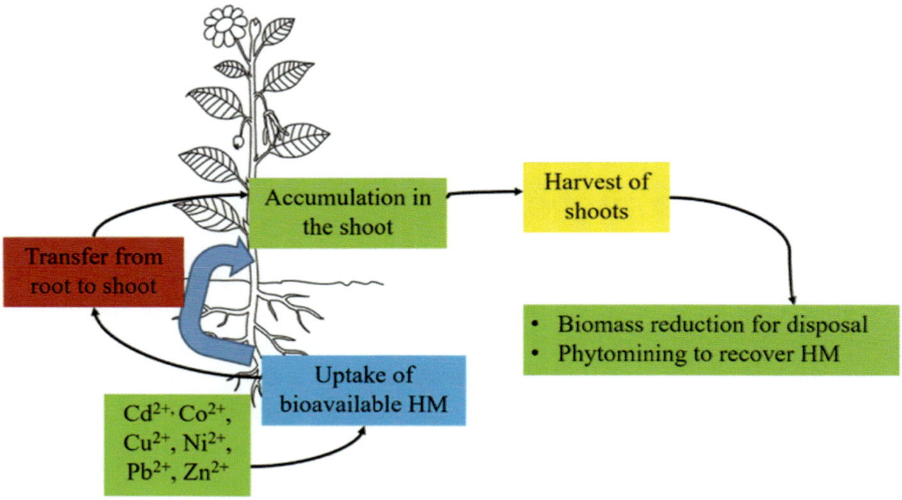

FIGURE 3.1 Schematic representation of phytoextraction process in metal bioremediation.

TABLE 3.1 Common Heavy Metals (HM): Sources, Effects, and Plants Used in Their Remediation

HM and Metalloids	Sources	Harmful Effects	Phytoextraction Plant Species
	[3]	[55]	[16]
As	Wood preservatives and pesticides.	Interferes with cellular processes.	*Corrigiola telephiifolia*
Cd	Paints, cement, and metal industries, plastic stabilizers, and rock phosphate.	Bone damage, kidney stones, carcinogenesis, and mutagenesis.	*Eleocharis acicularis, Azolla pinnata, Rogippa globose, Solanum photenocarpum*
Cr	Fly ash, manures, contaminated biosolids, leather, and steel industry.	Hair loss.	*Pteris vittata*
Cu	Zinc fertilizers and rock phosphate.	Anemia, intestinal irritation, dizziness, fatigue, kidney, and brain damage.	*Eleocharis acicularis*
Hg	Medical waste, coal combustion, surgical instruments, mining.	Ulcers, kidney damage, restlessness, fatigue, hair loss, depression.	–

TABLE 3.1 *(Continued)*

HM and Metalloids	Sources	Harmful Effects	Phytoextraction Plant Species
	[3]	[55]	[16]
Ni	Surgical instruments, kitchen appliances, metal industry, steel alloys, batteries.	Hair fall, cancers of the liver, stomach, throat lungs, genotoxicity.	*Allyssum caricum, A. murale, A. corsicum, A. bertolonii*
Pb	Electric batteries, agrochemicals, and lead-fueled urban traffic lights.	Memory loss, poor learning ability, coordination problem, renal failure, and cardiovascular disease.	*Euphorbia cheiradenia*
Zn	Fertilizers and pesticides.	Fatigue and dizziness.	*Eleocharis acicularis*

3.2.1.2 PHYTODEGRADATION

Phytodegradation (also known as phytotransformation) method involves uptake of inorganic and organic pollutants by plants and their degradation inside the plant cells using enzymes, such as laccases, dehalogenase, nitro-reductases. This vegetal degradation is applicable in cleansing pollutants off underground waters, soils, and sediments, since reduction and eventual metabolism occurs as a physiological process independent of microorganisms [2].

The process depends on the time by which the contaminant has resided in soil, the plant type, and the physicochemical characteristics of soils. Favas et al. [14] presented *Myriophyllum spicatum* and *Populus species* as examples of plants that are used in phytodegradation using enzymatic systems. Phytodegradation is, however, not much used for HMs due to their non-biodegradable nature [3].

3.2.1.3 PHYTOSTABILIZATION

Phytostabilization (also known as phytoimmobilization) is the process of incorporation of contaminants into the cell wall of roots, particularly in the lignin. HMs precipitate to insoluble compounds through root exudation, and contaminants are released to soil matrix, where they are trapped [2].

Their diffusion is also limited by inhibiting contaminant mobilization using surface adsorption, complexation, sorption, sedimentation processes, reduction of contaminant valences and accumulation in the roots, and rhizospheric precipitation [2].

Examples of plant species useful in phytostabilization are: *Alyssum, Gladiolus, Ascolepis, Eragrostis, and Haumaniastrum* species [3]; *Stipa austroitalica, Sinapis arvensis, Silybum marianum, Dasypyrum villosum,* and *Carduus pycnocephalus* [47]. *Athyrium wardii, Alnus rugosa,* and *Jatropha curcas* have also phytoimmobilized HMs from contaminated soils [70]. These plant species can exude high levels of chelating compounds, while reducing soil mobility. The plants must be fast-growing, poor translocators of contaminants above roots, possess dense roots and have tolerance to the contaminants, they take up [54].

Excretion of redox enzymes, which reduce the toxicity of HMs, is the case with Cr (VI) reduced to Cr (III) that the latter is less toxic, and it is a classic example of phytoimmobilization [68]. Etim [13] commended phytostabilization for the decontamination of sediments and soils off HMs, such as, As, Cd, Cr, Cu, Pb, and Zn. Although the method is effective in inactivating toxic metals, yet it is not a permanent solution since the contaminants are retained in the soils, and hence, these should be frequently monitored.

Metalliferous soils occur as restrictive environs for plant growth due to their toxicity, a phenomenon resulting to selection pressures. The habitats also have low fertility limiting normal plant growth unless the plants are HM tolerant [10]. Phytoimmobilization of metal-contaminated sites depends on the natural regrowth of native plants from seed banks in soils or the transplanting or seeding of areas with plant species that are HM resistant. The former is influenced by the quality and abundance of seed at the topsoil and is slow and cost-effective compared to the latter, which is faster and more effective [6]. To enhance the effectiveness of this technique, a combination of plants (grasses and trees) should be used in these metalliferous habitats.

3.2.1.4 PHYTOVOLATILIZATION

Phytovolatilization (also known as vegetal evaporation) utilizes plants that take up and volatilize metalloids and metals. Group 12 (11B), group 15 (VA) and group 16 (VIA) elements (such as Se, As, and Hg) are taken up by plant roots and converted to non-toxic gases that are released to the atmosphere via transpiration or evaporation [4]. Unlike phytoextraction and phytostabilization that sequester and immobilize a variety of HM pollutants, phytovolatilization is less applicable as pollutants should be taken up by roots, be translocated through the xylem to shoots, be transformed to volatile chemicals, and be vaporized to the atmosphere [39, 47].

The most essential aspect of phytovolatilization is transforming hazardous to non-hazardous pollutants, as is the case with Hg. Mercury is a priority hazardous substance since it is very toxic, can move very fast and can reside for long in the atmosphere. According to Wang et al. [65], once Hg is freed to the atmosphere, its retention time ranges from half to two years within which it can be transported to >10,000 km prior to deposition. For these reasons, its phytovolatilization and subsequent uptake by plants can deter its associated extensive pollution.

Examples of plant species used in phytovolatilization of Hg are: *Brassica napus, Liriodendron tulipifera, Nicotiana tabacum, and Arabidopsis thaliana*; while *Stanleya pinnata* and *Astragalus bisulcatus* are used in Se and As phytoremediation, respectively [14]. Wang et al. [66] suggested *Populus* and *Salix* trees as suitable for phytovolatilization while Ghosh and Singh [18] suggested the use of *Arabidopsis thaliana* and *Brassica juncea* plants for this strategy.

Pinto et al. [47] also suggested the use of *Eucalyptus globulus labill, Fragaria vesca L., Caulanthus* sp. and *Lepidium latifolium* for phytovolatilization of Hg. However, the possible release of these contaminants as they are carried from the roots to the vascular system is a potential hazard according to Adiloglu [2]. Padmavathiamma and Li [43] agreed with the opinions of Adiloglu [2] noting that phytovolatilization is a controversial phytoremediation technique since it involves the transfer of pollutants from one environment (soil to atmosphere to soil re-deposition) to another.

3.2.1.5 RHIZODEGRADATION AND RHIZOFILTRATION

Rhizodegradation (also referred to as phytostimulation) refers to the decomposition of pollutants including HM and organic compounds to the root zone of plants through microbial processes. Microorganisms in this process use carbon from plant metabolites and exudates as energy sources. The contaminants are released from the roots and influence microbial processes of that environment by dissolving them. Examples of plants useful in this process include [5]: reedmace (*Typha latifolia* L.), red berry (*Morus rubra* L.), mint

(*Mentha spicata* L.) and Lucerne (*Medicago sativa* L.). The technique is limited to organic contaminants [15].

In rhizofiltration, pollutants adhere to the roots or are absorbed during abiotic and biotic processes. The contaminants are transported to the plant parts for immobilization or even taken out of the plant using different approaches. Rhizofiltration is useful in phytoremediation of HMs in contaminated waters. Plants used for this purpose are planted in polluted waters directly or can be raised at their early stages in clean water and are then transplanted to contaminated waters, where they can adapt to the new environs. Once their roots are saturated, the plants are harvested. The technique is useful in phytoremediation of polluted natural and artificial water sources, and it uses both aquatic and terrestrial plants [40]. A schematic representation of these five main phytoremediation processes is shown in Figure 3.2.

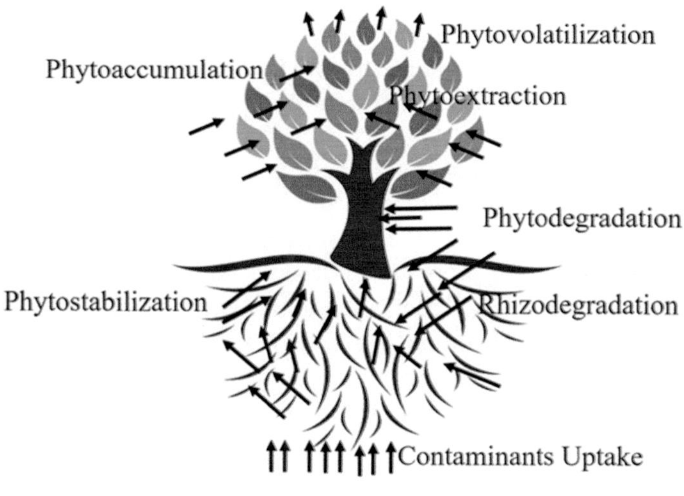

FIGURE 3.2 Schematic representation of conventional methods of phytoremediation.

3.2.1.6 PHYTOHYDRAULIC CONTROL

Other processes such as phytohydraulic control, phytodesalination, vegetative cover systems and buffer strips are conventional phytoremediation processes with some of the aspects of the techniques highlighted above.

The phytohydraulic technique uses plants to control the accumulation and movement of pollutants in surface and underground waters. During the transport of these pollutants, their metabolism and subsequent vaporization occur with the help of plant enzymes [50]. The impact area using this technique can be enlarged through root expansion using artificial systems despite the unstable absorption of water due to climatic and seasonal changes [2]. Examples of plants used in this technique include: *Populous, Eucalyptus, and Salix* tree species due to their high absorption and perspiration level [49]. The method is however limited for phytoremediation of HMs.

3.2.1.7 PHYTODESALINATION

The technique uses halophytes to free saline soils off some salts, some of which are metallic compounds [53, 69]. Compared to glycophytic plants, halophytes are better hyperaccumulators of HMs. According to Ali et al. [3], *Sesuvium portulacastrum* and *Suaeda maritima* have the capacity to sequester approximately 500 kg of sodium chloride from saline soils in four months. Although unique, yet the process is closely related to phytoextraction.

3.2.1.8 VEGETATIVE COVER SYSTEMS

Grasses and shrubs that grow on tailings or landfills can reduce rainwater infiltration and the eventual migration of pollutants. Their roots can also enhance soil aeration and improve transpiration, evaporation, and biodegradation [25]. Vegetative cover serves as barriers, which prevent and slow down pollution and are applied to contaminated habitats. Tailings are not appropriate for the development of plant roots, which presents a challenge in this technique. The problem can be alleviated by artificial fertilization of these environs and their constant control to prevent dominance of some crops compared to others [26].

3.2.1.9 RIPARIAN CORRIDORS/BUFFER STRIPS

Riparian corridors and buffer strips are plant systems that are often grown along stream ranks in strips to absorb pollutants from surface and underground waters. The aim is to slow down pollutant migration and prevent its mixing with aquifer systems [2]. Although the method has not been evaluated for cleansing of HMs from polluted environs, yet it has been shown to remove 8–100, 27–97, 67–96 and 71–90% of fecal coliforms, pesticides, phosphorous, and nitrogen from sediments, respectively [26]. *Populus* plant species is the common plant species used for this purpose. Contaminant removal using these plant systems uses some principles of rhizodegradation, phytodegradation, and phytovolatilization.

3.2.2 MODIFIED PHYTOREMEDIATION TECHNIQUES

Phytoremediation using green plants and the techniques discussed under Section 3.2.1 in this chapter are promising in alleviation of HMs. However, such approaches have some limitations for large-scale applications. Usually, plant species that are hyperaccumulators are slow-growing, rarely adapt to polluted environs and have low cumulative biomass according to Saifullah et al. [52]. Some of the limitations of the conventional phytoremediation strategies according to Ali et al. [3] include:

- Limited knowledge on breeding potential, diseases, insects, and pests spread among other agronomical aspects hinders full exploitation of these plant remediators.
- Mishandling of the above processes could lead to entry of contaminants to food chains and the extensive spread of their negative effects.
- Only a small portion of heavy metals are bioavailable and this aspect varies based on a variety of factors.
- The conventional phytoremediation techniques are only effective in environs with low to moderate pollution.
- The presence of low above ground biomass is a limiting factor to hyperaccumulator plants that remediate pollutants via phytoextraction.
- They require a long period to cleanse off pollutants from contaminated environs.

These limitations necessitate the modifications of conventional methods to enhance the applicability of phytoremediation on a large scale. The following sections discuss the recent advances towards improved phytoremediation.

3.2.2.1 MICROBIAL ASSISTED SYSTEMS

Soil microbes to some extent influence the availability of HMs around the plant rhizosphere. Some of the microorganisms live independently or in association with plant roots. For instance, mycorrhizal fungi live in association with higher plants and in varied forms, such as ericaceous, arbuscular, orchid, and ectomycorrhizas [55]. Associations of plants and such fungi can enhance the availability of plant nutrients (such as Ca, Co, Cu, K, N, Ni, P, S, and Zn) through hyphal network of this association. The phytoremediation ability of these microbes is in their ability to transform the chemical characteristics of root exudates and soil pH, leading to enhanced HM bioavailability and subsequent uptake by plant remediators [48].

Plant growth-promoting bacteria (PGPB) can facilitate the remediation of HMs from contaminated sediments and soils. They are classified as either free living rhizobacteria or symbiotic [57]. Through reduced enzyme activity, nitrogen fixation and ethylene generation during stress, the bacteria can promote the growth of biomass necessary for HM phytoextraction. The microbes can increase the ability of plants to withstand HM toxicity and stress, making them more effective pollutant hyperaccumulators [57]. Examples of microbes used in phytoremediation, their host plants, the metals they decontaminate and their effects in plants with HM stress are presented in Table 3.2.

TABLE 3.2 Microbes Used in Phytoremediation, Their Host Plants and Response to Heavy Metal (HM) Stress

Microbes	Host Plant	HM	Effects in HM Stressed Plants	References
Glomus mosseae	Sorghum	Cd, Cu, Pb	Immobilizes and accumulates the contaminants, transforms them to low toxic forms and promotes biomass growth.	[20]
Amanita muscaria, Scleroderma citrinum, Lactarius rufus L.	*Pinus sylvestris L.*	Cd, Pb, Zn	Reduces translocation of the contaminants to the shoot from roots.	[29]
Glomus mosseae	*Cajanus cajan L.*	Cd, Pb	Reduces HM translocation through the vascular system and promotes their accumulation in the roots.	[17]
Glomus etunicatum	*Calopogomium mucunoides*	Pb	Promotes Fe, P, and S uptake and retention of Pb in plant membranes. Prevents the production of reactive oxygen species.	[11]

TABLE 3.2 *(Continued)*

Microbes	Host Plant	HM	Effects in HM Stressed Plants	References
Glomus intraradices, Glomus viscosum, Glomus constrictum, Rhizophagus intraradices	*Phytolacca americana*	Mn	Accumulation of high Mn concentrations.	[67]
Funneliformis mosseae	*Cynodon dactylon*	Sb	Increased plant biomass, contaminant translocation and bioconcentration.	
Funneliformis mosseae, Claroideoglomus claroideum	*Calendula officinalis L.*	Cd, Pb	Improved antioxidant capacity and accumulation of secondary metabolic products, e.g., flavonoids and phenols.	[22]
Rhizophagus, Glomus and Acaulospora species	*Ricinus communis, Vetiveria zizanioides*	Pb	Uptake of high levels of Pb by roots and its transport to shoots.	[37]
Glomus fasciculatam	*Helianthus annuus L.*	Pb	Accumulation of HM in roots and shoots.	[56]
Rhizophagus irregularis	*Trifolium repens L.*	Pb	Distribution and accumulation of Pb in the roots and hyphae.	[36]
	Phragmites australis	Cd	Improved tolerance to the contaminant by the fungi.	[23]
G, intraradices, G. mosseae, G. claroideum, G. etunicatum, G. microaggregatum, G. geosporum	*Zea mays L.*	Hg	Biogeochemical cycling of the pollutant to transform it to less toxic products.	[28]
Phomopsis fukushii PDL-10, Glomerella truncata	*Solanum nigrum*	Cd	Improved uptake of the pollutant and its accumulation at the roots, shoots, and leaves.	[27]
Rhizophagus intraradices	*Oryza sativa*		Accumulation of the contaminant by roots, shoots, and in biomass.	[35]

TABLE 3.2 *(Continued)*

Microbes	Host Plant	HM	Effects in HM Stressed Plants	References
Arbuscular mycorrhiza (AM) fungi	*Solanum melongena*	As, Cd, Pb	Anti-oxidative defensive response against HM stress, enhanced plant growth for hyperaccumulation.	[9]
Rhizophagus intraradices, Funneliformis mosseae	*S. nigrum*	Cd	Concentration of the pollutant at the shoot.	[31]
Claroideoglomus, Funneliformis, Rhizophagus, and Glomerales species	*Daucus carota, Lactuca sativa*	Sb	Hyperaccumulation in the roots of the plants.	[46]
C. lamellosum, R. irregularis, F. mosseae	*Ricinus communis*	Cr (III) and (VI)	Decreased Cr (VI) levels and its subsequent decontamination to Cr (III).	[19]

3.2.2.2 CHEMICAL ASSISTED SYSTEMS

High HM concentrations in plant remediators and their biomass is an essential factor in successful phytoremediation since the pollutants must be available despite their low solubility naturally. Chemical modifications play a crucial role in enhancing the availability of HMs in polluted environs and their subsequent uptake and transport to the shoot [4, 39]. Different chelating agents and chemical amendments (such as sodium dodecyl sulfate (SDS), ethylene glycol tetraacetic acid (EGTA), ethylenediamine tetraacetic acid (EDTA)) can increase bioavailability of HMs in plants to enhance their hyperaccumulation in shoots and roots [21]. Figure 3.3 shows the chemical structure of three agents.

EDTA promotes metal absorption by enhancing xylem loading while EGTA acts as a chelating agent useful in phytoextraction of HMs. SDS, on the other hand, serves as a surfactant to enhance the solubility capacity of HMs such as Cd, Cu, Pb, and Zn, which makes them easily taken up by herbaceous species [21]. Selected plant remediators, the metals they take up and the chemical amendments used to enhance pollutant solubility are shown in Table 3.3.

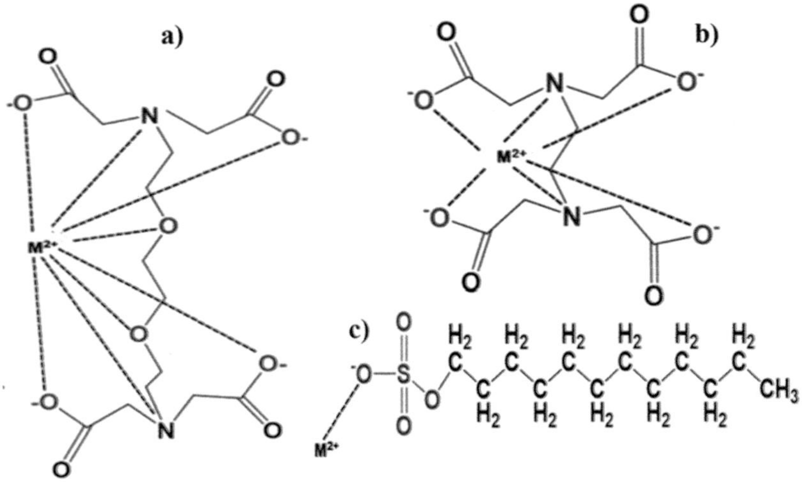

FIGURE 3.3 Chemical structures: (a) EGTA; (b) EDTA; and (c) SDS while binding to metals with a valence of 2 (M^{2+}), such as, Cd^{2+} and Pb^{2+}.

Although these chemical amendments enhance the uptake of HMs, yet they have several limitations. They pose some toxic effects on cultivated plant species, enzymatic activities in the soil and on soil microorganisms [21]. EDTA, for instance is non-biodegradable, persist in soils for long and as an exogenous chemical, has negative effects on the environment [41]. Metal chelators (MCs) have also been shown to result in stunted growth among plants [55]. These phenomena are associated with improved ability to uptake HMs by plant remediators despite their inability to tolerate such pollutants [21]. The chemicals are also very expensive considering that phytoremediation is done at a large scale in most cases.

Souza et al. [60] noted that the expensive nature of these chemical amendments during phytoremediation suggesting the use of organic acids such as oxalic, malic, citric, and acetic acids that are cost-effective, biodegradable, and result to minimal environmental pollution to form soluble HMs complexes as an alternative chemical modification approach.

TABLE 3.3 Examples of Plant Remediators

Plant Remediator	Metals	Chemical Amendments
Populus alba	Zn	SDS
Halimione portulacoides	Cu	SDS
Sesbania exaltata	Pb	EGTA, EDTA

TABLE 3.3 *(Continued)*

Plant Remediator	Metals	Chemical Amendments
Althaea roseas	Cd	EGTA, SDS
Calendula officinalis	Cd	EGTA, SDS
Cicer arietinum	Cd	EGTA
Mirabilis jalapa	Cd	EGTA, EDTA
Triticum aestivum	Pb, Zn	EDTA
Brassica juncea L.	Pb, Zn	
Zea maya L. cv.	Cu, Pb, Zn	
Phaseolus vulgaris L.	Cu, Pb, Zn	
Brassica juncea	Pb	
	Cd, Cr, Pb	
Brassica carinata	Cd, Cr, Pb	
Canavalia ensiformis L.	Pb	
Vetiver zizanioides	Cd, Pb, Zn	
Dianthus chinensis	Cd, Pb, Zn	
Helianthus annuus	Cd, Ni	
Tribulus terrestris	Cd, Pb	
Medicago sativa	Cr	
Calendula officinalis	Cd	EGTA, SDS
Targetes erecta	Pb	EDTA
Impatiens walleriana	Cd	EDTA

3.2.2.3 BIOCHAR ASSISTED SYSTEMS

Biochar describes a porous carbonaceous material produced through pyrolysis of sludges, organic manure, and plant materials, among other organic feedstocks [45]. It possesses some peculiar characteristics including the ability to stabilize HMs from its large surface area, alkaline nature, and high pH. This substance forms metal complexes with Ca, K, Mg, and Na ions through physical adsorption [34]. In combination with conventional phytoremediation techniques, biochar enhances the effectiveness of HM uptake by promoting plant growth and biomass production by 10% [3]. Biomass increment effects are associated with enhanced water holding capacity (WHC), cation exchange capacity and high nutrient content. Biochar also promotes the activity of microorganisms useful in phytoremediation among non-accumulators and hyperaccumulators, which enhances the resistance capacity of the plant remediators against HM pollutants [12].

3.2.2.4 GENETIC ENGINEERED SYSTEMS

Genetic engineering system in phytoremediation of HMs is based on over-expression of genes associated with HM tolerance, sequestration, translocation, and uptake in transgenic plants [3]. The introduction of specific genes to enhance HM phytoremediation is either direct via DNA transfer or using *Agrobacterium tumafaciens* or *Arabidopsis thaliana* mediated modifications resulting to transgenic plants [57]. These transgenic plants work by immobilizing HMs (such as Cd and Hg) and enhancing the tolerance of plants towards these pollutants.

Jagtap and Bapat [24] suggested that two strategies are applied in gene-enhanced phytoremediation: (i) in planta metabolism improvement; and (ii) ex-planta metabolism improvement. The latter involves genetic engineering of plants purposely to metal-specific ligands and reactive enzymes extracellularly. In planta strategies entail tree-metabolic modifications to enhance HM diffusion, transformation, and/or sequestration via the leaves, shoots, and roots. Genetic engineering models for phytoremediation use *in vitro* cultures and/or whole plants [24].

Metal absorption and transport across the plasma membrane is a function of transport proteins, whose binding sites have a high affinity to such pollutants. Transgenic plants are engineered to improve this transport process and to enhance HM tolerance to accumulate the pollutants in cell organelles that have low activity, such as, vacuoles. A case example is the ability of arsenic to complex with glutathione (-GSH) and phytochelatins (PCs) in vacuoles. PCs chelate HM with their thiol (-SH) group from cysteine to form complexes that are sequestered into vacuoles thereafter.

Metallothioneins (MTs) are also used in genetic modification of plant remediators. They bind HM ions to form mercaptide bonds with many cysteine residues, which enhance the tolerance to high levels of these pollutants [30]. MCs overexpression has also enhanced HM tolerance through their ability to sequester, transport, and absorb ions from plant fluids such as cytosol and compartmentalize them in the vacuole. Examples of these MCs include amino acids such as histidine and organic acids such as citrate and malate [24].

Enzymes responsible for chelation, stress tolerance, oxidative response and detoxification of HMs have been overexpressed in plants to enhance phytoremediation effectiveness. Examples of these enzymes are: GSH synthetase, aldehyde reductase, GSH synthetase, ACC deaminase and Hg^{2+} reductase, which are shown to enhance phytoextraction of HM pollutants [24, 30].

3.3 FACTORS AFFECTING PHYTOREMEDIATION OF HEAVY METALS (HMS)

A number of factors influence the uptake of HM pollutants as summarized in Figure 3.4.

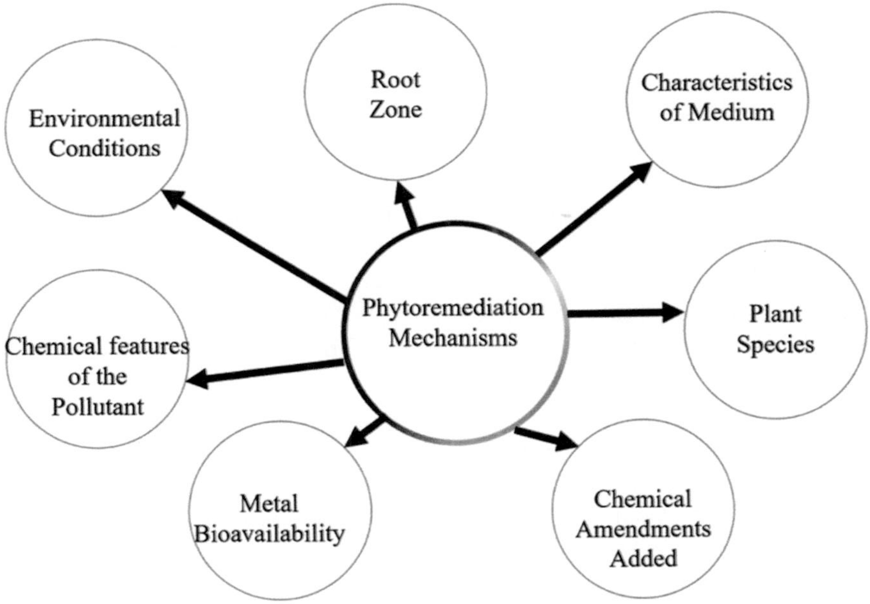

FIGURE 3.4 Factors that influence phytoremediation techniques.

The uptake of HMs depends on the plant varieties and species, and only those of superior remediation capacity are selected [50]. Singh and Prasad [59] claimed that the success of phytoremediation processes particularly phytoextraction is based on the selection of a suitable HM hyperaccumulator, which produces large biomass and can be easily be managed. The characteristics of medium of HM uptake, such as, its pH influences the effectiveness of remediation processes.

For example, the amount of Pb absorbed by plants is influenced by phosphorous, organic matter and soil pH according to Tangahu et al. [61], who noted that a reduction of pH between 6.5 and 7 using additives such as lime reduces the uptake of Pb ions from soils. The root zone is an essential determinant of phytoremediation owing to its capacity to absorb, accumulate, store, and metabolize contaminants within the plant tissue. The roots of plant remediators exude plant enzymes useful in the degradation of pollutants [35, 46].

Environmental conditions influence the vegetative uptake of HM pollutants. For example, temperature influences the growth and length of roots evident from examinations of phytoremediation plants grown in greenhouses compared to those under field conditions [38]. Phytoextraction processes are dependent on pollutant-specific hyperaccumulators. These factors advice for the need to conduct a mass balanced assessment on the fate of HM pollutants following metabolism to justify the applicability of phytoremediation compared to other conventional remediation techniques [62].

The bioavailability of HMs in the aqueous phase depends on a specific metal's retention time and the reaction capability with other substances; and elements in the water influence the uptake by plants. Once pollutants are bound to soil, their tendency to exist in ionic forms available for plant uptake is dependent on redox potential, organic matter content and soil pH. Plants usually oxygenate sediments and lower the pH, thus influencing HM availability. These activities can be enhanced by adding micronutrients and chelating agent on the plants [63].

The use of biodegradable physicochemical aspects (such as micronutrients and chelating agents) can increase HM bioavailability and enhance the activity of microorganisms involved in the phytoremediation. The use of synthetic chemical additives curtails remediation processes but also increases the risk of their leaching [61]. The chelating agents can also lower pH and enhance metal translocation to plant tissues.

3.4 EFFECTIVENESS OF PHYTOREMEDIATION MECHANISMS

Research studies have demonstrated the performance of plants for HM uptake [61]. Selected studies are summarized in Table 3.4, which confirms that phytoremediation is a promising alternative to HM treatment method that is eco-friendly. The process has been applied to remediate variety of HMs and uses variety of plants at different environs, such as, soils, and water of different characteristics. Furthermore, the use of chemical amendments (such as organic acids and EDTA) is shown to enhance the effectiveness of phytoremediation.

TABLE 3.4 Phytoremediation Effectiveness in Various Soil and Water Studies

Media and Uptake Mechanism	Contaminant and Concentration	Plant Used	Results
Phytoextraction in soils	0.1 g of sodium arsenate heptahydrate with a mixture of 50 ppm of As.	*Leersia oryzoides*	An increase in the size of the plant and a decrease of As concentrations in the shoot.
Soil and fly ash mixture	1,000 ppm of Pb, Zn, Ni, Mn, and Cu as lead nitrate, zinc sulfate, manganese chloride and copper sulfate, respectively.	*Scirpus littoralis*	The metals were accumulated in the plants below ground organs within 3 months.
Soil	Cd, Cr, Cu, Fe, Mn, Ni, Pb, and Zn.	*Brassica campestris* L., *Triticum aestivum*	The metals were accumulated in the plants' roots, shoots, leaves, and seeds in a period of 30 to 60 days. Pb was the most accumulated, followed by Fe, Mn, and Zn.
Phytoextraction	Standard Pb solution of 75 ppm	*Portulaca grandiflora, Sanvitalia procumbens,*	The plants accumulated Pb in their tissues with an efficacy process of 30 to 80%.
Soil	Cd, Cr, Cu, Ni, Pb, and Zn	*Brassica juncea, B. rapa, B. napa*	All plants showed tolerance to the HM, especially *B. rapa* for Cd and Pb accumulation in the leaves, stems, roots, fruit shell and seed in reducing order.
Phytoextraction	Cd, Zn	*Salix virminalis*	The HM were concentrated in the plant's leaves.

TABLE 3.4 *(Continued)*

Media and Uptake Mechanism	Contaminant and Concentration	Plant Used	Results
Soil	Zn in the form of $ZnSO_4$ at 50, 1,500, and 2,000 ppm	*Populus sp.*	A steady decrease in Zn concentrations in soil was observed as plants absorbed it as a mineral. At the roots, the concentration of the HM was higher compared to other above ground parts.
Soil	As (200 μg/g), Pb (1,000 μg/g)	*Salix sp. and Populus sp.*	Willows sequestered 40% of As and Pb, respectively from contaminated soil.
Phytoextraction	Cu, Pb, Zn	*B. juncea, S. virminalis* L. *Festuca arundinacea Schreb*	The metals were translocated to plants' tissues. Translocation of HM was enhanced by using EDTA.
Phytoextraction	Hg	*Triticum aestivum, Hordeum vulgare, Lupinus luteus*	Hg concentration decreased in soil to a level of 0.5 μg/g from 29.2 and 20.3 μg/g at 1–10 and 10–40 cm, respectively.
Lead polluted soils	Pb	*Agrostis capillaris*	Pb was translocated from the roots to the shoots, where it was concentrated.
Phytoextraction	As, Pb	*Pteris vittate, B. juncea*	Pb concentrations reduced with uptake by plants. Remediation process was enhanced by using EDTA.
Phyto-immobilization	Hg	*F. rubra, Poa pratensis, Armoracia lapathifolia, H. tuberosus, S. virminalis*	Translocation of Hg from soil to roots and aerial plant parts occurred.

TABLE 3.4 *(Continued)*

Media and Uptake Mechanism	Contaminant and Concentration	Plant Used	Results
Phytostabilization and phytoextraction	Cd, Cr, Cu and Zn	*Populus deltoides*	The four HM were accumulated in leaves, stem, roots, and woody cuttings.
	As, Co, Cu, Pb and Zn	*P. alba, P. tremula, P. nigra, Salix alba*	Concentrations of pollutants were higher in roots compared to above the ground parts.
	Cd	*S. fragilis, S. virminalis*	Cd was accumulated in above the ground tissues of these plants.
Rhizodegradation	Cd, Cu, and Zn	*Salix spp.*	The plants took up the HM, and the process was enhanced in the presence of ammonium acetate.
Hyrdoponic	As and Se	*Pteris vittata* L.	The pollutants were translocated to the fronds from the roots.
Hydroponic	Pb as lead nitrate	*B. juncea*	The plant accumulated the contaminant in the roots and above the ground tissues.
Phytofiltration	Hg as $HgCl_2$	*B. juncea*	The plant accumulated the pollutant 100 to 270 times more compared to initial concentrations.
Phytoextraction	Hg as $HgSO_4$	*Colocasia esculenta, Scirpus tabernaemontami, Eichornia crassipes*	Hg was taken up by these plants.
Hydroponic	Cu, Ni	*S. virminalis, S. triandra*	The metals were found concentrated in the plants above-ground tissues.

3.5 ECONOMIC EVALUATION OF PHYTOREMEDIATION IN CLEANING UP OF HEAVY METALS (HMS)

The unsustainable use of natural resources, including water, land, and air for economic advantage has an important input in manufacturing processes and has resulted in negative effects, such as environmental degradation and pollution [55]. The relationship between economic development and environmental quality has been proved to be inverse through empirical studies, with the tendencies being predominant in developing countries. The countries capitalize on cheap technologies, particularly at industries and urban areas, though they have severe environmental effects.

Soil and water contamination are the negative externalities of the economic activities. These forms of contamination require specialized techniques and strategies to evaluate their potential damage and mitigation measures. These undertakings are essential because the short-term effects of contamination are invisible while the long-term, are irreversible and have negative effects on income-generating activities, human health, and the environment [47, 48].

Soil and water pollution resulting from HM release to the environment emanates from agricultural and industrial activities and policy flaws that do not regulate these practices. Limitations of existent policies to subsidize phytoremediation initiatives hinder positive gains in the form of enhanced soil quality, higher land productivity, and improved water quality, which could result if the technology was adopted. The difficulty to quantify land degradation and restoration as well as its associated economic costs encourages persistent inaction on the socio-economic development damages, which result from the pollution [3].

Pollution of soil and water is a cost, whose quantification is based on a preference method that uses the environment as a production process input. The environment also acts as a feature of market goods depicted in changes of land prices influenced by soil health. Land degradation is a conventional market failure, which is influenced by both economic and political factors [7]. Different approaches to address the market failure are evident in economics and derive their logic from valorizing of non-market costs and benefits.

Total economic value (TEV) is a systematic method for quantifying the advantages of land restoration compared to its absence [44]. The direct valuation method as part of TEV considers variation in productivity of pristine and degraded environs. Production function method is also a common direct approach to assess market failure. The process compares the outputs to the inputs of a production process. Unlike direct methods, the indirect method considers environmental quality as a parameter influencing production processes since it affects the price and quality of resulting outputs. Using these approaches, phytoremediation value is based on crops to be grown and used in the cleaning process of polluted environs, the speed of bioremediation and the value of resultant biomass [32]. Returns on investment in phytoremediation are evaluated using a variety of methods, and it is essential for policymakers and researchers to analyze the economic factors involved in the process and its practicality [55].

Just like other pollutants, it is difficult to control individual behavior of HMs in soil and water. This is because contamination occurs due to both off and on-site practices and run-off from municipalities, industries, and neighboring fields. Costs affiliated with remediating polluted soils and water could be raised through tax imposition on responsible agencies. Pollution can also be stopped through intensive awareness on its long-term effects and through stringent enforcement of regulation on municipalities and industries to pre-treat wastes prior to their release [61]. Economic measures to solve the negative externalities include: taxing pollution or degradation units and introducing beneficial market measures for involved stakeholders. Apart from economic evaluation of phytoremediation, factors, such as its acceptance among end-users, its environmental sustainability and multi-functionality should be given precedence prior to its adoption [55].

3.6 SUMMARY

Phytoremediation, in this book chapter, entails strategies, such as, phytoextraction, phytodegradation, phytoimmobilization, phytovolatilization, and rhizodegradation. The performance of these conventional techniques can be enhanced using genetic engineering, microorganisms, biochar, and chemical amendments, which collectively serve to enhance HM bioavailability and subsequent uptake. The effectiveness of phytoremediation has been proven using a number of empirical studies. However, to validate the rationality of phytoremediation compared to preexistent techniques, which do not use plants, intensive research is required to understand and to optimize the economics and effectiveness of the approach towards sustainable environmental management.

KEYWORDS

- **bioremediation**
- **biotechnology**
- **contaminants**
- **heavy metal**
- **phytoremediation**
- **pollutants**

REFERENCES

1. Abhilash, C., Powell, R., Singh, B., & Singh, B., (2012). Plant-microbe interactions: Novel applications for exploitation in multipurpose remediation technologies. *Trends in Biotechnology, 30,* 416–420.
2. Adiloglu, S., (2018). Heavy metal removal with phytoremediation. In: Shiomi, N., (ed.), *Advances in Bioremediation and Phytoremediation* (pp. 123–141). London, UK: IntechOpen Publisher.
3. Ali, H., Khan, E., & Sajad, M., (2013). Phytoremediation of heavy metals-concepts and applications. *Chemosphere, 91*, 869–881.
4. Arshad, M., Silvestre, J., Merlina, G., Dumat, C., Pinelli, E., & Kallerhoff, J., (2012). Thidiazuron induces shoot organogenesis from mature leaf explants of scented *Pelargonium capitatum* cultivars. *Plant Cell Tissue Organic Culture, 108*, 315–322.
5. Aybar, M., Bilgin, A., & Sağlam, B., (2015). Removing heavy metals from the soil with phytoremediation. *Artvin Çoruh University; Journal of Natural Hazards and Environment, 1,* 59–65.
6. Bolan, N., Park, J., Robinson, B., Naidu, R., & Huh, K., (2011). Phytostabilization: Green approach to contaminants. *Advances in Agronomy, 112*, 145–204.
7. Champ, P. A., Boyle, K. J., & Brown, T. C., (2003). *Primer on Nonmarket Valuation* (p. 180). Boston, USA: Kluwer Academic Publishers.
8. Chaney, L., (1983). Plant uptake of inorganic waste constituents. In: Parr, J. F. E. A., (ed.), *Land Treatment of Hazardous Wastes* (pp. 50–76). Park Ridge, NJ: Noyes Data Corp.
9. Chaturvedi, R., Favas, P., Pratas, J., Varun, M., & Paul, M., (2018). Assessment of edibility and effect of arbuscular mycorrhizal fungi on T *Solanum melongena* L. Grown under heavy metal contaminated soil. *Ecotoxicology and Environmental Safety, 148*, 318–326.
10. Cordova, S., Neaman, A., Gonzalez, I., Ginocchio, R., & Fine, P., (2011). The effect of lime and compost amendments on the potential for the revegetation of metal-polluted, acidic soils. *Geoderma, 166*, 135–144.
11. De Souza, L., Andrade, S., De Souza, S., & Schiavinato, M., (2012). Arbuscular mycorrhiza confers Pb tolerance in *Calopogonium mucunoides*. *Acta Physiologiea Plantarum, 34*, 523–531.

12. Elad, Y., Cytryn, E., Harel, Y. M., Lew, B., & Graber, E. R., (2012). The biochar effect: Plant resistance to biotic stresses. *Phytopathologia Mediterranea, 50*, 335–349.
13. Etim, E., (2012). Phytoremediation and its mechanisms: A review. *International Journal of Environment and Bioenergy, 2*(3), 120–136.
14. Favas, P., Pratas, J., Varun, M., D'Souza, R., & Paul, M., (2014). Phytoremediation of soils contaminated with metals and metalloids at mining areas: Potential of native flora. In: Maria, C., & Hernandez, S., (eds.), *Environmental Risk Assessment of Soil Contamination* (p. 24). Shanghai, China: InTech. online.
15. Frers, C., (2008). El uso de plantas acuáticas en el tratamiento de aguas residuales (Use of aquatic plants to treat waste water). *Observatorio Medioambiental (Environmental Observatory), 11*, 301–305.
16. Garcia-Salgado, S., & Garcia-Casillas, D., (2012). Arsenic and heavy metal uptake and accumulation in native plant species from soils polluted by mining activities. *Water Air Soil Pollution, 223*, 559–572.
17. Garg, N., & Aggarwal, N., (2011). Effects of interactions between cadmium and lead on growth, nitrogen fixation, phytochelatin, and glutathione production in mycorrhizal *Cajanus cajan* (L.) millsp. *Journal of Plant Growth Regulation, 30*, 286–300.
18. Ghosh, M., & Singh, S., (2005). Review on phytoremediation of heavy metals and utilization of its byproducts. *Applied Ecology and Environmental Research, 3*, 1–18.
19. Gil-Cardeza, M., Müller, D., Amaya-Martin, S., Viassolo, R., & Gómez, E., (2018). Differential responses to high soil chromium of two arbuscular mycorrhizal fungi communities isolated from Cr-polluted and non-polluted rhizospheres of *Ricinus communis*. *Science of Total Environment, 625*, 1113–1121.
20. Gonzalez-Chavez, M., Carrillo-Gonzalez, R., Wright, S., & Nichols, K., (2004). The role of glomalin, a protein produced by *Arbuscular Mycorrhizal* Fungi, in sequestering potentially toxic elements. *Environmental Pollution, 130*, 317–323.
21. Hasan, M., Uddin, N., & Ara-Sharmeen, I., (2019). Assisting phytoremediation of heavy metals using chemical amendments. *Plants, 8*, 295–305.
22. Hristozkova, M., Geneva, M., & Stancheva, I., (2016). Contribution of *Arbuscular Mycorrhizal* fungi in attenuation of heavy metal impact on *Calendula officinalis* development. *Applied Soil Ecology, 101*, 57–63.
23. Huang, X., Ho, S., Zhu, S., Ma, F., Wu, J., Yang, J., & Wang, L., (2017). Adaptive response of *Arbuscular Mycorrhizal* symbiosis to accumulation of elements and translocation in *Phragmites australis* affected by cadmium stress. *Journal of Environmental Management, 197*, 448–455.
24. Jagtap, U., & Bapat, V., (2015). Genetic engineering of plants for heavy metal removal from soil. *Heavy Metal Contamination of Soils, 18*, 433–470.
25. Jorba, M., & Vallejo, R., (2008). Ecological restoration of quarries: a case with application of organic amendments and irrigation. *Ecosistemas, 17*(3), 119–132.
26. Kalkan, H., Orman, Ş., & Kaplan, M., (2011). Phytoremediation Technique in Reclamation of Contaminated Lands. *Journal of Agriculture and Food Science, 25*(4), 103–108.
27. Khan, A., Waqas, M., Ullah, I., Khan, A., Khan, M., Lee, I., & Shin, J., (2017). Culturable endophytic fungal diversity in the cadmium hyperaccumulator *Solanum nigrum* L. and their role in enhancing phytoremediation. *Environmental and Experimental Botany, 135*, 126–135.

28. Kodre, A., Arcon, I., Vogel-Mikuš, K., Debeljak, M., Potisek, M., & Likar, M., (2017). *Arbuscular Mycorrhizal* fungi alter hg root uptake and ligand environment as studied by x-ray absorption fine structure. *Environmental and Experimental Botany, 133*, 12–23.
29. Krupa, P., & Kozdrój, J., (2007). Ectomycorrhizal fungi and associated bacteria provide protection against heavy metals in inoculated pine (*Pinus sylvestris* L.) seedlings. *Water, Air and Soil Pollution, 182*, 83–90.
30. Leszczyszyn, O., Imam, H., & Blindauer, C., (2013). Diversity and distribution of plant metallothioneins: Review of structure, properties and functions. *Metallomics, 5*, 1146–1169.
31. Li, H., Li, X., Xiang, L., Zhao, H., Li, Y., & Cai, Q., (2018). Phytoremediation of soil co-contaminated with Cd and BDE-209 using hyperaccumulator enhanced by AM fungi and surfactant. *Science of Total Environment, 613–614*, 447–455.
32. Liwandowski, I., Schmidt, U., Londo, M., & Faaij, A., (2006). The economic value of the phytoremediation function assessed by the example of cadmium remediation by willow (*Salix ssp*). *Agricultural Systems, 89*, 68–89.
33. Lorestani, B., Cheraghi, M., & Yousefi, N., (2012). The potential of phytoremediation using hyperaccumulator plants: A case study at a lead-zinc mine site. *International Journal of Phytoremediation, 14,* 786–795.
34. Lu, H., Zhang, W., Yang, Y., Huang, X., Wang, S., & Qiu, R., (2012). Relative distribution of pb^{2+} sorption mechanisms by sludge-derived biochar. *Water Research, 46,* 854–862.
35. Luo, J., Cai, L., Qi, S., Wu, J., & Gu, X., (2017). Improvement effects of cytokinin on EDTA assisted phytoremediation and the associated environmental risks. *Chemosphere, 180*, 386–393.
36. Malekzadeh, E., Aliasgharzad, N., & Majidi, J., (2016). Contribution of glomalin to Pb sequestration by arbuscular mycorrhizal fungus in a sand culture system with clover plant. *Eurasian Journal of Soil Biology, 74*, 45–51.
37. Mani, D., Kumar, C., & Patel, N., (2016). Integrated micro-biochemical approach for phytoremediation of cadmium and lead-contaminated soils using *Gladiolus grandiflorus* L. flower. *Ecotoxicology and Environmental Safety, 124*, 435–446.
38. Merkl, N., Schultze-Kraft, R., & Infante, C., (2005). Phytoremediation in the tropics: Influence of heavy crude oil on root morphological characteristics of graminoids. *Environmental Pollution, 138*(1), 86–91.
39. Mignardi, S., Corami, A., & Ferrini, V., (2012). Evaluation of the effectiveness of phosphate treatment for the remediation of mine waste soils contaminated with Cd, Cu, Pb, and Zn. *Chemosphere, 86*, 354–360.
40. Mirsal, I., (2004). *Soil Pollution: Origin, Monitoring and Remediation* (p. 440). Berlin Heidelberg, Germany: Springer–Verlag.
41. Muhlbachov, A. G., (2011). Soil microbial activities and heavy metal mobility in long-term contaminated soils after addition of EDTA and EDDS. *Ecological Engineering, 37*, 1064–1071.
42. Nascimento, C., & Xing, B., (2006). Phytoextraction: A review on enhanced metal availability and plant accumulation. *Sciential Agricola, 63*(3), 299–311.
43. Padmavathiamma, P., & Li, Y., (2017). Phytoremediation technology: Hyper-accumulation of metals in plants. *Water, Air and Soil Pollution, 184*, 105–126.
44. Pagiola, S., Ritter, K., & Bishop, T., (2004). *How Much is an Ecosystem Worth? Assessing the Economic Value of Conservation* (p. 234). Washington, DC: The International Bank for Reconstruction.

45. Paz-Ferreiro, J., Lu, H., Fu, S., Mendez, A., & Gasco, G., (2014). Use of Phytoremediation and biochar to remediate heavy metal polluted soils: A review. *Solid Earth, 5*, 65–75.
46. Pierart, A., Dumat, C., Maes, A., & Sejalon-Delmas, N., (2018). Influence of *Arbuscular Mycorrhizal* fungi on antimony Phyto-uptake and compartmentation in vegetables cultivated in urban gardens. *Chemosphere, 191*, 272–279.
47. Pinto, A., De Varennes, A., Fonseca, R., & Martins, T. D., (2015). Phytoremediation of soils contaminated with heavy metals: Techniques and strategies. In: Ansari, A., (ed.), *Phytoremediation: Management of Environmental Contaminants* (pp. 133–155). Switzerland: Springer International Publishing.
48. Pinto, A., De Varennes, C., Dias, C., & Lopes, M., (2018). Microbial assisted phytoremediation: A convenient use of plant and microbes to clean up soils. In: Ansari, A., Gill, S., Gill, R., Lanza, G. R., & Newman, L., (eds.), *Phytoremediation* (pp. 213–237). Cham: Springer.
49. Pivetz, B., (2001). *Ground Water Issue: Phytoremediation of Contaminated Soil and Ground Water at Hazardous Waste Sites* (p. 36). Washington - DC: United States Environmental Protection Agency (US EPA).
50. Prasad, M., (2004). Phytoremediation of metals and radionuclides in the environment: The case for natural hyperaccumulators, metal transporters, soil-amending chelators and transgenic plants. In: Prasad, M., (ed.), *Heavy Metal Stress in Plants: From Biomolecules to Ecosystems* (pp. 345–391, 2nd edn.). Berlin: Springer.
51. Rafati, M., Khorasani, N., & Moattar, F., (2011). Phytoremediation potential of *Populus alba* and *Morus alba* for cadmium, chromium and nickel absorption from polluted soil. *International Journal of Environmental Research, 5*, 961–970.
52. Saifullah, S., Bibi, S., Ahmad, M., & Ok, Y. S., (2014). Effectiveness of zinc application to minimize cadmium toxicity and accumulation in wheat *Triticum aestivum* L. *Environment and Earth Sciences, 71*, 1663–1672.
53. Sakai, Y., Ma, Y., Xu, C., Wu, H., Zhu, W., & Yang, J., (2012). Phytodesalination of a salt-affected soil with four halophytes in China. *Journal of Arid Land Studies, 22*, 17–20.
54. Santos, E., Abreu, M., Varennes, A., Macías, F., Leitão, S., & Cerejeira, M., (2013). Evaluation of chemical parameters and ecotoxicity of a soil developed on gossan following application of polyacrylates and growth of *Spergularia purpurea*. *Science of Total Environment, 461–462*, 360–370.
55. Sarwar, N., Imran, M., & Shaheen, M., (2016). Phytoremediation strategies for soils contaminated with heavy metals: Modifications and future perspectives. *Chemosphere, 171*, 710–721.
56. Schneider, J., Bundschuh, J., & Nascimento, C., (2016). Arbuscular mycorrhizal fungi-assisted phytoremediation of a lead-contaminated site. *Science of the Total Environment, 572*, 86–97.
57. Seth, C., (2012). Review on mechanisms of plant tolerance and role of transgenic plants in environmental clean-up. *Botanical Review, 78*, 32–62.
58. Shabani, N., & Sayadi, M. H., (2012). Evaluation of heavy metals accumulation by two emergent macrophytes from the polluted soil: An experimental study. *Environmentalist, 32*, 91–98.
59. Singh, A., & Prasad, S. M., (2011). Reduction of heavy metal load in food chain: Technology assessment. *Reviews in Environmental Science and Biotechnology, 10*, 199–214.

60. Souza, L. A., & Piotto, F. A., (2013). Use of non-hyperaccumulator plant species for the phytoextraction of heavy metals using chelating agents. *Scientia Agricola, 70*, 290–295.
61. Tangahu, B., Abdullah, S., Basri, H., Idris, M., Anuar, N., & Mukhlisin, M., (2011). Review on heavy metals (As, Pb, and Hg) uptake by plants through phytoremediation. *International Journal of Chemical Engineering, 12*, 1–32.
62. Tu, L., Ma, A., Fayiga, O., & Zillioux, E. J., (2004). Phytoremediation of arsenic-contaminated groundwater by the arsenic hyperaccumulating fern *Pteris vittata* L. *International Journal of Phytoremediation, 6*(1), 35–47.
63. Van, G. L., Meers, E., & Guisson, R., (2007). Phytoremediation for heavy metal-contaminated soils combined with bioenergy production. *Journal of Environmental Engineering and Landscape Management, 15*(4), 227–236.
64. Vithanage, M., Dabrowska, B., Mukherjee, B., Sandhi, A., & Bhattacharya, P., (2012). Arsenic uptake by plants and possible phytoremediation applications: Brief overview. *Environmental Chemistry Letters, 10*, 217–224.
65. Wang, F., Lin, X., & Yin, R., (2007). Inoculation with *Arbuscular Mycorrhizal* fungus *Acaulospora mellea* decreases Cu phytoextraction by Maize from Cu-contaminated soil. *Pedobiologia, 51*, 99–109.
66. Wang, J., Feng, X., Anderson, C., Xing, Y., & Shang, L., (2012). Remediation of mercury-contaminated sites: Review. *Journal of Hazardous Material, 221–222*, 1–18.
67. Wei, Y., Hou, H., & Li, J., (2014). Molecular diversity of arbuscular mycorrhizal fungi associated with an mn hyperaccumulator, *Phytolacca americana*, in Mn mining area. *Applied Soil Ecology, 82*, 11–17.
68. Wu, G., Kang, H., Zhang, X., Shao, H., Chu, L., & Ruan, C., (2010). Critical review on the bio-removal of hazardous heavy metals from contaminated soils: Issues, progress, eco-environmental concerns and opportunities. *Journal of Hazardous Materials, 174*, 1–8.
69. Zorrig, W., Rabhi, M., Ferchichi, S., Smaoui, A., & Abdelly, C., (2012). Phytodesalination: A solution for salt-affected soils in arid and semi-arid regions. *Journal of Arid Land Studies, 22*, 299–302.
70. Zou, T., Li, T., Zhang, X., Yu, H., & Luo, H., (2011). Lead accumulation and tolerance characteristics of *Athyrium wardii* (Hook.) as a potential phytostabilizer. *Journal of Hazardous Material, 186*, 683–689.

CHAPTER 4

PHYTOREMEDIATION TECHNIQUE FOR PAH-POLLUTED SOILS: APPLICATIONS, CHALLENGES, AND OPPORTUNITIES

ARUNIMA NAYAK, BRIJ BHUSHAN, and SHREYA KOTNALA

ABSTRACT

Soil contamination by polycyclic aromatic hydrocarbons (PAH) is carcinogenic, recalcitrant, and can have detrimental biological and environmental effects. In this context, phytoremediation has emerged as a cost-effective, green bioremediation technology that uses plants and microorganisms to remediate PAH-contaminated soils. This chapter introduces basic concepts, mechanisms, and underlying factors; and addresses: (i) research status in the remediation of soil contaminated sites; (ii) hurdles during large-scale usage of phytoremediation; (iii) restoration of damaged ecosystems; and (iv) future prospects.

4.1 INTRODUCTION

Polycyclic aromatic hydrocarbons (PAHs) having four to seven fused aromatic rings (chrysene, coronenes, etc.), have been identified as hazardous micropollutants by the US-EPA [17, 60], due to: (i) their toxic nature; (ii) tendency to bio-accumulate; (iii) resistance towards biodegradation; (iv) carcinogenic nature; and (v) persistent organic nature [15, 31, 32]. Although PAH are universally present in the environment, but on a global level, approximately 90% are found in soil [59]. Due to rapid industrialization in the past decade, the concentration of PAHs in the environment has been increasing at an alarming rate. Because of their toxicity, the PAHs have gained significant attention in the scientific and commercial communities.

The common sources of PAHs are either natural (such as forest fires, volcanic eruptions, oil spills, etc.), or anthropogenic (activities such as burning of fossil fuels, incineration of municipal solid wastes (MSW), processing of petroleum products, etc.) [22]. Various technologies (such as (i) incineration and thermal desorption; (ii) solvent extraction/soil washing; (iii) chemical oxidation; and (iv) bioremediation) have been used for the remediation of PAH contaminated soils to minimize the potential risks associated with its use.

Remediation using biological methods have gained importance because of their environmental friendliness and cost-effectiveness, according to Kuppusamy et al. [33–35]. The process of bioremediation of PAH contaminated soils includes both *in-situ* and *ex-situ* methods. While, land farming, biostimulation, bioaugmentation, composting, and phytoremediation are identified as *in-situ* methods, the *ex-situ* treatment involves the use of bioreactors [33–35]. Phytoremediation is an *in-situ* process that involves the use of plants for gradual and safe decontamination of PAH-contaminated soils. Plants having deep, long penetrating and large surface area roots require low maintenance costs and are generally preferred [1].

As compared to other treatment options, phytoremediation has been employed very effectively in the treatment of large areas of PAHs contaminated soils. Studies have shown that various species of grass and legumes together can enhance the removal of PAHs. Research studies by Reilley [47] have indicated that soils planted with legume alfalfa and grasses had significantly lower incidences of PAHs. This led to an enhancement of 30–40% degradation as compared to unplanted soils. In another study by Aprill and Sims [2], degradation of four PAHs was found to be higher in soils planted with grasses as compared to unplanted soils. Such studies illustrate the importance of phytoremediation in enhancing the degradation of PAHs.

Various strategies have been used to enhance the rate of phytoremediation, e.g., via the use of plant growth-promoting rhizobacteria. Studies have shown that bacteria are known to enhance the growth and development of roots of plants to improve phytoremediation [24, 25].

The aim of this chapter is to provide: (i) an up-to-date review on various aspects of phytoremediation method; (ii) technological background behind the technique; (iii) advantages, disadvantages, and challenges; and (iv) assessment of process parameters that affect the efficiency of this method during reduction of PAHs from the polluted soil.

4.2 MECHANISMS FOR PHYTOREMEDIATION OF PAH POLLUTED SOIL

Phytoremediation has served as a very effective, cheap, and green method for the degradation of soil contaminated with various toxic pollutants, such as PAHs [64]. The concept of using plants for the decontamination of soil was recognized as a mature technology in 1991 [12, 46]. The method of decontamination of soil via degradation of the toxic contaminants varies among different plant species. Based on the mechanism adopted to decontaminate or degrade the PAHs, the phytoremediation technique has been classified into phytoextraction, phytovolatilization, phytodegradation, and rhizodegradation.

The process of phytoextraction via absorption of pollutants from soil is best exhibited by high biomass producing plant species, such as *Brassica sp.*, *Arundo donax* and *Typha spp.* The roots of such plants in the rhizosphere region extract the contaminants from the soil. Such contaminants are then transported to stems and leaves and get accumulated there. The biomass is later harvested for their safe disposal in the form of incineration and/or landfilling. Factors that accelerate such mechanism are: microbial activities thriving in the rhizosphere region, physical, and chemical properties of the contaminant, and that of the soil.

During the process of phytovolatilization, the organic-based contaminants (such, PAHs) are absorbed by the roots, get transported to the roots and shoots via xylem and finally get released into the environment via transpiration in their modified lesser toxic forms [20]. The disadvantage is that there is a possibility that the modified form of the pollutant may be toxic and may resettle in the environment after its release from the plant.

Phytodegradation process involves the uptake of PAHs by the roots of plant species followed by their degradation via enzymes that are involved in the metabolic activities of the plant [7, 52]. Such enzymes are: dehalogenase, peroxidase, nitroreductase, nitrilase, and phosphatase [6, 61]. Each enzyme has been identified to be specific for a particular degradation activity.

The process of rhizodegradation involves the degradation of PAHs in the rhizosphere zone via synergistic activity with different microbial classes, such as yeast, bacteria, fungi, etc. [14, 53]. The microbial concentration in the rhizosphere is usually 10- to 1,000-folds greater than that in the bulk soil. Oxygen released from the roots to the microbial community near the rhizosphere region along with enzyme exudates can enhance the rhizodegradation process as well as can help in the uptake by the plants via phytoextraction.

The size and intensity of root propagation in the rhizosphere region have a positive influence on this degradation mechanism. Because of the tendency of the soil contaminant to get mineralized, rhizodegradation favors the improvement of soil characteristics.

4.3 FACTORS AFFECTING PHYTOREMEDIATION OF PAHs

4.3.1 CHARACTERISTICS OF PAH

The physical and chemical properties of PAHs that can influence its degradation are: molecular weight, solubility in aqueous medium, lipophilicity, volatilizing capacity, availability of functional groups, etc. [4]. Low molecular weight (LMW) PAHs (e.g., acenaphthene, and phenanthrene) when present in soil are more mobile; hence are easily available for uptake by plants and can undergo rapid phytoremediation [8]. Whereas, high molecular weight (HMW) PAHs having a greater number of fused benzene rings are more recalcitrant. Studies have shown that stable PAHs, whose origin is due to decades of industrial activity, demonstrated better sorption to soil organic carbon and thus exhibited lower bioavailability for uptake via plant roots. This is in comparison to PAHs added experimentally to a specific study site [27].

According to Juhasz and Naidu [28], PAH hydrophobicity (K_{ow}: octanol-water partitioning constant) has a direct relationship with organic matter partitioning and hence on its bioavailability and degradation. Consequently, such PAHs will have a greater tendency to exhibit minimum phytoremediation. Various other studies have demonstrated that degradation, transport, and sequestration of PAHs are dependent on their bioavailability in the environment [3, 51], which in turn is dependent on their water solubility, dissolution rate and adsorption capacity onto the soil matrix. Biodegradation half-life values of different PAHs [58] have been found to be linked to the molecular weights of PAH. PAHs having more than four fused rings have a longer half-life of biodegradation as compared to their LMW counterparts.

4.3.2 PHYSICOCHEMICAL CHARACTERISTICS OF SOIL

The degree of degradation of PAH in soil is dependent on soil properties, such as pH, organic matter content, and clay content. In soils having a higher concentration of organic matter, there is increased competitive binding of

PAHs for plant lipids and for soil organic carbon leading to lesser bioavailability for plants. Consequently, the tendency for the degradation of such PAH decreases [2, 13]. In a study by Briggs et al. [9] to test the uptake potential by plants cultivated in hydroponic conditions, organic molecules having a high log K_{ow} could be bounded more strongly to roots. Gauthier et al. [19] stated that the degree of aromaticity in organic matter increases with time, which in turn can result in greater sorption of hydrophobic PAHs.

Karickhoff et al. [29] reported that pyrene showed greater adsorption onto soil with increased soil clay content. A similar study showed that the binding potential of anthracene was increased by four times compared to a 2-fold increase of fluoranthene, benz[a]anthracene, and benzo[a]pyrene, when the soil composition consisted of clayey marsh sediment as compared to sandy soil [18]. The contribution of soil to adsorb and finally degrade PAHs thus depends on its physical and chemical characteristics of soil, which ultimately is dependent on the geographical location and geology.

In a study by Tucker et al. [56], soils with different compositions (bentonite and kaolinite) demonstrated different rates of degradation of pyrene. In this study, the greater amount of pyrene got adsorbed to bentonite as compared to that for kaolinite; and this resulted in its low bioavailability. Transfer of PAH from soil to plants is dependent on soil moisture content [2]. High moisture content with low organic matter content of soil has favored the uptake of PAHs by plants. In a study by Beck et al. [8], PAHs are strongly bounded by soil, thus retarding uptake by roots in soils having low moisture content but with high organic matter content.

4.3.3 ROOTS

Roots constitute the contact point between the contaminants and the soil and hence serve as a nodal role in phytoremediation [43, 62]. The role of roots is multi-functional, ranging from providing a natural habitat for the microbial community in the rhizosphere region, modifying, and strengthening soil composition, regulating plant growth, absorbing water and nutrients, providing mechanical support for the plants, and finally contributing towards sustainable environment during stress. Thus, maximization of root-soil contact can promote the phytoremediation of PAHs in contaminated soils. Longer root lengths and higher root density in grasses allow the maximum contact with the PAHs in the soil, thus accelerating the phytoremediation of PAHs. The high rate of growth of grasses is highly favored for bringing about higher degradation rates for PAHs in surface soil. For PAHs at greater depth below the topsoil, deep penetrating roots (as in the case of poplar) are preferred.

4.3.4 PLANT SPECIES

Greater efficiency of phytoremediation can be achieved via judicious selection of a plant species [20, 37, 40]. The first important criteria for the selection are the ability of such plants to survive the climatic conditions at the contaminated site. The other criteria are that plants should have tolerance to the contaminant concentration, high biomass, and long and high-density roots for ensuring maximum soil-contaminant interaction.

Selection of native plants is desirable since such plants have greater adaptability to both climate and soil conditions, especially at the contaminated site. Besides, native plants can facilitate restoration of habitat after the phytoremediation. The ecological risks associated with planting a non-native plant to an ecosystem are thereby eliminated.

4.4 APPLICATIONS OF PHYTOREMEDIATION FOR REMOVAL OF PAHS FROM CONTAMINATED SOIL

Plants are known to influence the fate of PAHs in soil by inducing phytoremediation via different mechanisms through their varied root parameters (such as morphology, root exudates) and associated microbial communities. Nutrients and contaminants are adsorbed via roots from the soil. Different enzymes are secreted from the plants that act as surfactants for enabling the bioavailability of the contaminants. Plants have the capacity to store the contaminants and finally sequester them. Via their metabolic pathway, plants have the capacity to degrade the contaminants to less toxic forms [39, 57].

Lastly, plants along with their interaction with the microbial community near the rhizosphere have brought about the degradation of recalcitrant organic contaminants [45]. A detailed literature study carried out on phytoremediation of PAH contaminated soils reveals that the plants for phytoremediation fall under three broad categories, such as grass, field crops, and trees.

Various research studies have shown that grasses and annual herbs have been used extensively for the treatment of PAH-contaminated sites. The reason for the selection of such species for phytoremediation was because of the fibrous root system, which could ensure greater plant-soil interaction.

Also, grasses have the capacity to adapt to different climatic and soil conditions. Grasses also require less care and hence are favorable candidates for phytoremediation.

Tall fescue (*Festuca arundinacea*), ryegrass (*Lolium multiflorum*) and clover (*Melilotus officinalis*) were used to evaluate the effectiveness of phytoremediation of a contaminated site with total PAH concentration of 400 to 5,000 mg/kg [45]. The soil was pretreated via biological methods of composting and the addition of biosolids. Plant growth recorded after 12 months showed a significant reduction in the concentrations of several PAH compounds in the soil. The study showed that an elongated time and presence of well fibrous roots contributed to better efficiency of phytoremediation.

Rezek et al. [48] conducted a study on the contaminated soil having 15 PAHs, on which ryegrass (*Lolium perenne*) was cultivated for 18 months in a greenhouse environment. Results revealed that the concentrations of the studied PAHs were decreased by 50%. Maximum degradation took place in fluoranthene and pyrene, which were major contaminants in original soil. The authors concluded that aging of the soil could have been the contributing factor for the phytodegradation. Proper selection of plants is one of the necessary conditions to ensure high PAH degradation and to ensure a stable vegetation cover at the contaminated site.

A test was conducted by Smith et al. [54] to assess the relevance of germination trials on the growth of 7 types of grass and legume species that were cultivated in 3 different types of contaminated soils. Soil spiked separately with PAH, coal tar and finally soil extracted from a coking plant site (found to be contaminated with aged PAHs) were selected for the study. Spiked coking plant-soil demonstrated poor growth characteristics in all plant species. Four out of seven and 3 out of seven exhibited stunted growth in the coal tar spiked soil and in the soil spiked with a mixture of 7 PAHs. Toxicity due to PAHs brought about reduced vegetation in the contaminated site. Tall fescue (*Festuca arundinacea*) and switchgrass (*Panicum virgatum*) were cultivated on PAH-contaminated soil.

The study conducted by Cofield et al. [11] revealed that approximately 40% of the PAHs were removed. The study highlighted the fact that there is a probability for soils contaminated with high concentration of PAHs that residual toxicity may exist. In yet another study [37], higher degradation rates (>99% and 77–94%) for two studied PAHs (phenanthrene and pyrene) were obtained by using native grass and legume families in an 80-day experimental trial. Results revealed that legumes had higher pyrene removal efficiency as compared to that for grasses. No difference was observed in their performance for the removal of phenanthrene. Literature study has shown less usage of trees with regard to PAH phytodegradation. Trees have added features like extended and penetrating root systems, which are desirable for their use in phytoremediation. In one study, red mulberry trees were used to enhance the degradation of PAH in an industrial sludge basin [44].

Mueller and Shann [42] planted native trees in soil that was artificially contaminated with three PAHs (anthracene, phenanthrene, and pyrene). In one year, dissipation of PAHs was observed extensively. The primary reason cited for enhanced PAH loss from soil was microbial degradation due to higher bioavailability and extractability that was observed throughout soil aging experiments. HMW PAH (such as pyrene) undergoes minimum degradation in soil.

In a study by Liste and Alexander [38], nine plant species were investigated to assess their capability to degrade pyrene spiked soil. The plants under study were field crops, horticultural crops, and trees. Within 8 weeks of experimental observation, it was found that almost all plant species caused increased pyrene degradation. Vegetated soil caused approximately 74% of the pyrene degradation as compared to 40% in unplanted soil. In 28 days, pepper, and radish resulted in the fastest pyrene degradation. The results have enlightened the necessity of conducting a preliminary plant screening test for a specific pollutant.

Cultivation of contaminated soil with mixed plants showed better PAH degradation performance as evident from a study conducted by Xu et al. [63]. The maize, ryegrass, and white clover was used separately and in combination to assess the degradation of phenanthrene and pyrene in contaminated soil. At the end of 60 days, degradation extents brought about from maize planted soils was to the tune of 92.10% and 88.36% for phenanthrene and pyrene, respectively. In contrast, combined plant cultivation (maize and ryegrass) showed enhanced degradation of phenanthrene (98.2%) and pyrene (95.8%) in soils.

Alfalfa was cultivated in a soil that was spiked with benzo[a]pyrene by Hamdi et al. [21]. After six months, PAH reduction was approximately 64%. In addition to PAH degradation, there was significant improvement in the soil characteristics. Limited period of treatment and the choice of plant species may have resulted in the presence of residual toxicity in the studied soil site (Table 4.1).

TABLE 4.1 Studies on Phytoremediation of Soils Contaminated with PAH

Plant Type	Soil Type	PAH Study	Results	References
Field crops, horticultural crops, and trees	PAH spiked soil	Pyrene	After 8 weeks all studied plant enhanced pyrene degradation. Planted soil brought about 74% degradation as compared to unplanted soil (40%) pepper and radish resulted in fastest pyrene degradation.	[38]
Grass	Manufactured gas plant site in Bedford, Indiana, and further pretreated via composting and biosolids addition	Total PAH conc. (400–5,000 mg/kg); high molecular weight PAHs (5- and 6-rings)	All plants reduced concentrations of several PAHs.	[45]
Grass	Aged contaminated soil	15 high molecular weight PAHs	Concentrations of the studied PAHs decreased by 50%. Maximum degradation took place in fluoranthene and pyrene.	[48]
Grass	PAH contaminated soil	16 PAHs at conc. of 100–500 mg/kg	Approximately 40% of the PAHs were removed. Soil had residual toxicity.	[11]
Grass (maize, ryegrass and clover)	Artificial creosote-contaminated soil	Phenanthrene and pyrene	Soil planted with maize: removal of phenanthrene (92.1%) and pyrene (88.36%) soil planted with maize and ryegrass: removal of phenanthrene (98.2%) and pyrene (95.8%).	[63]
Grasses and legumes	PAH spiked soil	Phenanthrene, pyrene	Higher degradation rates (>99% and 77–94%) for phenanthrene and pyrene were obtained in an 80-day experimental trial.	[37]

TABLE 4.1 *(Continued)*

Plant Type	Soil Type	PAH Study	Results	References
Grasses, legumes	Mixed PAH spiked soil, coal tar spiked soil and soil extracted from coking plant site that was contaminated with aged PAHs	7 high molecular weight and aged PAHs	Germination was ok in all soil types, but all plant species demonstrated reduced growth in the coking plant soil. Four out of seven and 3 out of seven exhibited stunted growth in the coal tar spiked soil and in the soil spiked with a mixture of 7 PAHs.	[54]
Legumes	Soil spiked with aged PAH	Benzo[a]pyrene	64% PAH removal with improved soil characteristics.	[21]
Trees	PAH in an industrial sludge basin	Anthracene, phenanthrene, pyrene	Rapid and extensive loss of PAHs was observed after one year of growth.	[42]

4.5 STRATEGIES FOR ENHANCED PHYTOREMEDIATION

Phytoremediation has many advantages over other technologies, some of which have been experimentally documented as evident from literature studies. Studies have shown that the growth of plant species can get hindered by the presence of PAHs [55, 65]. The growth of such plants is very slow, and it becomes difficult to achieve sufficient biomass to bring a significant remediation of soil. Also, in some contaminated sites, soil microorganisms are low, which can neither facilitate plant growth nor favor contaminant degradation [26]. Plant growth and phytoremediation efficiency can be enhanced by the external addition of rhizobacteria (bioaugmentation) and by the use of plant growth regulators [50].

4.5.1 USE OF MICROBES

It is the symbiotic relationship among plants, rhizospheric microorganisms and the environment that cause enhanced phytoremediation efficiency [50]. It is basically the bacterial population in the soil rhizosphere that has the capacity to degrade PAH into simpler molecules via their secreting enzymes. However, the rate of degradation is very slow. Bioaugmentation is the process of adding degrading microbes or basically bacteria to the contaminated site to augment the rate of phytoremediation [30].

Various studies have assessed the role of bioaugmentation to enhance the process of phytoremediation of PAH contaminated soils. According to Eskandary et al. [16], two rhizospheric bacterial species were inoculated onto *Festuca arundinacea* seeds that were planted in a PAH contaminated site. The results showed that the inoculated plant sample demonstrated higher biomass as compared to the untreated counterparts. The PAH concentration in the contaminated site was decreased significantly as a result of the combined action of plant and bacterial inoculation.

Effect of a bacterial consortium was studied on the phytoremediation capacity of two plant species (*Sorghum* and *Onobrychis sativa*) for dissipation of two PAHs (phenanthrene and pyrene) from contaminated soil [5]. The dissipation efficiency was tested under three conditions: natural, phytoremediation, and bioaugmentation, separately, and combined. After 120 days of treatment, both plant species exhibited good removal efficiency for the studied PAH. Combination of phytoremediation with bioaugmentation resulted in enhanced removal efficiency of 22 and 16%, respectively, for pyrene and phenanthrene from contaminated soil.

4.5.2 PLANT GROWTH REGULATORS

The plant growth regulators [10, 23] (such as auxins, cytokines, gibberellins, etc.), are known to increase the plant biomass and decrease the negative effects due to the presence of PAH in soil. The efficiency of their activity depends on their usage concentration, environmental factors that can affect their absorption and physiological state of the plant [50].

Rostami et al. [49] evaluated the effects of Indole Acetic Acid-IAA and externally added microorganism (*Pseudomonas aeruginosa*) on the phytoremediation efficiency of *sorghum bicolor* for pyrene removal from the soil. Soil treatments spiked with pyrene were: (T0) unplanted soil, (T1) soil planted with sorghum, (T2) soil planted with sorghum and application of IAA, (T3) soil planted with sorghum and application of *Pseudomonas sp.*, and (T4) soil planted with sorghum and simultaneous application of IAA and *Pseudomonas sp.* Observations after 90 days of treatment showed that planted soils demonstrated higher pyrene removal rates as compared to unplanted soil. Highest pyrene removal rate (92%) was observed in the planted soil (T4) with simultaneous application of IAA and *Pseudomonas sp.* Plant biomass, bacterial population in the rhizosphere and pyrene removal were increased in T2, T3 and T4 as compared to that in T1. The results therefore demonstrated the positive influence of the addition of plant growth regulator and bioaugmentation on the phytoremediation efficiency of *Sorghum* for removal of pyrene from soil.

4.6 CHALLENGES AND OPPORTUNITIES

The growing acceptance of phytoremediation as a technology for the treatment of PAH contaminated soil is mainly because of its cost-effectiveness, easy application, and eco-friendliness. The inherent characteristics of using plants and soil microorganisms for reducing soil contaminants with simultaneous restoration of ecosystem have made the technology green and sustainable. The technology has the potential to dissipate not only PAH but also inorganic metal ions via diverse mechanisms, such as phytodegradation, phytoaccumulation, phytoextraction, phytotransformation, and phytovolatilization.

In fact, in some instances, depending on the severity of contaminated site, multiple plant cycles are required. Enhancing the rate of phytoremediation and growth of plant biomass can be achieved via the addition of soil amendments, plant growth regulators or via bioaugmentation.

Another important challenge is that phytoremediation of PAH is dependent on abiotic factors, such as soil properties, local climatic conditions, and availability of nutrients. Highly toxic contaminants, such as HMW PAHs and highly contaminated site can put a risk to the plant growth and the microbial community. Also, certain PAHs being hydrophobic are less bioavailable for phytoremediation. Under such conditions, the application of a single process for dissipating PAH from the soil may be difficult and time-consuming.

In a study conducted by Huang et al. [24], a phytoremediation system comprising of volatilization, photooxidation, microbial remediation and phytoremediation processes were adopted for the removal of 16-toxic PAHs from the contaminated soil. Approximately 78% removal efficiency was demonstrated, which was 23% more than the phytoremediation treatment alone. Another significant achievement was the capability of the multi-process systems to remove the higher molecular weight PAHs (especially the 4- and 5-ringed PAHs).

4.7 SUMMARY

The use of plants for degradation of PAH with simultaneous restoration of ecosystem is a sustainable and green approach. Real time contaminated soil has PAH, many organic and inorganic contaminants that have diverse physicochemical characteristics. Many plant species (grasses, legumes, and few trees) having favorable physiologies need to be identified and assessed for their phytoremediation capacities. Studies undertaken on a laboratory scale need to focus on optimization of the process parameters during the uptake of contaminants by the plants so that there is consistency of the experimental results with real time applications. Lastly, more studies are required to assess the optimal use of plant growth regulators and bioaugmentation processes to enhance the phytoremediation effects.

KEYWORDS

- **Arundo donax**
- **bioaugmentation**
- **phytoremediation**
- **plant growth regulators**
- **polycyclic aromatic hydrocarbon (PAH)**
- **soil contamination**

REFERENCES

1. Alagic, S. C., & Jovanovic, V. P. S., (2016). Bioaccumulation of HMW PAHs in roots of *blackberry* from Serbia: Phytoremediation and biomonitoring aspects. *Science of the Total Environment, 562*, 561–570.
2. Aprill, W., & Sims, R. C., (1990). Evaluation of the use of *prairie* grasses for stimulating polycyclic aromatic hydrocarbon treatment in soil. *Chemosphere, 20*(1, 2), 253–265.
3. Aronstein, B. N., & Alexander, M., (1992). Surfactants at low concentrations stimulate biodegradation of sorbed hydrocarbons in samples of aquifer sands and soil slurries. *Environmental Toxicology Chemistry, 11*(9), 1227–1233.
4. Atlas, R. M., (1981). Microbial degradation of petroleum hydrocarbons: An environmental perspective. *Microbiological Reviews, 45*(1), 180–209.
5. Baneshi, M. M., Kalantary, R. R., & Jafari, A. J., (2014). Effect of bioaugmentation to enhance phytoremediation for removal of phenanthrene and pyrene from soil

with *Sorghum* and *Onobrychis Sativa*. *Journal of Environmental Health Sciences & Engineering, 12*, 24–28.

6. Bamforth, S. M., & Singleton, I., (2005). Bioremediation of polycyclic aromatic hydrocarbons: Current knowledge and future directions: Review. *Journal of Chemical Technology & Biotechnology, 80*, 723–736.
7. Barceló, J., & Poschenrieder, C., (2003). Phytoremediation: Principles and perspectives. *Contributions to Science, Institut d'Estudis Catalans, Barcelona, 2*(3), 333–344.
8. Beck, A. J., Johnson, D. L., & Jones, K. C., (1996). The form and bioavailability of non-ionic organic chemicals in sewage sludge-amended agricultural soils. *Science of the Total Environment, 185*(1–3), 125–149.
9. Briggs, G. G., Bromilow, R. H., & Evans, A. A., (1983). Relationships between lipophilicity and the distribution of non-ionized chemicals in barley shoots following uptake by the roots. *Pesticide Science, 14*, 492–500.
10. Cassina, L., Tassi, E., & Morelli, E., (2011). Exogenous cytokinin treatments of an Ni hyperaccumulator, *Alyssum Murale*, grown in a serpentine soil: Implications for phytoextraction. *International Journal of Phytoremediation, 13*, 90–101.
11. Cofield, N., Banks, M. K., & Schwab, A. P., (2008). Lability of polycyclic aromatic hydrocarbons in the rhizosphere. *Chemosphere, 70*, 1644–1652.
12. Cristaldi, A., Conti, G. O., & Jho, E. H., (2017). Phytoremediation of contaminated soils by heavy metals and PAHs: Brief review. *Environmental Technology & Innovation, 8*, 309–326.
13. Christman, R. F., & Pfaender, F. K., (2006). Molecular implications of hydrophobic organic partitioning theory. *Acta Hydrochimica et Hydrobiologica, 34*, 367–374.
14. Cunningham, S. D., & Ow, D. W., (1996). Promises and prospects for phytoremediation. *Plant Physiology, 110*, 715–719.
15. Duan, L., Naidu, R., Thavamani, P., Meaklim, J., & Megharaj, M., (2015). Managing long term polycyclic aromatic hydrocarbon contaminated soils: A risk-based approach. *Environmental Science and Pollution Research, 22*, 8927–8941.
16. Eskandary, S., Tahmourespour, A., Hoodaji, M., & Abdollahi, A., (2017). The synergistic use of plant and isolated bacteria to clean up polycyclic aromatic hydrocarbons from contaminated soil. *Journal of Environmental Health Science and Engineering, 15*, 12–17.
17. Gan, S., Lau, E. V., & Ng, H. K., (2009). Remediation of soils contaminated with polycyclic aromatic hydrocarbons (PAHs). *Journal of Hazardous Materials, 172*, 532–549.
18. Gardner, W. S., Lee, R. F., Tenore, K. R., & Smith, L. W., (1979). Degradation of selected polycyclic aromatic hydrocarbons in coastal sediments: Importance of microbes and polychaete worms. *Water, Air, & Soil Pollution, 11*, 339–347.
19. Gauthier, T. D., Siitz, W. R., & Grant, C. L., (1987). Effects of structural and compositional variations of dissolved humic materials on pyrene K_{oc} values. *Environmental Science & Technology, 21*, 243–248.
20. Gerhardt, K. E., Huang, X. D., Glick, B. R., & Greenberg, B. M., (2009). Phytoremediation and rhizoremediation of organic soil contaminants: Potential and challenges. *Plant Science, 176*, 20–30.
21. Hamdi, H., Benzarti, S., Aoyama, I., & Jedidi, N., (2012). Rehabilitation of degraded soils containing aged PAHs based on phytoremediation with alfalfa (*Medicago Sativa L.*). *International Biodeterioration & Biodegradation, 67*, 40–47.

22. Haritash, A. K., & Kaushik, C. P., (2009). Biodegradation aspects of polycyclic aromatic hydrocarbons (PAHs): A review. *Journal of Hazardous Materials, 169*, 1–15.
23. Hopkins, W. G., (2008). *Introduction to Plant Physiology* (4th edn., p. 528).; New York: John Wiley & Sons, Inc.
24. Huang, X. D., El-Alawi, Y., & Penrose, D. M., (2004). Multi-process phytoremediation system for removal of polycyclic aromatic hydrocarbons from contaminated soils. *Environmental Pollution, 130*, 465–476.
25. Huang, X. D., El-Alawi, Y., & Penrose, D. M., (2004). Responses of three grass species to creosote during phytoremediation. *Environmental Pollution, 130*, 453–463.
26. Huang, X. D., Glick, B. R., & Greenberg, M. B., (2001). combining remediation technologies increases kinetics for the removal of persistent organic contaminants from soil. In: Greenberg, B. M., (ed.), *Environmental Toxicology and Risk Assessment* (pp. 271–278). ASTM.
27. Joner, E. J., Leyval, C., & Van, C. J., (2006). *Ectomycorrhizas* impede phytoremediation of polycyclic aromatic hydrocarbons (PAHs) both within and beyond the rhizosphere. *Environmental Pollution, 142*, 34–38.
28. Juhasz, A. L., & Naidu, R., (2000). Bioremediation of high molecular weight polycyclic aromatic hydrocarbons: A review of the microbial degradation of benzo[a] pyrene. *International Biodeterioration & Biodegradation, 45*, 57–88.
29. Karickhoff, S. W., Brown, D. S., & Scott, T. A., (1979). Sorption of hydrophobic pollutants on natural sediments. *Water Research, 13*, 241–248.
30. Kuiper, E. L., Lagendijk, G. V., & Lugtenberg, B., (2004). Rhizoremediation: A beneficial plant-microbe interaction. *Molecular Plant-Microbe Interactions, 17*, 6–15.
31. Kuppusamy, S., & Thavamani, P., (2016). Bioaugmentation with novel microbial formula vs. natural attenuation of a long-term mixed contaminated solid treatability studies in solid-and slurry-phase microcosms. *Water Air Soil Pollution, 227*, 1–15.
32. Kuppusamy, S., Thavamani, P., Megharaj, M., & Naidu, R., (2016). Biodegradation of polycyclic aromatic hydrocarbons (PAHs) by novel bacterial consortia tolerant to diverse physical settings-assessments in liquid-and slurry-phase systems. *International Biodeterioration & Biodegradation, 108*, 149–157.
33. Kuppusamy, S., Thavamani, P., & Venkateswarlu, K., (2016). Remediation approaches for polycyclic aromatic hydrocarbons (PAHs) contaminated soils: Technological constraints, emerging trends and future directions. *Chemosphere, 168*, 944–968.
34. Kuppusamy, S., Palanisami, T., & Megharaj, M., (2016). *In-Situ* remediation approaches for the management of contaminated sites: Comprehensive overview. *Reviews of Environmental Contamination and Toxicology, 236*, 1–115.
35. Kuppusamy, S., Palanisami, T., Megharaj, M., Venkateswarlu, K., & Naidu, R., (2016). *Ex-situ* remediation technologies for environmental pollutants: A critical perspective. *Reviews of Environmental Contamination and Toxicology, 236*, 117–192.
36. Lee, J. H., (2013). Overview of phytoremediation as a potentially promising technology for environmental pollution control. *Biotechnology and Bioprocess Engineering, 18*, 431–439.
37. Lee, S. H., Lee, W. S., Lee, C. H., & Kim, J. G., (2008). Degradation of phenanthrene and pyrene in rhizosphere of grasses and legumes. *Journal of Hazardous Materials, 153*, 892–898.
38. Liste, H. H., & Alexander, M., (2000). Plant-promoted pyrene degradation in soil. *Chemosphere, 40*, 7–10.

39. Meagher, R. B., (2000). Phytoremediation of toxic elemental and organic pollutants. *Current Opinion in Plant Biology, 3*, 153–162.
40. Mench, M., Lepp, N., & Bert, V., (2010). Successes and limitations of phytotechnologies at field scale: Outcomes, assessment and outlook from COST action 859. *Journal of Soils and Sediments, 10*, 1039–1070.
41. McDonald, J. P., (1995). *Proceedings of 10th Annual Conference on Hazardous Waste Research* (pp. 325–332). Kansas State University, Manhattan, Kansas, USA.
42. Mueller, K. E., & Shann, J. R., (2006). PAH dissipation in spiked soil: Impacts of bioavailability, microbial activity and trees. *Chemosphere, 64*, 1006–1014.
43. Neergaard, E., Lyshede, O. B., Gahoonia, T. S., Care, D., & Hooker, J. E., (2000). Anatomy and histology of roots and root-soil boundary. In: Smit, A. L., Bengough, A. G., Engels, C., Van, N. M., & Pellerin, S., & Van De, G. S. C., (eds.), *Root Methods. A Handbook* (pp. 33–73). Springer-Verlag: Berlin/ Heidelberg.
44. Olson, P. E., Fletcher, J. S., & Philp, P. R., (2001). Natural attenuation/ phytoremediation in the vadose zone of a former industrial sludge basin. *Environmental Science and Pollution Research, 8*, 243–249.
45. Parrish, Z. D., Banks, M. K., & Schwab, A. P., (2005). Assessment of contaminant liability during phytoremediation of polycyclic aromatic hydrocarbon impacted soil. *Environmental Pollution, 137*(2), 187–197.
46. Prabakarana, K., Li, J., Anand, K. A., Leng, Z., Zou, C. B., & Du, D., (2019). Managing environmental contamination through phytoremediation by invasive plants: A review. *Ecological Engineering, 138*, 28–37.
47. Reilley, K. A., Banks, M. K., & Schwab, A. P., (1996). Organic chemicals in the environment: Dissipation of polycyclic aromatic hydrocarbons in the rhizosphere. *Journal of Environmental Quality, 25*, 212–219.
48. Rezek, J., Wiesche, C., Mackova, M., Zadrazil, F., & Macek, T., (2008). The effect of *Ryegrass* (*Lolium Perenne)* on decrease of PAH content in long-term contaminated soil. *Chemosphere, 70*, 1603–1608.
49. Rostami, S., Azhdarpoor, A., & Rostami, M., (2016). Effects of simultaneous application of plant growth regulators and bioaugmentation on improvement of phytoremediation of pyrene contaminated soils. *Chemosphere, 161*, 219–223.
50. Rostami, S., & Azhdarpoor, A., (2019). The application of plant growth regulators to improve phytoremediation of contaminated soils: Review. *Chemosphere, 220*, 818–827.
51. Semple, K. T., Doick, K. J., & Jones, K. C., (2004). Defining bioavailability and bioaccessibility of contaminated soil and sediment is complicated. *Environmental Science & Technology, 38*(12), 228A–231A.
52. Sharma, P., & Pandey, S., (2014). Status of phytoremediation in world scenario. *International Journal of Environmental Bioremediation & Biodegradation, 2*(4), 178–191.
53. Shimp, J. F., Tracy, J. C., & Davis, L. C., (1993). Beneficial effects of plants in the remediation of soil and groundwater contaminated with organic materials. *Critical Reviews in Environmental Science and Technology, 23*, 41–77.
54. Smith, M. J., Flowers, T. H., Duncan, H. J., & Alder, J., (2006). Effects of polycyclic aromatic hydrocarbons on germination and subsequent growth of grasses and legumes in freshly contaminated soil and soil with aged PAHs residues. *Environmental Pollution, 141*(3), 519–525.

55. Suthersan, S. S., (2002). *Natural and Enhanced Remediation Systems* (p. 239). Arcadis/ Lewis Publishers, London.
56. Tucker, B., McLean, J. E., Brar, J. S., & Anderson, A. J., (1995). Soil composition and suppression of pyrene mineralization by *Phanerochate chrysosporium*. In: Erickson, L. E., Tillison, D. E., Grant, S. C., & McDonald, J. P., (eds.), *Proceedings of 10th Conference on Hazardous Waste Research* (pp. 325–332). Kansas State University, Manhattan, Kansas, USA.
57. Vidali, M., (2001). Bioremediation: Overview. *Pure and Applied Chemistry, 73*(7), 1163–1172.
58. Wild, S. R., & Jones, K. C., (1992). Polynuclear aromatic hydrocarbons uptake by carrots grown in sludge-amended soil. *Journal of Environmental Quality, 21*, 217–225.
59. Wild, S. R., & Jones, K. C., (1995). Polynuclear aromatic hydrocarbons in the united kingdom environment: Preliminary source inventory and budget. *Environmental Pollution, 88*, 91–108.
60. Wilson, S. C., & Jones, K. C., (1993). Bioremediation of soil contaminated with polynuclear aromatic hydrocarbons: Review. *Environmental Pollution, 81*, 229–249.
61. Winquist, E., Björklöf, K., & Schultz, E., (2013). Bioremediation of PAH-contaminated soil with fungi-from laboratory to field scale. *International Biodeterioration & Biodegradation, 86*, 238–247.
62. Wu, G., Shao, H. B., Chu, L. Y., & Cai, J. W., (2007). Insights into molecular mechanisms of mutual effect between plants and the environment: A review. *Agrónomy for Sustainable Development, 27*, 69–78.
63. Xu, S. Y., Chen, Y. X., Wu, W. X., Wang, K. X., Lin, Q., & Liang, X. Q., (2006). Enhanced dissipation of phenanthrene and pyrene in spiked soils by combined plants cultivation. *Science of The Total Environment, 363*(1–3), 206–215.
64. Yadav, K. K., & Gupta, N., (2018). Mechanistic understanding and holistic approach of phytoremediation: A review on application and future prospects. *Ecological Engineering, 120*, 274–298.
65. Yanzheng, G., & Lizhong, Z., (2004). Plant uptake, accumulation and translocation of phenanthrene and pyrene in soils. *Chemosphere, 55*, 1169–1178.

CHAPTER 5

ASSESSMENT AND OVERVIEW OF PETROLEUM HYDROCARBON BIOREMEDIATION

NAGA RAJU MADDELA, M. SUBHOSH CHANDRA,
M. SRINIVASULU, and PALLAVAL V. BRAMHACHARI

ABSTRACT

Petroleum hydrocarbon (PH) pollutants cause huge ecological impacts after their anthropogenic entry into the ecosystem. Pollutant's nature, composition, physicochemical properties have a substantial impact on PHs removal from the contaminated sites. Serious human health risks posed by PHs pollution have attracted scientific communities to search for emerging and sustainable remediation strategies, such as biological remediation methods (*in situ* and *ex situ* conditions). This chapter addresses: (i) key issues in the bioremediation of PHs-contaminated sites; (ii) growth of bioremediation; (iii) bioavailability of PHs to microorganisms; (iv) PHs degrading bacteria; (v) biodegradation mechanisms of PHs; (vi) PHs bioremediation strategies and their practicalities; and (vii) an overview of restoration of PHs-contaminated soil.

5.1 INTRODUCTION

Restoration and cleaning of petroleum hydrocarbons (PHs) polluted sites require an eco-friendly, sustainable, and cost-efficient methods. In bioremediation strategies, microbial inoculants have provided removal efficiency of 99.9% in PHs polluted sites. The bioremediation can be achieved in two ways: (i) introduction of contaminant-degrading microflora to (so-called *bioaugmentation*) the media which to be cleaned; and (ii) altering the existing environmental conditions to stimulate local microflora (so-called *biostimulation*) for subsequent degradation of PHs. There is a growing interest in bioremediation for the restoration of PH-contaminated sites (Figure 5.1).

Research studies and corresponding publications on different strategies of bioremediation have significantly increased during 2010–2019 (Figure 5.1), indicating that the bioremediation is one of the active research areas in the field of restoration of PH-contaminated sites.

There are considerable benefits (Table 5.1) and contributions of bioremediation towards the environment, thus bioremediation has become a preferred choice for the restoration of contaminated sites around the world. For example, the area available for landfills has become scare due to urbanization and industrialization. Therefore, bioremediation has become an attractive and cost-effective strategy for the removal of PHs pollutants from the media [44].

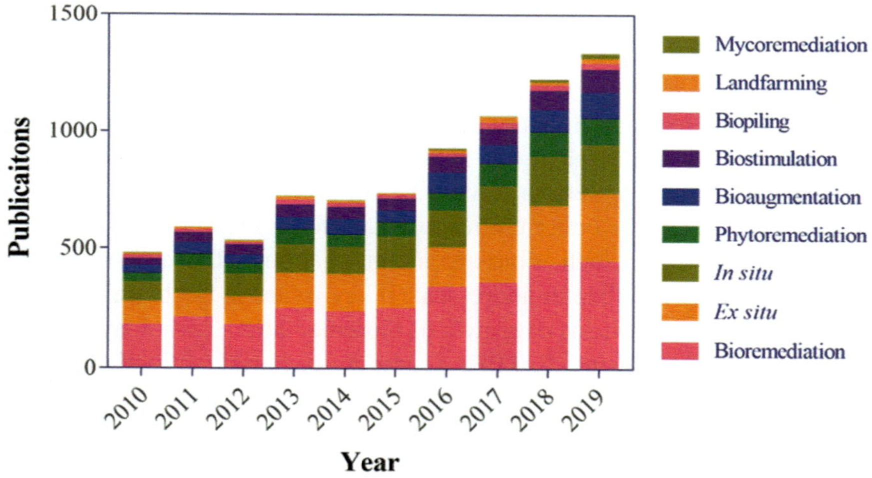

FIGURE 5.1 Number of technical publications on bioremediation of petroleum hydrocarbon (PH)-polluted sites.

Source: Data accessed (26 May 2020) from ScienceDirect using different keywords (shown in legend) with the combination of "petroleum hydrocarbons."

Bioremediation is one of the attractive approaches at several sites, such as petroleum stations, industrial sites, landfills, farms, on-site sanitation systems, mine site tailings, accidental chemical spills, etc. Genetic modification of microorganisms is also in practice for bioremediation of crude oil-polluted sites. Under certain incidences, skilled operators are needed to handle specialized bioremediation equipment, such as soil and groundwater remediation systems, fully integrated custom environmental remediation systems, dual-phase recovery systems, soil vapor extraction systems, air sparging systems, low-profile air strippers, air/water separators, enhanced oil/water separators, mobile dewatering aerators, etc. It should be noted that if PHs are not removed at the spilled site by a proper remediation strategy, then PHs are subjected to different kinds of weathering processes [40].

Major benefits of bioremediation are: Acceptability from the health and regulatory agencies; Can be executed in *in-situ* which avoids the transport of PHs pollutants; Contaminated media become useful in a short time; Less energy consumption; Less equipment is needed; Low-cost operation; Natural process and eco-friendly with no harmful effects; PHs are less likely to be escaped; and public acceptability due to less disturbance.

However, the possible outcomes of bioremediation are complete mineralization, partial transformation of pollutants and subsequently alterations in their mobility. The problem is, when there is no complete mineralization of pollutants by microorganisms, there is the high level possibility of the formation of toxic intermediates. Therefore, there is a great threat to the ecosystem by these toxic intermediate products. These intermediate substances may also be carcinogens [30], which implies that damage posed by intermediate compounds will be higher than that of parent pollutants [39, 41].

At the same time, there are certain microorganisms, which can detoxify these intermediates to non-toxic forms. Therefore, metabolic, and chemical pathways concerning pollutant versus degrader should be known completely to achieve sustainable restoration of contaminated site. Another major limitation of bioremediation is adaptation of allochthonous microflora after their introduction at the contaminated site, since environmental factors (such as temperature, pH, quality, and quantity of nutrients and contaminants, oxygen, and other molecules, solubility of pollutants, etc. [4]) play crucial role in the adaptability of microorganisms. For these reasons, results related to microbial degradation of contaminants at the laboratory level may not be expected at the field level, and there are always significant differences in the success rates between small-scale laboratory studies and field operations [74]. Furthermore, depending on the conditions, bioremediation may take even more time than other alternative remediation strategies (e.g., landfilling, incineration, etc.) [30].

Environmental pollution by PHs is mostly by accidental discharge of hazardous substances. According to International Tanker Owners Pollution Federation (ITOPF), oil spills from tankers (so-called accidental) has been decreased drastically over the past four and half decades (https://ourworldindata.org/oil-spills). For example, large oil spills (i.e., >700 tons) have been decreased by 93% between the 1970s and 2010s. However, oils spills have occurred on a regular basis, for example, very recently (29 May 2020), there was a flooding of 17,500 tons of diesel fuel (https://en.wikipedia.org/wiki/List_of_oil_spills) into the river in Russia due to the failure of fuel storage tank at Norilsk-Taimyr Energy's Thermal Power Plant. This is an ongoing and latest industrial disaster in Russia. Though many strategies are being used to restore such oil spill sites, yet bioremediation strategy has given successful results [8].

This chapter highlights certain key issues on bioavailability of PHs, potential microorganisms that degrade PHs, microbial degradation, bioremediation strategies and practices, phytoremediation, technological assessments, and case studies, and future directions.

5.2 BIOAVAILABILITY OF PETROLEUM HYDROCARBONS (PHS)

The amount of PHs that is readily taken up by microorganisms for further degradation is a bioavailable fraction; and the rate of removal of PHs from the contaminated sites is greatly affected by the nature of bioavailability of PHs [37]. In fact, lesser bioavailability of PHs is one of the principal limitations of bioremediation [73]. Bioavailable or bioaccessible fractions of PHs are very useful in the risk assessments, as actual exposure limits can be predicted through these fractions. It is noteworthy that the bioavailability of PHs can be manipulated by the addition of certain lipid molecules. For example, there is solubilization of hexadecane by rhamnolipids (biosurfactant), thus resulting in the enhanced bioavailability of hexadecane for bacteria (e.g., *Pseudomonas aeruginosa*) [48]. Likewise, Tween-80 is an another biosurfactant, which is frequently used in the removal of PHs by microorganisms either from medium or soil [51, 62].

Biosurfactants not only can improve the bioremediation process but also can improve the biodegradation rates by microorganisms. The other benefits offered by biosurfactants include: increased sorption of PHs onto microbial cell surfaces, increased trans-membrane transport through the induction of structural changes in PHs, and stimulation of co-metabolism [46], etc. Furthermore, the size of the PHs has considerable impact on their bioavailability. According to a biota (*Echinometra mathaei*)-sediment accumulation factors (BSAF), it has been found that both less and high molecular weight

(HMW) polycyclic aromatic hydrocarbons (PAHs) are known to be accumulated in the sea urchins; however, low molecular weight (LMW) PAHs were more bioavailable than the HMW PAHs [31].

The bioavailable propensities of PHs are seemingly different between *in vitro* and *in vivo* conditions, which demand greater attention while determining the bioavailability of pollutants. Supporting this, a very recent investigation [50] revealed that dimethyl sulfoxide (DMSO) extraction, interference with protein in cell culture medium, dilution during *in vitro* operations could have a significant impact on the bioavailability of aliphatic and aromatic fractions of PHs. These factors are commonly overlooked in the studies emphasizing on the characterization of PHs for bioavailability propensities. Temporal changes in the bioavailability of pollutants can also be predicted by using certain machine learning (ML) models (e.g., artificial neural network (ANN), random forest (RF)) [14], and such models can aid in the better understanding of bioavailable fractions of PHs in soil.

Furthermore, biometric extraction (BE) is also widely used in the characterization of bioavailable propensities of PHs [64]. Technically in BE method, there is a coating of solid-phase microextraction fibers with polydimethylsiloxane (PDMS), which create an appropriate passive sampling format for the characterization of bioavailability of PHs. Likewise, aquatic hazard assessments of PHs are made easy by BE methods; consequently, BE methods have wide implications in the hazard and risk evaluations of PHs and their mixtures. It is based on chemical extraction methods (e.g., mild organic solvent, supercritical fluid with pure CO_2, cyclodextrin, solid-phase with Tenax, surfactants, etc.), persulfate oxidation, and several bioassays (e.g., respirometry, bioluminescence, mRNA quantitation, toxicity tests based on earthworm and springtail, **Microtox®** toxicity test, plant growth tests, cytotoxicity to algae and Daphnia, etc.) [37]. In conclusion, insights into the bioavailability of PHs in the contaminated soil are important for a successful bioremediation.

5.3 ROLE OF MICROORGANISMS IN DEGRADATION OF PHS

Oleophilic microorganisms or oil-eating microorganisms (OEMs) are the perfect tools to combat persistent environmental pollutants, such as, PHs; and these OEMs can decompose PHs to their basic and non-toxic elements through a process of bioremediation. OEMs are ubiquitously distributed in the environment [62], and enzymatic activities of individual and/or consortium of OEMs can offer an eco-friendly and economic approach for the restoration of PHs-polluted soils. When crude oil containing environments were analyzed, predominant organisms were belonging to prokaryotes (bacteria and archaea). Figure 5.2 describes the abundance of three different taxonomic domains at the crude oil-polluted sites. Irrespective of the samples, prokaryotes were more dominant domains over eukaryotes in the crude oil containing environment [65].

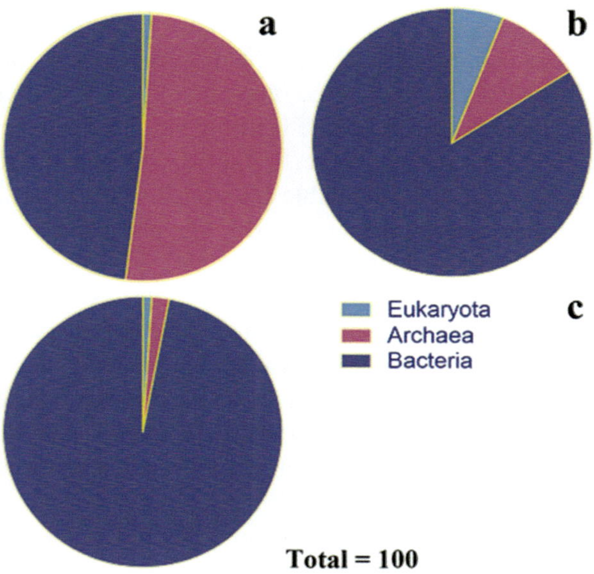

FIGURE 5.2 Relative abundance of taxa (domain level) in different crude oil containing environments: (a) sample of reservoir BA-1; (b) sample of reservoir BA-2 oil phase; and (c) water phase.

In one of recent investigations by authors of this chapter, they isolated two bacteria (e.g., *Bacillus cereus* isolate-1 and *B. thuringiensis* isolate-2) and two fungi (e.g., *Geomyces pannorum* isolate-1 and *Geomyces* sp. isolate-2) from the crude oil-polluted soil in Ecuadorian amazon rainforest [52]. Their degradation potential of *n*-alkanes (diesel in liquid medium) by selected strains reached up to 63–77% in 30 days. When sterile soil was spiked with 10% crude oil and treated with *Geomyces* spp., TPHs (total petroleum hydrocarbons) were removed up to 80% in 30 days *in vitro* [53]. When they tested a mixed culture of above 4 strains at field-level for 90 days, the 87% removal of TPHs was observed from the soil artificially contaminated with 5% heavy crude oil (21.4 API (American Petroleum Institute) gravity) [54]. Interestingly, the mixed culture showed a high adaptability to the environmental conditions at the experimental site, as the recorded rainfall and temperature range during 90 days of investigations in the low-land amazon rainforest were ~630 mm and 6–29°C, respectively [54].

Recent investigations revealed that several species of microorganism were known to degrade PHs [73]. For example, potential bacteria to degrade aliphatic hydrocarbons are:

- *Acinetobacter* sp.;
- *Alcanivorax* sp.;
- *Azoarcus* sp.;
- *Bacillus* sp.;
- *Brevibacterium;*
- *Desulfococcus* sp.;
- *Desulfosarcina* sp.;
- *Marinobacter* sp.;
- *Micrococcus* sp.;
- *Ochrobactrum* sp.;
- *Oleispira* sp.;
- *Pseudomonas* sp.;
- *Rhodococcus* sp.;
- *Stenotrophomonas* sp.;
- *Thalassolituus* sp., etc.

Fungal members include: *Aspergillus* sp., *Candida* sp., *Penicillium* sp., *Pseudozyma* sp., etc. Similarly, there are several bacteria that can degrade monoaromatics, e.g., *Acinetobacter* sp, *Archaeoglobus fulgidus*, *Aromatoleum aromaticum*, *Bacillus* sp., *Halomonas* sp., *Pseudomonas* sp., *Rhodococcus* sp., *Sphingobacterium* sp., etc. It is not surprising that several microorganisms were identified with polyaromatic degrading potential, e.g., *Achromobacter insolitus*, *Bacillus* sp., *Cycloclasticus* sp., *Phanaerochaete chrysporium*, *Pseudomonas* sp., *Vibrio* sp., *Penicillium janthinellum*, etc.

Key enzymes that do support microorganisms to degrade PHs are: methane monooxygenases, AlkB related alkane hydroxylases, eukaryotic P450, bacterial P450 oxygenase system, etc. [73]. Also, there are several environmental factors influencing the potential of microorganisms for PHs degradation, such as nutrients, microbial community, toxicity of end products, activity of water, bioavailability, temperature, oxygen, salinity, pH, etc. [3]. According to response surface methodology study, the temperature influences the degradation of PHs by microorganisms [69]. Nonetheless, the above insights clearly state that microorganisms have a full suite of prospective to decompose the components of PHs; and thus, there is a growing interest on the bioremediation for the restoration of PHs-contaminated soils (Figure 5.1).

5.4 MICROBIAL DEGRADATION

Microorganisms degrade PHs in series of steps, which are regulated by different enzymes. These metabolic reactions are highly specific, vary from species to species. At the same time, PHs are highly heterogeneous chemically in nature [43]. In the degradation of PHs, consortium is always effective than the individual species, due to metabolic cooperation between the strains in a consortium. Also, simple PHs (such as *n*-alkanes [72]) are always rapidly degraded as compared to complex PHs (e.g., PAHs), due to: prevalence of *n*-alkanes at the polluted sites, adaptability of pollutant degraders and the potential of microorganisms to degrade *n*-alkanes (Figure 5.3). Furthermore, information required for the degradation of PHs is located either in chromosomes or plasmids or both. Microorganisms can degrade the PHs either by phototrophically or chemotrophically (aerobic and anaerobic).

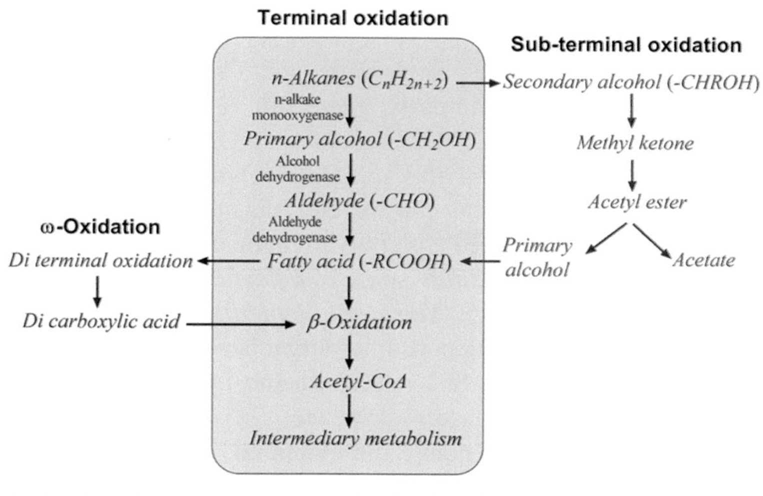

FIGURE 5.3 Possible pathways for aerobic microbial degradation of *n*-alkanes.

Degradation of PHs by phototrophic microorganisms is highly feasible in aquatic ecosystems due to the distribution of phototrophic bacteria in aquatic systems. Phototrophic microbial biofilms consist of diverse phototrophs (e.g., diatoms, green algae, cyanobacteria, other phototrophic bacteria) that play an important role in the treatment of wastewater containing PHs [60].

Similarly, *Desulfatiglans aniline* (SO_4^{2-}: reducing phototrophic purple bacterium) is able to decompose aniline either alone or in combination with *Thicapsa roseopersicina* (sulfide-consuming phototrophic bacterium) [79]. However, chemotrophic degradation of PHs by bacteria through aerobic [47] and anaerobic [68] pathways is quite common, and these pathways are well-mediated by certain key enzymes, such as hydroxylases, dioxygenases for ring-hydroxylation, alkyle/benzylsuccinate synthases, etc. It is worth noting that there are several analytical procedures for the detection identification of PHs and their products [42].

Saturated aliphatics and aromatics are principal fractions of PHs, and are subjected to decomposition by various hydroxylases. Initial steps during the aerobic decomposition of PHs vary depending on the type of substrate. For example, in one pathway, alkane monooxygenases (encoded by *AlkB* gene) can decompose aliphatic hydrocarbons aerobically by transferring electrons from NADPH to the substrates in the presence of certain cofactors (e.g., rubredoxin, rubredoxin reductase) [25]. Whereas, LMW PAHs are decomposed by either phenol monooxygenases (encoded by *Phe* gene) or naphthalene dioxygenases (encoded by *Nah* gene), where there is an addition of single or two atoms of O_2, respectively.

For these reasons, characterization of bacteria for genes encoding hydroxylases is an active research field for searching potential bacteria degrading of PHs [56], and for further implications of such potential bacteria in the bioremediation of PHs-polluted soils. During the aerobic metabolism, PHs are converted to intermediates of tricarboxylic acid (TCA), subsequently cell biomass is synthesized from the precursors (e.g., acetyl-CoA, succinate, and pyruvate). It is important to note that the bacteria utilize O_2 in two steps to decompose the PHs aerobically: (i) first step involves the initial attack of substrate (so-called PHs); and (ii) second step is of final respiratory chain [73].

Degradation scenarios related to complex PHs (such as aromatic hydrocarbons) are different compared to degradation of saturated PHs. In general, degradation of aromatic hydrocarbons occurs through meta-cleavage (ring cleavage between the hydroxyl groups) and ortho-cleavage (ring cleavage before the hydroxyl groups); and the metabolites formed are further converted to intermediates of TCA cycle. Overall, the microbial degradability of different PHs is in the following order: *n*-alkanes > branched alkanes > low molecular weight aromatics > cyclic alkanes.

Principal mechanism during the bioremediation is an oxidation-reduction process (Redox reactions) [29]. Evidently, there is frequent addition of either electron acceptors or electron donors during bioremediation for the degradation of pollutants, which are in either reduced form (e.g., hydrocarbons) or in oxidized form (e.g., chlorinated solvents), respectively. The most common type of electron acceptors of bioremediation are: O_2, NO_3^-, Mn^{3+}, Mn^{4+}, Fe^{3+}, SO_4^{2-}, CO_2, etc. Similarly, many organics (e.g., alcohols, fats, PHs, sugars, etc.), can act as electron donors during bioremediation. Nevertheless, the redox potential aerobic biotransformation is significantly higher than many other biotransformation reactions (e.g., anaerobic processes), and the order of redox potentials of different biotransformation reactions is as follows: aerobic (600 ~ 400 E_h in mV) > denitrification (400 ~ 200) > iron (Fe^{3+}) reduction (300 ~ 100) > SO_4^{2-} reduction (0 ~ –150) > fermentation (–150 ~ –220). Thus, aerobic process is a common process for the removal of pollutants from the contaminated media.

Anaerobic degradation of PHs has been reported under different anoxic conditions (e.g., iron-, manganese-, nitrate-, sulfate-reducing environments, etc.). There are four different mechanisms involved in the initial activation of anaerobic degradation of PHs, such as fumarate addition, methylation, hydroxylation, and direct carboxylation.

Once there is a successful initiation reaction, it leads to subsequent decomposition reactions, such as ring saturation, β-oxidation and/or ring cleavage. The net result of these decomposition reactions is the formation of intermediate metabolites (such as benzoyl-CoA (benzoyl coenzyme A)), which can be either converted into cellular biomass or completely oxidized [55].

5.5 BIOREMEDIATION STRATEGIES AND PRACTICES

PHs-polluted soils can be remediated by using natural attenuation (NA), land-farming, biostimulation, bioaugmentation, biopiling, bioventing, slurry-bioreactors, phytoremediation with high removal efficiency of 75–100% of TPHs [36, 63]. The main reasons for the attraction of biological based-remediation strategies are because these are: eco-friendly, cost-efficient, minimum site interruption, effective even at low concentrations of PHs [34].

On the other side, certain limitations are: longer treatment time, low predictability, highly reliant on environmental factors, etc. The main characteristics of conventional bioremediation methods are as follows [45]:

1. **Natural Attenuation (NA):** This uses indigenous microorganisms under natural environmental conditions. Though NA is cost-effective, it requires long-term observation and not effective sometimes.
2. **Bioaugmentation:** There is an addition of highly efficient pollutant-degrading microorganisms to the polluted sites, which facilitates the rapid removal of pollutants from the contaminated sites. However, *bioaugmentation* suffers from poor adaptability of augmented microflora.
3. **Biostimulation:** Environmental factors are managed by the addition of chemicals (such as fertilizers, biosurfactants, etc.), so that native microorganisms are stimulated. Thus, *biostimulation* is always better than NA. However, *biostimulation* is not always effective. Besides, there is a need to determine optimal C/N/P ratio and PHs bioavailability.
4. **Bioventing:** This is an *in-situ* remediation technology (Figure 5.4) and is used to decontaminate the soils at vadose or unsaturated zone by injecting air or pure oxygen. Therefore, *bioventing* is an aerobic biodegradation process, which relies completely on indigenous oleophilic microorganisms. Though *bioventing* has its own advantages, such as cost-effective (\$45–140 ton^{-1}) and requires short treatment time, yet high pollutant concentrations are in-turn toxic to microorganisms.
5. **Landfarming:** There is a spreading of excavated polluted soil on the ground surface in a thin layer, which accelerates the natural biodegradation process. Practically, *land farming* is easy to design and implement, with a shorter treatment time and cost-effective (\$30–60 ton^{-1}). However, *landfarming* suffers from certain limitations, such as difficulty in achieving >95% pollutant removal efficiency, not effective against sites containing TPHs >50,000 ppm, requires larger surface area, and generates substantial air pollution through the generation of dust and vapor.
6. **Slurry-Phase Reactors:** Whereas *slurry-phase reactors* are used to treat excavated soil in a bioreactor. Solids (10–30%) are maintained as suspensions and mixed with nutrients and oxygen in the reactors. *Slurry-phase reactors* have been used successfully to remediate the soils and sludge samples polluted by PHs. However, the main

limitations of reactor-based treatments include difficulty in the excavation of contaminated soil, and sizing of soil prior to placing them into reactor is difficult and expensive.

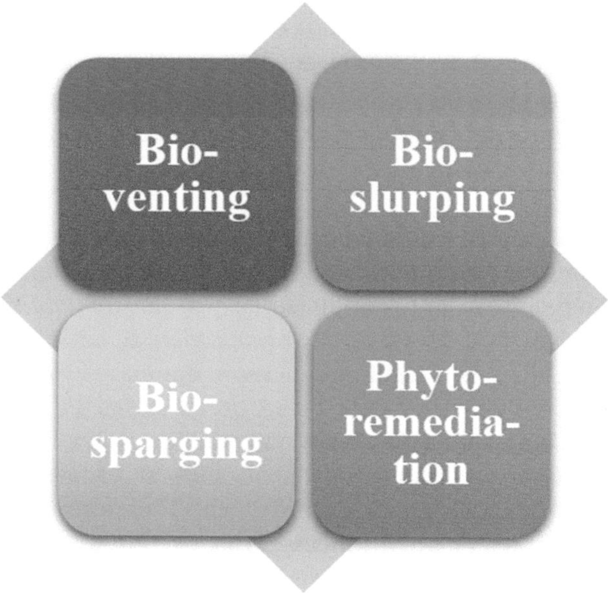

FIGURE 5.4 *In-situ* bioremediation strategies.

However, recently, several emerging approaches have been evolved for the enhanced removal of PHs from the polluted sites. These techniques include:

- Bacteria-fungi consortium [15];
- Earthworm-associated bioremediation [57];
- Genetically engineered microorganisms (GEMs) [11];
- Hybrid microbial electrochemical technologies [23];
- Microbial cells on humic acid-modified biofuel ash [76];
- Microbial electrochemical systems (MES) [78];
- Microbial enzymes [66];
- Mixed technique of composting and zero-valent iron nanoparticles (nZVI) [22];
- Phytoremediation [24]; and
- Rhizoremediation [35], etc.

Although different microbial-based tools have been tested for the successful restoration of PHs-polluted soils, yet there is a considerable gap between lab- and field-level results, attributable to limitations in bioremediation [67].

One of the most critical factors in the selection of potential microorganisms to decompose PHs in a particular habitat. The pollutant may not be homogeneous at the contaminated site, e.g., the man-made ecosystem (such as laboratory environment). Furthermore, depending on the threshold values, pollutants act as stressors rather than nutrients. There are also certain sensors on the surface of microorganisms, which do decide the movement of microbial cell either towards or away from the contaminant. Therefore, it is necessary to know the type of sensors present in the microbial cells. The other limiting factors of bioremediation include: low hydrophilicity of PHs, microbial interactions, soil physicochemical properties (e.g., temperature, pH, organic load, pollutant nature), etc.

5.6 PHYTOREMEDIATION

There are several complex interactions among various components during phytoremediation of polluted sites, where green plants associated microorganisms, soil amendments, and agronomic techniques could have a significant influence on the outcome of remediation. In principle, plants that are used in the phytoremediation are often harvested, therefore, this process provides the best way of eliminating the pollutants from the site [5]. There are many successful case studies of phytoremediation, where different plant species have proved their potential in the restoration of PHs-contaminated soils (Table 5.1).

TABLE 5.1 Phytoremediation of PHs-Contaminated Soils

Plant Species	Removal Potential	References
Azolla filiculoides Lam	Aliphatic and aromatic hydrocarbons were degraded by 73–94% and 77–81%, respectively in 15 days	[33]
Calotropis procera L.	62% reduction in PHs in the presence of biochar and urban waste compost	[28]
Cynodon dactylon (L.) Pers.	37–48% reduction in crude oil in the presence of NPK, after 180 days	[10]
Helianthus annus	69% reduction of TPHs on 90 days	[16]

TABLE 5.1 *(Continued)*

Plant Species	Removal Potential	References
Iris dichotoma Pall.	20–30% degradation of PHs	[13]
Iris pseudacorus L.	31–42% degradation of TPHs	[77]
Lemna paucicostata	98% reduction in TPHs after 120 days	[17]
Polypogon monspoliensis	85% reduction in PHs in 45 days	[20]
Ricinus communis L.	92% degradation of TPHs in 20 weeks	[26]
Ryegrass	85% degradation of crude oil after 75 days in the presence of biochar, compost, and bacterial consortium	[27]
Salicornia spp.	40–60% reduction in PHs in 100 days	[1]
Spartina anglica	9–36% reduction in TPHs in 60 days in the presence of biochar and rhamnolipids	[80]
Vetiveria zizanioides	88.5% degradation of aliphatic hydrocarbons in 15 months	[61]
Zea mays L. with plant growth-promoting bacteria	43% degradation of PHs in 60 days	[6]
Zea mays with *Streptomyces* association	70% reduction of PHs from soil	[9]

Several mechanisms are involved during the process of phytoremediation, such as:

1. **Phytodegradation (Phytotransformation):** There is a degradation (either in soil or with in the plant system) of pollutants by secreted plant enzymes, later degraded products are reabsorbed by plants and subjected to transpiration;
2. **Phytoextraction:** Pollutants are accumulated into harvestable plant biomass, and if the plants show high rates of accumulation of pollutants, then these plants are referred to as 'hyperaccumulators;' and
3. **Phytostabilization:** There is an immobilization of pollutants by soil particles with the help of plants, called phytostabilization, this makes pollutant less bioavailable;
4. **Phytostimulation (Rizodegradation):** In this process root exudates (e.g., carbohydrates, and acids) stimulate the rhizosphere microorganisms to degrade the pollutants [7];

5. **Phytovolatilization:** Pollutants are taken up by plants and are released into the atmosphere through the process of transpiration [18].
6. **Rhizofiltration:** There is a removal of pollutants during the filtration of water by the mass of roots, called rhizofiltration, in this case, pollutants are always adsorbed in or adsorbed to the root surface;

Several factors need to be considered for the successful implications of phytoremediation of PHs-contaminated soil at *in situ* level [2]. Key aspects include: nature of contaminant(s), conditions of polluted sites, intensity of contamination, time frame for clean-up, and type of plant species to be used. In fact, selection of suitable plant species for phytoremediation is uneconomical; therefore, economic methods should be implemented strictly in finding the suitable phytoremediators. For example, combination of analytic hierarchy process (AHP), technique of order of preference by similarity to ideal solutions (TOPSIS), triangular fuzzy numbers (TFNs) [75] can be a suitable choice for the effective selection of plants for the remediation of PHs-contaminated soils [32, 49].

5.7 TECHNOLOGICAL ASSESSMENTS AND CASE STUDIES

Several parameters need to be considered during the cost analysis of bioremediation to make the strategy viable, such as Initial investment (Fixed cost), operation costs (including maintenance, personnel, supplies, and services), marketing, and administration. According to a recent investigation, the treatment cost for hydrocarbon contaminated soil by biostimulation and bioaugmentation strategies under aerobic conditions was found to be US$ 59 and 77 per m^3, respectively [59].

Interestingly, 90% of the treatment cost by biostimulation was due to the amendment of vermicompost (40% w/w), which demands to search for low-cost alternative substrates. For different bioremediation strategies (https://frtr.gov/matrix2/section4/4-3.html), details of cost analysis as determined by remedial action cost engineering and requirements (RACER) software are presented in Table 5.2.

It can be observed that the cost of one strategy invariably differs between treatment of small sites and large sites. In addition, the treatment cost depends on the nature of the contaminated site. U.S. Department of Defense (DoD) and Federal Agencies (e.g., federal remediation technology roundtable:

FRTR) have jointly assessed the cost for each bioremediation strategy by considering the above parameters [21].

For example, through an *in-situ* soil technology (i.e., Bioventing), the unit treatment cost for small site (contaminated volume = 83 cubic yards) was 9–12 times higher than that for large site (1,667 cubic yards). There was a significant increase in the cost (small site = 4%; large site = 38%) between the treatment of easy and difficult sites by bioventing. Importantly, there is a higher extraction well spacing (ft) (37% ↑) and average vapor flow rate per well (50% ↑) at easy site than at difficult site.

TABLE 5.2 Different Bioremediation Strategies and Cost Analysis

Bioremediation Strategy	Cost Range (US$/m^3)
Biopiles	130–260
Bioventing	79–109
Compositing	365–422
Enhanced bioremediation	30–100
Landfarming	<100
Phytoremediation	147–483
Slurry phase biological treatment	130–200
Slurry phase biological treatment including off-gas treatment	160–210

As a result, a greater number of vapor extraction wells (50–58% ↑) are to be installed at the difficult site, and this will be responsible for higher bioventing marked-up costs for the treatment of difficult site than the easy site. In addition, remedial design cost is also higher for difficult site than the easy site. Nevertheless, for the successful and cost-effective remediation of contaminated site, other parameters should also be taken into consideration, as listed in Figure 5.5. There are two types of costs associated with a full-scale bioremediation:

1. **Principal Capital Cost:** This can be from installation of equipment like blowers, off-gas treatment, blowers, pumps, tanks, equipment related to material-sizing and process-monitoring; and
2. **Operating Costs:** These include labor, energy, chemical, and microbiological analysis, water, health, and safety supplies, nutrients, and consumable supplies, equipment repair and maintenance, site preparation, mobilization, and demobilization, and disposable of waste. Data and report preparation costs are also part of operating costs.

FIGURE 5.5 Pre-evaluations of site clean-up.
Source: Modified after Ref. [19].

It should also be noted that there may be certain unseen difficulties during the clean-up process, which causes an additional cost and can be considered as a contingency cost, which can be added to capital and operating costs. In simple words, a higher degree of uncertainty in remediation causes higher contingency costs. There are several case studies, which have successfully restored the PHs-contaminated sites and estimated the treatment costs [38]. The details related to bioremediation strategies are shown in Table 5.3.

TABLE 5.3 Bioremediation of PHs-Contaminated Sites: Case Studies

Place	Site and Treatment	Strategy	Cost (US$)
Alaska, USA [45]	Leakage of JP-4 jet fuel at Eielson air force base (AFB). PHs concentrated >5.25 ft below ground surface with 6.1 m thickness of saturated zone with contaminant. Treatment was carried out during 1991–1994.	Bioventing plus three soil warming technologies (i.e., active, passive, and surface warming)	758,077
Bogota, Colombia [45]	PHs-contaminated soil (12 plots of 20 m^2 each)	Landfarming and Bioaugmentation (*Lysinibacillus sphaericus*)	–
Central Alberta, Canada [12]	4,000 m^3 of PHs contaminated soil treated during 1999–2000	Biopiles (passive aeration with a Trackhoe and mixing head unit	–
Colorado, USA [70]	Treatment of underground storage tanks (35–40 ft below ground level) at Lowry AFB, Denver. Remediation operator was Engineering Science, Inc., Denver	Bioventing	28,650
Florida, USA [70]	8,100 yds^3 of creosote-contaminated soil treated at Brown Wood Preserving Superfund Site, Live Oak during 1989–1990. Operator was John Ryan Remediation Technologies, Inc (ReTeC, Seattle, WA, USA)	Land treatment, bioaugmentation	565,400
Gujarat, India [58]	14,694 m^3 of oil-polluted soil excavated at the accidental spill site and remediated by ONGC (Oil and Natural Gas Corporation Limited, Delhi, India) and ONGC-TERI (The Energy and Resources Institute) Biotech Ltd. (Delhi, India) in June 2008	Bioaugmentation and biostimulation	–
Missouri, USA [71]	PAHs contaminated site at Scott Lumber Company Superfund Site, Alton. Site was remediated by Remediation Technologies, Inc. (ReTeC), MT, USA	Biostimulation	54 million

TABLE 5.3 *(Continued)*

Place	Site and Treatment	Strategy	Cost (US$)
Texas, USA [70]	300,000 tons of PHs polluted soil/sludge collected at French Limited Superfund Site, Crosby, and was treated in 1983. Operator was French Limited Task Group.	Slurry-phase bioremediation by MixFlo™ aeration system	49 million
Utah, USA [70]	5,000 yds^3 area at JP-4 Fuel spill site at Hill AFB treated in 1985	Soil vapor extraction, Bioventing	335,000

5.8 RECOMMENDATIONS AND FUTURE DIRECTIONS

- There should be more emphasis on the use of combination of techniques, where there should be proper consideration of sustainable factors like economic, social, and environmental factors. For example, for the remediation of sites with very heavy contamination, conventional clean-up can be used first, then phytoremediation can be used as a final or polishing step.
- It is necessary to search for novel synthetic dispersants with less toxicity and density. As we know, dispersant treated oil will be more available to the degrader, and novel biocompatible surfactants enhance the physical contacts between contaminant and degrader.
- Regarding biostimulation, fertilizers with less toxicity, less solubility and carbon-neutral are essential to make the biostimulation strategy more viable. Uric acid is a promising fertilizer, which has less solubility, less toxicity, high affinity for PHs, abundantly available in the environment.
- There is also a necessity to couple the synthetic biology with computational biology and whole-cell simulations in the field of bioremediation studies to find the degraders with a more effective metabolic potential. By this way, it is feasibly possible to move beyond the traditional biostimulation strategy where existing bacteria are stimulated by addition chemical additives.
- Still, there are certain gaps in understanding the microbial uptake of PHs. Therefore, emphasis on theoretical basis is obligatory for in

depth understanding of the mechanisms behind the interfacial interactions between bacteria and contaminants.

- It is also necessary to explore more and undiscovered resources for PHs-degrading bacteria by using biotechnological approaches (such as high-throughput screening).
- Also, future studies should be focused on finding novel functional genes regulating the pathways of PH's degradation in different bacteria; this will provide additional insights into the molecular mechanism of microbial remediation strategies.
- Use of technological advances of synthetic biology should not be neglected in constructing genetically engineered bacteria, this will enhance the implications of microbial degradation of PHs, subsequently bioremediation of PHs-contaminated sites.

5.9 SUMMARY

This chapter has summarized the importance of microbial degradation of PHs in the restoration of the contaminated sites. The chapter discusses key issues, such as, bioavailability of PHs, potential microorganisms degrade PHs, mechanism of microbial degradation of PHs, bioremediation strategies and practicalities, and phytoremediation. Finally, a special emphasis has been paid on technological assessments of different bioremediation strategies and case studies regarding bioremediation of PHs-contaminated sites.

Bioavailability of PHs is one of the important factors, which influence the rate of degradation of PHs by bacteria, and the bioavailability of PHs is highly influenced by the environmental factors. It is highly advisable to use a combination of strategies for the optimum removal of contaminant during the clean-up process. The bioremediation is achievable in the range of US$ < 50 to ~ 500 per m^3.

ACKNOWLEDGMENT

NRM greatly acknowledges the Universidad Técnica de Manabí, Portoviejo, Ecuador, for the financial assistance (No. PYT1184-CONV2019-FCS0035).

KEYWORDS

- **bioavailability**
- **bioremediation**
- **cost analysis**
- **microbial electrochemical systems**
- **petroleum hydrocarbon (PH)**
- **polycyclic aromatic hydrocarbons**

REFERENCES

1. Abdollahzadeh, T., Niazi, A., & Moghadam, A., (2019). Phytoremediation of petroleum-contaminated soil by *Salicornia*: From PSY activity to physiological and morphological communications. *Environmental Technology, 40*(21), 2789–2801.
2. Adesipo, A. A., Freese, D., & Nwadinigwe, A. O., (2020). Prospects of in-situ remediation of crude oil contaminated lands in Nigeria. *Scientific African, 8*. Article ID: e00403.
3. Al-Hawash, A. B., Dragh, M. A., & Li, S., (2018). Principles of microbial degradation of petroleum hydrocarbons in the environment. *The Egyptian Journal of Aquatic Research, 44*(2), 71–76.
4. Al Disi, Z., Jaoua, S., Al-Thani, D., Al-Meer, S., & Zouari, N., (2017). Considering the specific impact of harsh conditions and oil weathering on diversity, adaptation, and activity of hydrocarbon-degrading bacteria in strategies of bioremediation of harsh oily-polluted soils. *BioMed Research International, 2017*. Article ID: 8649350.
5. Ali, H., Khan, E., & Sajad, M. A., (2013). Phytoremediation of heavy metals: Concepts and applications. *Chemosphere, 91*(7), 869–881.
6. Asghar, H. N., Rafique, H. M., Khan, M. Y., & Zahir, Z. A., (2017). Phytoremediation of light crude oil by maize (*Zea mays* L.) bio-augmented with plant growth-promoting bacteria. *Soil and Sediment Contamination: An International Journal, 26*(7–8), 749–763.
7. Attah-Olottah, R., & Adeniyi, K. A., (2018). Application of phytoremediation in the management of oil spillage: Review. *Global Journal of Earth and Environmental Science, 3*(3), 16–22.
8. Baniasadi, M., & Mousavi, S. M., (2018). A comprehensive review on the bioremediation of oil spills. Chapter 10. In: Kumar, V., Kumar, M., & Prasad, R., (eds.), *Microbial Action on Hydrocarbons* (pp. 223–254). Singapore: Springer Nature.
9. Baoune, H., Aparicio, J. D., & Acuña, A., (2019). Effectiveness of the *Zea mays-Streptomyces* association for the phytoremediation of petroleum hydrocarbons impacted soils. *Ecotoxicology and Environmental Safety, 184*. Article ID: 109591.
10. Basumatary, B., & Bordoloi, S., (2016). Phytoremediation of crude oil-contaminated soil using *Cynodon dactylon* (L.) pers. Chapter 3. In: Abid, A. A., Sarvajeet, S. G., Ritu, G., Guy, R. L., & Lee, N., (eds.), *Phytoremediation* (pp. 41–51). Cham. – Switzerland: Springer Nature.

11. Benjamin, S. R., (2019). De. Lima, F., & Rathoure, A. K., (2019). Genetically engineered microorganisms for bioremediation processes: GEMs for bioremediation. In: *Biotechnology: Concepts, Methodologies, Tools, and Applications* (pp. 1607–1634). IGI Global.
12. Carcamo, A., & Powers, J., (2019). *Bioremediation: Case Studies in Central Alberta.* https://www.scribd.com/document/247203848/Bioremediation-Case-Studies-in-Central-Alberta (accessed on 1 August 2021).
13. Cheng, L., Wang, Y., Cai, Z., Liu, J., Yu, B., & Zhou, Q., (2017). Phytoremediation of petroleum hydrocarbon-contaminated saline-alkali soil by wild ornamental *Iridaceae* species. *International Journal of Phytoremediation, 19*(3), 300–308.
14. Cipullo, S., Snapir, B., Prpich, G., & Campo, P., (2019). Prediction of bioavailability and toxicity of complex chemical mixtures through machine learning models. *Chemosphere, 215*, 388–395.
15. Cocarta, D. M., Dumitru, D. M., & Pesciaroli, L., (2019). Cultivable hydrocarbonoclastic microbial community from historically polluted soil: Tests for consortium development. *Soil and Sediment Contamination: An International Journal, 28*(3), 334–345.
16. Dickson, U. J., Coffey, M., & Mortimer, R. J. G., (2020). Investigating the potential of sunflower species, fermented palm wine and *Pleurotus ostreatus* for treatment of petroleum-contaminated soil. *Chemosphere, 240*. Article ID: 124881.
17. Ekperusi, A. O., Nwachukwu, E. O., & Sikoki, F. D., (2020). Assessing and modelling the efficacy of *Lemna paucicostata* for the phytoremediation of petroleum hydrocarbons in crude oil-contaminated wetlands. *Scientific Reports, 10*(1), 1–9.
18. Eskander, S., & Saleh, H., (2017). Phytoremediation: Overview. *Environmental Science and Engineering, Soil Pollution and Phytoremediation, 11*, 124–161.
19. Faris, B., & Vlassopoulos, D., (2003). Systematic approach to *in situ* bioremediation in groundwater. *Bioremediation: The Journal of Environmental Cleanup Costs, Technologies & Techniques, 13*, (2), 27–52.
20. Farzamisepehr, M., & Nourozi, F., (2018). Physiological responses of *Polypogon monspeliensis* L. in petroleum-contaminated soils. *Plant Physiology, 8*(2), 2391–2401.
21. FRTR (Federal Remediation Technology Roundtable)., (2002). *Remediation Technologies Screening Matrix and Reference Guide,* Version 4.0. https://frtr.gov/matrix2/Preface/REPORT_DOC_PAGE.html (accessed on 1 August 2021).
22. Galdames, A., Mendoza, A., & Orueta, M., (2017). Development of new remediation technologies for contaminated soils based on the application of zero-valent iron nanoparticles and bioremediation with compost. *Resource-Efficient Technologies, 3*(2), 166–176.
23. Ghangrekar, M. M., Sathe, S. M., & Chakraborty, I., (2020). *In situ* bioremediation techniques for the removal of emerging contaminants and heavy metals using hybrid microbial electrochemical technologies. Chapter 9. In: Maulin, P. S., Rodriguez-Couto, S., & Sevinç, Ş. S., (eds.), *Emerging Technologies in Environmental Bioremediation* (pp. 233–255). New York: Elsevier Inc.
24. Gupta, P. K., Ranjan, S., & Gupta, S. K., (2019). Phytoremediation of petroleum hydrocarbon-polluted sites: Application, challenges, and future prospects. Chapter 8. In: Sanjay, K. G., & Faizal, B., (eds.), *Application of Microalgae in Wastewater Treatment* (Vol. 1, pp. 145–162). Cham – Switzerland: Springer Nature.
25. Hamamura, N., Fukui, M., Ward, D. M., & Inskeep, W. P., (2008). Assessing soil microbial populations responding to crude-oil amendment at different temperatures

using phylogenetic, functional gene (*Alkb*) and physiological analyses. *Environmental Science and Technology, 42*(20), 7580–7586.
26. Hasibuan, S. Y., Yani, M., & Mansur, I., (2019). The effectiveness of oil spill dispersant addition for phytoremediation of petroleum-contaminated soil using *Ricinus communis* L. *Journal of Degraded and Mining Lands Management, 6*(3), 1811–1814.
27. Hussain, F., Hussain, I., & Khan, A. H., (2018). A combined application of biochar, compost, and bacterial consortia with Italian ryegrass enhanced phytoremediation of petroleum hydrocarbon contaminated soil. *Environmental and Experimental Botany, 153*, 80–88.
28. Jahantab, E., Jafari, M., Motesharezadeh, B., Tavili, A., & Zargham, N., (2018) Remediation of petroleum-contaminated soils using '*Stipagrostis plumosa, Calotropis procera* L and *Medicago sativa*' under different organic amendment treatments. *Ecopersia, 6*(2), 101–109.
29. Jugnia, L. B., Manno, D., Dodard, S., Greer, C. W., & Hendry, M., (2019). Manipulating redox conditions to enhance *in situ* bioremediation of RDX in groundwater at a contaminated site. *Science of the Total Environment, 676,* 368–377.
30. Juwarkar, A. A., Singh, S. K., & Mudhoo, A. A., (2010). Comprehensive overview of elements in bioremediation. *Reviews in Environmental Science and Bio/Technology, 9*(3), 215–288.
31. Keshavarzifard, M., Moore, F., Keshavarzi, B., & Sharifi, R., (2017). Polycyclic aromatic hydrocarbons (PAHs) in sediment and sea urchin (*Echinometra mathaei*) from the intertidal ecosystem of the Northern Persian Gulf: Distribution, sources, and bioavailability. *Marine Pollution Bulletin, 123*(1, 2), 373–380.
32. Kirkham, M. B., (2006). Cadmium in plants on polluted soils: Effects of soil factors, hyperaccumulation, and amendments. *Geoderma, 137*(1), 19–32.
33. Kösesakal, T., Ünal, M., Kulen, O., Memon, A., & Yüksel, B., (2016). Phytoremediation of petroleum hydrocarbons by using a freshwater fern species *Azolla filiculoides* Lam. *International Journal of Phytoremediation, 18*(5), 467–476.
34. Koshlaf, E., & Ball, A. S., (2017). Soil bioremediation approaches for petroleum hydrocarbon polluted environments. *AIMS Microbiology, 3*(1), 25–31.
35. Kotoky, R., Rajkumari, J., & Pandey, P., (2018). The rhizosphere microbiome: Significance in rhizoremediation of polyaromatic hydrocarbon contaminated soil. *Journal of Environmental Management, 217*, 858–870.
36. Kuppusamy, S., Maddela, N. R., Megharaj, M., & Venkateswarlu, K., (2020). Approaches for remediation of sites contaminated with total petroleum hydrocarbons. Chapter 7. *Total Petroleum Hydrocarbons: Environmental Fate, Toxicity, and Remediation* (1st edn., pp. 167–205). Cham – Switzerland: Springer Nature.
37. Kuppusamy, S., Maddela, N. R., Megharaj, M., & Venkateswarlu, K., (2020). Bioavailability of total petroleum hydrocarbons. Chapter 4. In: *Total Petroleum Hydrocarbons: Environmental Fate, Toxicity, and Remediation* (1st edn., pp. 79–94). Cham – Switzerland: Springer Nature.
38. Kuppusamy, S., Maddela, N. R., Megharaj, M., & Venkateswarlu, K., (2020). Case studies on remediation of sites contaminated with total petroleum hydrocarbons. Chapter 9. In: *Total Petroleum Hydrocarbons: Environmental Fate, Toxicity, and Remediation* (1st edn., pp. 225–256). Cham – Switzerland: Springer Nature.
39. Kuppusamy, S., Maddela, N. R., Megharaj, M., & Venkateswarlu, K., (2020). Ecological impacts of total petroleum hydrocarbons. Chapter 5. In: *Total Petroleum*

Hydrocarbons: Environmental Fate, Toxicity, and Remediation (1st edn., pp. 95–138). Cham – Switzerland: Springer Nature.

40. Kuppusamy, S., Maddela, N. R., Megharaj, M., & Venkateswarlu, K., (2020). Fate of total petroleum hydrocarbons in the environment. Chapter 3. In: *Total Petroleum Hydrocarbons: Environmental Fate, Toxicity, and Remediation* (1st edn., pp. 57–77). Cham – Switzerland: Springer Nature.
41. Kuppusamy, S., Maddela, N. R., Megharaj, M., & Venkateswarlu, K., (2020). Impact of total petroleum hydrocarbons on human health. Chapter 6. In: *Total Petroleum Hydrocarbons: Environmental Fate, Toxicity, and Remediation* (1st edn., pp. 139–165). Cham – Switzerland: Springer Nature.
42. Kuppusamy, S., Maddela, N. R., Megharaj, M., & Venkateswarlu, K., (2020). Methodologies for analysis and identification of total petroleum hydrocarbons. Chapter 2. In: *Total Petroleum Hydrocarbons: Environmental Fate, Toxicity, and Remediation* (1st edn., pp. 29–55). Cham – Switzerland: Springer Nature.
43. Kuppusamy, S., Maddela, N. R., Megharaj, M., & Venkateswarlu, K., (2020). An overview of total petroleum hydrocarbons. Chapter 1. In: *Total Petroleum Hydrocarbons: Environmental Fate, Toxicity, and Remediation* (1st edn., pp. 1–27). Cham – Switzerland: Springer Nature, (2020).
44. Kuppusamy, S., Maddela, N. R., Megharaj, M., & Venkateswarlu, K., (2020). Regulatory guidelines for total petroleum hydrocarbon contamination. Chapter 8 In: *Total Petroleum Hydrocarbons: Environmental Fate, Toxicity, and Remediation* (1st edn., pp. 207–224). Cham – Switzerland: Springer Nature.
45. Kuppusamy, S., Maddela, N. R., Megharaj, M., & Venkateswarlu, K., (2019). *Total Petroleum Hydrocarbons: Environmental Fate, Toxicity, and Remediation* (1st edn., p. 264). Cham – Switzerland: Springer Nature.
46. Li, Q., Huang, Y., Wen, D., Fu, R., & Feng, L., (2020). Application of alkyl polyglycosides for enhanced bioremediation of petroleum hydrocarbon-contaminated soil using *Sphingomonas changbaiensis* and *Pseudomonas stutzeri*. *Science of The Total Environment, 719*. Article ID: 137456.
47. Liu, Q., Tang, J., Gao, K., Gurav, R., & Giesy, J. P., (2017). Aerobic degradation of crude oil by microorganisms in soils from four geographic regions of China. *Scientific Reports, 7*(1), 1–12.
48. Liu, Y., Zeng, G., Zhong, H., & Wang, Z., (2017). Effect of rhamnolipid solubilization on hexadecane bioavailability: Enhancement or reduction? *Journal of Hazardous Materials, 322*, 394–401.
49. Lone, M. I., He, Z., Stoffella, P. J., & Yang, X., (2008). Phytoremediation of heavy metal polluted soils and water: Progresses and perspectives. *Journal of Zhejiang University Science B, 9*(3), 210–220.
50. Luo, Y. S., Ferguson, K. C., Rusyn, I., & Chiu, W. A., (2020). *In vitro* bioavailability of the hydrocarbon fractions of dimethyl sulfoxide extracts of petroleum substances. *Toxicological Sciences, 174*(2), 168–177.
51. Maddela, N. R., Burgos, R., & Kadiyala, V., (2016). Removal of petroleum hydrocarbons from crude oil in solid and slurry phase by mixed soil microorganisms isolated from Ecuadorian oil fields. *International Biodeterioration & Biodegradation, 108*, 85–90.
52. Maddela, N. R., Masabanda, M., & Leiva-Mora, M., (2015). Novel diesel-oil-degrading bacteria and fungi from the Ecuadorian Amazon rainforest. *Water Science and Technology, 71*(10), 1554–1561.

53. Maddela, N. R., Scalvenzi, L., Pérez, M., Montero, C., & Gooty, J. M., (2015). Efficiency of indigenous filamentous fungi for biodegradation of petroleum hydrocarbons in medium and soil: Laboratory study from Ecuador. *Bulletin of Environmental Contamination and Toxicology, 95*(3), 385–394.
54. Maddela, N. R., Scalvenzi, L., & Venkateswarlu, K., (2017). Microbial degradation of total petroleum hydrocarbons in crude oil: Field-scale study at the low-land rainforest of Ecuador. *Environmental Technology, 38*(20), 2543–2550.
55. Meckenstock, R. U., Boll, M., & Mouttaki, H., (2016). anaerobic degradation of benzene and polycyclic aromatic hydrocarbons. *Journal of Molecular Microbiology and Biotechnology, 26*(1–3), 92–118.
56. Meynet, P., Hale, S. E., & Davenport, R. J., (2012). Effect of activated carbon amendment on bacterial community structure and functions in a PAH impacted urban soil. *Environmental Science and Technology, 46*(9), 5057–5066.
57. Nneji, L. M., Somade, O. T., & Adeyi, A. O., (2016). Earthworm - assisted bioremediation of petroleum hydrocarbon-contaminated soils from motorcar mechanic workshops in Ibadan, Oyo State, Southwestern Nigeria. *Bioremediation Journal, 20*(4), 263–285.
58. Oilzapper, (2017). *Bioremediation Technology: Bioremediation of Oil Spill in Gujarat Oil Field in India (Western India)*. http://www.otbl.co.in/Bioremediation.php (accessed on 1 August 2021).
59. Orellana, R., Cumsille, A. G. M., & Rojas, C., (2017). Assessing technical and economic feasibility of complete bioremediation for soils chronically polluted with petroleum hydrocarbons. *Journal of Bioremediation and Biodegradation, 2017*, 1–7.
60. Paniagua-Michel, J., (2017). Wastewater treatment using phototrophic-heterotrophic biofilms and microbial mats. Chapter 9. In: Bhumi, N. T., & Dhananjay, K., (eds.), *Prospects and Challenges in Algal Biotechnology* (pp. 257–275). Singapore: Springer Nature.
61. Rajaei, S., & Seyedi, S. M., (2018). Phytoremediation of petroleum-contaminated soils by *Vetiveria zizanioides* (L.) Nash. *Clean – Soil, Air, Water, 46*(8). Article ID: 1800244.
62. Raju, M. N., Leo, R., & Herminia, S. S., (2017). Biodegradation of diesel, crude oil and spent lubricating oil by soil isolates of *Bacillus* spp. *Bulletin of Environmental Contamination and Toxicology, 98*(5), 698–705.
63. Raju, M. N., & Scalvenzi, L., (2017). Petroleum degradation: Promising biotechnological tools for bioremediation. Chapter 17. In: Zoveidavianpoor, M., (ed.), *Recent Insights in Petroleum Science and Engineering* (pp. 351–369). Rijeka – Croatia; IntechOpen.
64. Redman, A. D., Butler, J. D., & Letinski, D. J., (2018). Technical basis for using passive sampling as a biomimetic extraction procedure to assess bioavailability and predict toxicity of petroleum substances. *Chemosphere, 199*, 585–594.
65. Sierra-Garcia, I. N., & Belgini, D. R. B., (2020). In depth metagenomic analysis in contrasting oil wells reveals syntrophic bacterial and archaeal associations for oil biodegradation in petroleum reservoirs. *Science of The Total Environment, 715*. Article ID: 136646.
66. Singh, J. S., & Singh, D. P., (2017). Methanotrophs: Emerging bioremediation tool with unique broad-spectrum methane monooxygenase (MMO) enzyme. Chapter 1. In: Singh, J. S., & Seneviratne, G., (eds.), *Agro-Environmental Sustainability* (pp. 1–18). Cham – Swiezerland: Springer Nature.

67. Singh, P., Jain, R., & Srivastava, N., (2017). Current and emerging trends in bioremediation of petrochemical waste: A review. *Critical Reviews in Environmental Science and Technology, 47*(3), 155–201.
68. Song, B., Tang, J., Zhen, M., & Liu, X., (2019). Influence of graphene oxide and biochar on anaerobic degradation of petroleum hydrocarbons. *Journal of Bioscience and Bioengineering, 128*(1), 72–79.
69. Sun, S., Liu, Q., Chen, S., Yu, W., Zhao, C., & Chen, H., (2019). Optimization for microbial degradation of petroleum hydrocarbon (TPH) By *Enterobacter* Sp. S-1 using response surface methodology. *Petroleum Science and Technology, 37*(7), 821–828.
70. US-EPA, (1995). *Abstracts of Remediation Case Studies.* Prepared by Member Agencies of the Federal Remediation Technologies Roundtable; https://www.epa.gov/sites/production/files/2015-04/documents/frtr-abs.pdf (accessed on 1 August 2021).
71. US-EPA, (1995). *Cost and Performance Report: Land Treatment at the Scott Lumber Company Superfund Site Alton. Missouri.* https://clu-in.org/products/costperf/LNDTREAT/Scott.htm (accessed on 1 August 2021).
72. Varjani, S. J., (2017). Microbial degradation of petroleum hydrocarbons. *Bioresource Technology, 223*, 277–286.
73. Varjani, S. J., & Upasani, V. N., (2017). New look on factors affecting microbial degradation of petroleum hydrocarbon pollutants. *International Biodeterioration & Biodegradation, 120*, 71–83.
74. Vidali, M., (2001). Bioremediation: An overview. *Pure and Applied Chemistry, 73*(7), 1163–1172.
75. Wang, B., Xie, H. L., Ren, H. Y., Li, X., Chen, L., & Wu, B. C., (2019). Application of AHP, TOPSIS, and TFNs to plant selection for phytoremediation of petroleum-contaminated soils in shale gas and oil fields. *Journal of Cleaner Production, 233*, 13–22.
76. Wang, X., Zheng, J., & Han, Z., (2019). Bioremediation of crude oil - contaminated soil by hydrocarbon-degrading microorganisms immobilized on humic acid-modified biofuel ash. *Journal of Chemical Technology & Biotechnology, 94*(6), 1904–1912.
77. Wang, Y. N., Cheng, L. J., & Zhou, Q. X., (2016). Phytoremediation of petroleum contaminated soils with *Iris pseudacorus* L. and the metabolic analysis in roots. *Huan Jing Ke Xue Huanjing Kexue, 37*(4), 1531–1538.
78. Wu, Y., Jing, X., Gao, C., Huang, Q., & Cai, P., (2018). Recent advances in microbial electrochemical system for soil bioremediation. *Chemosphere, 211*, 156–163.
79. Xie, X., & Müller, N., (2019). Enhanced aniline degradation by *Desulfatiglans anilini*: Synthetic microbial community with the phototrophic purple sulfur bacterium *Thiocapsa roseopersicina. Systematic and Applied Microbiology, 42*(5). Article ID: 125998.
80. Zhen, M., Chen, H., Liu, Q., Song, B., Wang, Y., & Tang, J., (2019). Combination of rhamnolipid and biochar in assisting phytoremediation of petroleum hydrocarbon contaminated soil using *Spartina anglica. Journal of Environmental Sciences, 85*, 107–118.

PART III
Contaminated Sites and Xenobiotics

CHAPTER 6

PHYTOREMEDIATION OF SOILS CONTAMINATED BY HEAVY METALS (MERCURY, CADMIUM, AND LEAD)

MOHD. YOUSUF RATHER, ASHAQ AHMAD DAR, and MOHAMMED LATIF KHAN

ABSTRACT

Heavy metals (HMs) are non-degradable pollutants that are naturally present in the soil but toxic to humans and the environment at higher concentrations. Phytoremediation has been used to remediate HMs from the contaminated soils for deterring impacts on human health. The chapter outlines the general aspects of phytoremediation and various techniques used in phytoremediation. It also focuses on their use in the remediation of the soils contaminated with mercury, cadmium, and lead.

6.1 INTRODUCTION

Globally, environmental pollution is regarded as one of the grave concerns at present. Increasing environmental pollution is massively contributing to human health degradation as it is perceptible in the air, water, and on land. The anthropogenic activities since the introduction of industries have led to the augmentation of either deliberate or accidental chemical release of pollutants into the environment [89]. Hence, substantial consideration has been paid towards chemical pollutants, specifically heavy metals (HMs) [108].

The HMs are toxic and they tend to bioaccumulate in the environment [140]. They can pose a threat to humans in the long run, as was the case in Minamata and itai-itai diseases. The HM pollution has increased due to the growth and development of metal plating industries, mining, fertilizer industries, tanneries, battery works, pesticide industries, paper industries, etc. These pollutants, directly and indirectly, have led to soil contamination [32]. HMs are naturally present in the soil, but they are regarded as contaminants if their concentration is more than the acceptable limits. Some of the HMs are non-essential for plant growth, and when they are taken up by plants along with other metals, these get accumulated in different tissues in their toxic forms. Therefore, these HMs can easily enter the food chain (Figure 6.1).

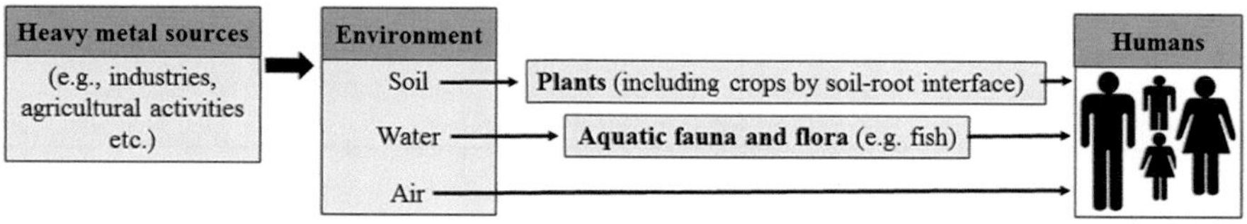

FIGURE 6.1 Heavy metal contamination to humans via food chain.

Some of the effects of HMs on human are [6]:

- Arsenic is carcinogenic and causes cellular damage and death;
- Barium damages cardiovascular and endocrine systems;
- Cadmium damages the renal tubules that are active in nutrient reabsorption process;
- Chromium causes skin diseases and ulcerations;
- Cobalt causes neurological and endocrine disorders;
- Copper causes gastrointestinal bleeding, hemolytic anemia;
- Lead causes respiratory diseases;
- Manganese causes neuro-toxicity;
- Mercury is a neurotoxin and damages lungs and kidneys;
- Nickel causes lung inflammation and irritation;
- Selenium causes respiratory and cardiovascular diseases;
- Strontium has general toxic effects on whole-body;
- Zinc causes liver and kidney failure, and stomach and abdominal cramps.

This is a serious health issue because humans can also get affected as they can be exposed to these contaminants by several routes like direct contact, inhalation, and ingestion [30]. They can also be accumulated in crop plants, which probably might result in food contamination [3]. Thus, removal of HMs from contaminated soils becomes an important area of remediation research to avoid or to minimize their impact on humans and the environment.

Before the development of advanced technologies for soil remediation, the HM contaminated soils were excavated and abandoned in a landfill. This did not solve the problem of HM contamination, but instead, the process relocates the HM contaminated soil from the polluted area to landfill site [95]. Afterwards, several physicochemical techniques were developed for the remediation of soils contaminated with HMs, however, their disadvantages are technical complications in application and associated high cost [136]. Besides, these methods might also disturb the native micro-flora and micro-fauna of the soil.

Phytoremediation is one of the developed alternatives to these physical and chemical methods for decontamination of HMs from polluted soils. Phytoremediation is an *in-situ* process in which plant species is used to remove and reduce the contaminants from the environment. The process has attracted a good deal of interest from soil remediation research community because the method is effective, economical, and simple. Additionally, the plants for the remediation process have no adverse effects on the topsoil, and hence the process is eco-friendly [32, 128].

There have been several advancements in the phytoremediation methods, such as phytodegradation [8, 35, 153], phytovolatilization [32, 119, 123, 142, 146], rhizofilitration [67, 98, 104, 107, 162] and phytoextraction [52, 91]. The improvements have made the technique more competent, and therefore it has also been used for the treatment of heavily contaminated industrial sites.

Therefore, considering the need of acquiring knowledge on concepts and advancements of phytoremediation technique and its application in the treatment of HM contaminated soils, this chapter reviews the process of phytoremediation with its background, and also discusses various techniques, factors affecting the process. The chapter also assesses the usefulness of phytoremediation technology by discussing the remediation of mercury, cadmium, and lead contaminated soils.

6.2 HEAVY METALS (HMS) IN SOILS

The contaminants and pollutants released into the environment as a result of various anthropogenic activities get finally accumulated in soil [32].

These substances because of various physicochemical properties affect the soil equilibrium and can also enter the food chain. The HMs among these contaminants are the most hazardous and toxic ones. HMs are metallic chemical elements with relatively high density and are harmful even at low concentrations. They are also defined as the elements that have a density of >5 g/cm^{-3} and weight ranging from 63.5 to 200.6 [127]. HMs, that are naturally present in the soil, include: cadmium, chromium, copper, mercury, lead, nickel, and zinc. Among these, zinc, copper, manganese, nickel, and cobalt are regarded as essential elements (micronutrients), which are important for plant growth and development, but they can be toxic at higher concentrations [148]. Some of the HMs, such as lead, cadmium, chromium, mercury, and arsenic are toxic even at low concentrations.

The concentration of HMs in the soil has increased due to increased industrial activities [20, 61], which are a most important anthropogenic source of HM pollution. A number of HMs (e.g., copper, tin, chromium, lead, mercury, zinc, nickel, and arsenic) are released into the environment as they are widely used in agriculture and industries (tanneries petrochemicals) [138]. The sources of some HMs are presented in Table 6.1.

TABLE 6.1 Sources of Some Heavy Metals

Heavy Metal	Sources	References
Arsenic	Antipyretics, antiseptics, caustics, antispasmodics, herbicides, fungicides, insecticides, paint, and cosmetics.	[25]
Barium	Ore mining and refining.	[81]
Cadmium	Both natural and anthropogenic sources: Batteries, ceramics, electronic, and metal finishing industries, electroplating industries, pigments, petroleum products, textiles, insecticides, solders, television sets, metallurgical industries.	[113]
Chromium	Mostly industrial emissions: Leather tanning, metallurgy, electroplating, and refractory.	[5]
Cobalt	Metal industry, construction industry, cosmetics, and jewelry, medical exposure.	[68]
Copper	Electronic chips, batteries, cellphones, semi-conductors, water pipes, fertilizer industry, pulp, and paper industry, fungicides, insecticides, catalysts, and metal processing products.	[144]

TABLE 6.1 *(Continued)*

Heavy Metal	Sources	References
Lead	Used in domestic, agricultural, and domestic applications: Melting and smelting of ores, exhaust from automobiles, fertilizer, pesticide, additive in pigment and gasoline, urban soil waste.	[170]
Manganese	Mining and metallurgical industries.	[92]
Mercury	Anthropogenic sources (volcanic eruptions and hot-springs) as well as anthropogenic activities: metal mining, fossil fuel combustion, medical waste, fungicides, fertilizers.	[94]
Nickel	Various industrial and technological applications: Mining, smelting, emissions from vehicles, disposal of household, municipal, and industrial waste, steel manufacturing and cement industry.	[121]
Selenium	Electronics, paint industry, glass industry, ceramics, metallurgy, chemical industry and pharmaceutics and in agriculture.	[40]
Strontium	Thermal power-plants, ferrous, and non-ferrous metallurgy enterprises, and chemical industry.	[90]
Zinc	Mostly anthropogenic sources: Galvanizing iron and steel, construction, siding, apparatus housings, office hardware, heating, and ventilation conduits, vehicle, and building enterprises for roofing.	[144]

In soil, these HMs can be found either as free ions or as soluble metal complexes. Also, they can be present in the form of carbonates, oxides, hydroxides, and silicates and are associated with soil organic matter [66]. For example:

- Cadmium is present in the form of insoluble complexes [138];
- The mobility of zinc is decided by the soil properties (such as pH, other chemical elements and cation exchange) [14];
- The bioavailability of nickel is restricted by soil texture, pH, organic matter content, presence of hydroxides [1];
- Copper in the form of cupric ions binds with soil organic as well as inorganic compounds [143];
- The bioavailability of arsenic depends on its ionic form [33], etc.

The availability of the metals to the plants is important for their absorption and the bioavailability of these metals will depend on their solubility in soil.

6.3 REMEDIATION METHODS FOR HEAVY METAL (HM) CONTAMINATED SOILS

The increased levels of HMs pollution in soil will also adversely affect food safety; therefore, it becomes essential to remediate the HM contaminated soils in the interest of avoiding adverse impacts on the environment and health [32]. Some of the techniques used for the remediation of HM contaminated soils are discussed briefly here:

1. **Natural Attenuation (NA):** These processes which uses the natural mechanisms to remove and reduce the contaminant quantities from the polluted sites. These processes are carried out by the environment itself and do not involve any sort of human intervention.
2. **Soil Replacement:** In this method, the contaminant concentration in the soil is first diluted and then the capacity of the soil is increased. The methods involve three concepts which are soil spading (a process of naturally degrading and reducing the contaminant by plowing and then spreading it), soil replacement (the process of replacing the contaminated soil with fresh and clean soil. This process is applied to small regions and new soil is treated before importing to avoid the pollution) and new soil importing (the method involves the addition of new and clean soil to the contaminated area and thereby diluting the contaminant. This method is expensive, laborious, and might disturb the soil ecosystem).
3. **Thermal Desorption:** The process of remediation of soil HMs by the application of heat using infrared radiation, steam, and microwave is known as *thermal desorption*. The heat volatilizes the metals which are then collected and removed by a carrier gas or vacuum negative pressure. The thermal desorption process may be low-temperature desorption (90–320°C) and high-temperature desorption (320–560°C). This is very simple but requires a long time for desorption, and the devices used are expensive.

4. **Air-Stripping:** This method in which air is injected to stimulate volatilization and then the pollutant is removed by subsequent adsorption.
5. **Incineration:** This is the method of soil contaminant remediation, in which the contaminants are combusted.
6. **Electro-Kinetic Remediation:** This method uses direct electric current to generate an electric field gradient that is used to remove HMs from contaminated soils. The method uses two electrodes to generate a low electric field, and the contaminants are transferred from soil to these electrodes by electro-migration flow. Although the method is considered as environment friendly and is cost-effective and easy to install but it has low efficiency.
7. **Soil Flushing:** It is another method that uses various aquatic solutions to solubilize the HM contaminants for their removal.
8. **Chemical Washing:** The technique that uses the chemicals for washing and scrubbing of contaminated soil and then separates the cleaned soil from HMs is called *chemical washing* or *soil washing*. The technique requires the multiple mechanical processes to wash the soil by chemicals liquids which comprises of freshwater, chelating agents, and other solvents.
9. **Oxidation:** Chemical oxidants are also applied to the contaminated soil to oxidize and subsequently degrade the contaminants, and the process is called *oxidation*.
10. **Solidification:** The contaminants from soil are also removed by tallying additives that can diminish their mobility, and the technique is known as *solidification*.

Bioremediation technology uses biological processes to remove the contaminants and minimize their impacts on the environment. The microorganisms, in particular, besides higher plants have been used for the removal of contaminants. These processes are readily accepted by the general public as they have the tendency to maintain the general ecological balance of nature. The principal objective of bioremediation is to clean up the contaminated and polluted sites by exploiting the biodegradative processes of nature. The processes are least expensive and moreover are a permanent solution to the contamination problems.

The remediation processes carried out without human intervention are referred to as *intrinsic bioremediation*, and the processes in which the properties of contaminants are manipulated to enhance the remediation rate are known as *enhanced bioremediation*. The enhanced bioremediation has two major approaches, which are biostimulation and bioaugmentation. *Biostimulation,* predominantly an alteration of environment, is the stimulation of bacterial communities present in the soil to enhance their proficiency to remediate the contaminants, while as b*ioaugmentation* is the addition of supplementary bacterial species to achieve improvements in the rate of contaminant degradation. Commonly, both these processes are used in combined form. Some other bioremediation technologies currently in use include:

1. **Air Spraying Technique:** which injects air into the contaminated area to stimulate aerobic degradation.
2. **Bioleaching:** A bioremediation process that uses microorganisms for contaminant extraction from the soil.
3. **Biosorption:** The process of using biological materials for adsorption of contaminants from polluted soils is called *Biosorption*.
4. **Biofiltration:** This method uses air filters of high surface area with microorganisms of contaminant degradation ability supported on it.
5. **Biopiling:** The contaminant to be removed is collected through the introduction of nutrients in an aerated environment.
6. **Biofuffing:** This is a process of contaminant removal from soil by increasing its porosity by augering.
7. **Biotransformation:** This process converts contaminant to less toxic form using biological processes.
8. **Rhizoremediation:** Rhizoremediation is the breakdown of organic contaminants in the soil zone around plant roots (the rhizosphere), generally as a result of plant roots stimulating microorganisms' catalytic activity.
9. **Composting:** The process of stimulation of microbial growth by the addition of biodegradable substrate for biodegradation is called *composting*.

All these remediation methods have some disadvantages, such as efficiency, cost-effectiveness, etc., and hence phytoremediation has been favored over them because the method is perceived as economical, efficient, novel, eco-friendly, and comprehensible. However, for the uptake of these HMs by plants through roots, their bioavailability is very important, and the bioavailability of a metal depends on its solubility [32]. Hence, the HMs present in the soil solution in turn control and regulate the bioavailability of other metals to plants.

6.3.1 PHYTOREMEDIATION

The first studies on phytoremediation were carried out in the 1950s and the term phytoremediation was coined in 1989 [104, 123]. The plant species act as "solar-driven pumping and filtering systems" and either absorb or degrade the contaminants. The method exploits the plants and the accompanying soil microbes for the remediation process [3].

Most studies reveal that the soils irrigated with HM contaminated wastewaters have a metal concentration of beyond permissible levels and plants readily uptake these metals [29]. This remediation process has good public acceptance as it protects the original identity of soil micro-fauna and micro-flora and is regarded as an eco-friendly method. The plant species having the ability to detoxify toxic metals do so by several mechanisms and strategies, such as chelating the metals at various cellular and sub-cellular levels resulting in their sequestration [47].

The phytoremediation of HMs from soils starts from its absorption though soil-root interface to its transport through phloem/xylem and finally its accumulation in various plant parts. Soon after the absorption of HMs by plant roots, its primary bulk flow and secondary bulk flow occur through xylem and phloem, respectively [88]. The uptake and transport can occur passively or actively. The HM ions in this process bind with special proteins of the plasma membrane for transportation, for example, *the Arabidopsis thaliana* plant has about 150 different proteins that can transport metal cations [59]. The final sequestration of these HMs takes place in the cell vacuoles [58, 96].

6.4 PLANTS USED FOR PHYTOREMEDIATION OF HEAVY METALS (HMS) FROM SOIL

For growth, plants require nutrients, and they are readily absorbed from the soil through their root systems and subsequently taken up either through active or passive mechanisms. The nutrients are transported either via apoplast or via symplast through roots and thereafter by xylem to other aerial parts [6, 32].

Some of the HMs are also taken by the plants along with the nutrients, but their uptake in higher concentrations can have adverse effects on them as these metals are known to be phytotoxic [66, 108]. However, *metallophytes* (woody or herbaceous plant species) can accumulate these HMs at low concentrations without showing any signs of damage, and so they are used for the phytoremediation process [9]. The efficiency of the process of phytoremediation relies on plant species, contaminants, and environment. Several plant species are known for their phytoremediation capability, and they are selected for remediation process based on their ability to absorb specific contaminants. Metallophytes are classified into three types, e.g.:

1. **Metal Indicators:** These are plants with the ability to uptake the HMs of higher concentrations from soil and accumulate them in their aerial parts. So, the shoots and leaves of metal indicator plants can reflect the possible concentrations of HMs present in soil. On continuous exposure to higher metal concentration, the plants show toxicity symptoms and eventually die off. An example of metal indicator species is *Coccotylus truncates* [161].
2. **Metal Excluders** These can accumulate the HMs present in the soil on their roots and cease their transport to above-ground tissues. These plants are efficiently used for phytostabilization techniques. An example of metal excluder species is *Silene paradoxa* [10, 103, 157, 169].
3. **Metal Hyperaccumulators** These are plants that have the ability to accumulate and transfer the high concentration of metals in foliage are classified as metal hyperaccumulator. Hyperaccumulation can be described as the potential of plants to accumulate metals in 100 times higher concentrations in their shoots in comparison to non-accumulator plants. Such plants remain healthy and do not show any toxicity signs even after accumulating toxic metals at higher concentrations. An example of metal hyperaccumulator species is *Ricinus communis* [17]. Around 400 metal hyperaccumulator species have been identified. The hyperaccumulator plants should be able to accumulate higher concentrations of metals; they should have higher growth and biomass production and resistant to pests and diseases [149]. Mostly metal hyperaccumulator plants are widely preferred for the phytoremediation process because these plants use these accumulated metals in various ecological and physiological functions, and hence, they halt the entry of these toxic metals into the food chain. The bioavailability of metals to plants or their stabilization could be enhanced by manipulating the soil parameters and the

normal plants in association with this could be used as an alternative to hyperaccumulator plants. Both herbs as well as woody plants have been used for the phytoremediation process and are reported to be the suitable remediators for certain specific habitats.

6.5 FACTORS AFFECTING PHYTOREMEDIATION

The mechanism of contaminant uptake and phytoremediation process efficiency depends on the type of plant or plant species, bioavailability of the contaminant, nature of the contaminant and soil properties [126]. The plant species must be capable of reducing or removing the toxic contaminants, and the contaminants must not be too toxic to the plant species neither it shall be in too less concentration so that it is not bioavailable to plant species and the environment also must be suitable for supporting the plant species. These factors are discussed briefly in subsections.

6.5.1 PLANT SPECIES

Selection of a plant for remediation of HMs is important as different plant species will have a tendency to accumulate different concentrations of HMs. The two basic requirements for the selection of a plant species for phytoremediation process are: the ability to accumulate more concentration of metals and to yield large biomass [136]. Plants also aid microorganisms in the augmentation process of contaminant degradation by secreting some factors into the soil. Additionally, the plants provide nutrients required for microorganism growth to support co-metabolism in rhizosphere which also enhances the contaminant degradation process.

6.5.2 PROPERTIES OF MEDIUM

The environment must support the phytoremediation process by having an appropriate pH, temperature, nutrient/fertilizer supply, moisture content, organic matter content, etc.; because these parameters affect the growth of plants and subsequently their metal absorption capacity. For example, fertilizers improve the health of plants as well as improve their efficiency of remediation [63]. The soil type and its characteristics is also an important parameter. Soil additives such as zeolites or coal fly ash decrease the bioavailability of metals.

6.5.3 BIOAVAILABILITY OF METALS

The bioavailability of metals to plants plays a significant role in their remediation, and the metals which are readily available to plants will be easier to remediate. Their bioavailability itself depends on pH, organic matter content, redox potential, etc. Some metal species can form complexes with organic C-based ligands, which influence their solubility in the soil [3]. It has also been observed that metals with higher oxidation states are less soluble, and the amount of organic matter also results in a decrease in their bioavailability in soil [37]. The contaminant must be soluble in water to such an extent so that it can be easily transported by the plant to its tissues. Microbes have also been reported to catalyze redox reactions, and by doing so, they alter the metal mobility and their bioavailability.

6.5.4 ADDITION OF CHELATING AGENTS

The chelating agents also affect the phytoremediation process as their addition can increase the HM uptake by the plants and can also stimulate the microorganisms present in and around the plant and thereby enhancing the uptake capacity. The improved rate of HM uptake will result in a reduction in the time and cost of the remediation process. The chelating agents are added to alkaline soils because the HM bioavailability is very low at alkaline pH [42]. The common effective chelating agent used for soil washing technique is ethylene-di-amine-tetra-acetic acid (EDTA), but the problem associated with EDTA is its expensive nature and slow biodegradability. Some chelating agents can significantly ease the HM uptake by plants; for example, it has been reported that EDTA enhances the lead uptake by *Zea mays* and *Brassica juncea* [13, 56].

6.6 PHYTOREMEDIATION OF MERCURY, CADMIUM, AND LEAD IN CONTAMINATED SOILS

Cadmium, chromium, copper, lead, mercury, and nickel are the common HMs found in the contaminated sites [45]. Among them, mercury, cadmium, and lead are the most significant and poisonous HMs even at very low concentrations (0.0001–0.1 mg/L) and are not known for any biological importance [4, 145]. Hence, the remediation of mercury, cadmium, and lead-contaminated soils using various plant species are discussed briefly in subsections.

6.6.1 MERCURY

Mercury and its compounds are the most hazardous environmental pollutants having strong bioaccumulation, carcinogenic, genotoxic, neurotoxic, and teratogenic effects [78]. Other major effects of mercury toxicity include: brain damage, depression, drowsiness, fatigue, hair loss, insomnia, memory loss, restlessness, lung, and kidney failure [44]. The soils contaminated with mercury had caused dreadful kidney and neurological damages and even death [83]. Anthropogenic global emissions of mercury to air from 20 key sectors were approximately 2,220 tons in 2015 [139]. Highly toxic methylmercury tends to bioaccumulate in the food chain and the environment, and therefore the remediation of mercury-contaminated soils is a global concern. Phytoremediation, because of its advantages than the physical and chemical methods has been adapted for the treatment of mercury-contaminated soils. The research on the exploration and examination of mercury hyperaccumulator plant species has been the research focus.

More than 200 plant species, mostly herbs have been found to accumulate and transfer mercury in the past 20 years [75]. As such, there are no accepted mercury hyperaccumulator plant species so far, though *Eremochloa ciliaris* and *Erato polymnioides* are regarded as "potential mercury hyperaccumulators" [105]. Some plant species for mercury remediation in contaminated soils are presented in Table 6.2.

TABLE 6.2 Plant Species Used for Remediation of Mercury Contaminated Soils

Plant Species	Family	References
Abelmoschus esculentus	Malvaceae	[48]
Aster ageratoides	Asteraceae	[105]
Axonopus compressus	Poaceae	[18]
Brassica ampestris	Brassicaceae	[105]
Brassica juncea	Brassicaceae	[16]
Buddleja officinalis	Scrophulariaceae	[105]
Chenopodium glaucum L.	Amaranthaceae	[147]

TABLE 6.2 *(Continued)*

Plant Species	Family	References
Cibotium barometz	Orchidaceae	[105]
Corrydalisedulis maxim	Papaveraceae	[105]
Cyclosorus acuminatus	Thelypteridaceae	[105]
Cyperus eragrostis	Cyperaceae	[85]
Cyrtomium macrophyllum	Dryopteridaceae	[156]
Datura stramonium	Solanaceae	[85]
Eremochkoa ciliaris	Poaceae	[105]
Fallopia multiflora	Polygonaceae	[105]
Gynura bicolor	Asteraceae	[105]
Houttuynia crdata	Saururaceae	[105]
Ixeris sonchifolia	Asteraceae	[105]
Jatropha curcas	Euphorbiaceae	[84]
Macleaya cordata	Papaveraceae	[105]
Medicago sativa L.	Fabaceae	[166]
Miscanthus sinensis	Poaceae	[167]
Morus nigra	Moraceae	[50]
Onchus brachyotus	Asteraceae	[105]
Oryza sativa L.	Poaceae	[72]
Oxalis corniculata L.	Oxalidaceae	[77]
Paspalum distichum L.	Poaceae	[36]
Phaseolus vulgaris	Fabaceae	[111]
Polypogon monospeliensis	Poaceae	[129]
Portulaca oleracea	Portulacaceae	[105]
Pratia nummularia	Campanulaceae	[105]
Primula sikkimensis	Primulaceae	[105]
Petrus vittata	Pteridaceae	[26]
Rumex acetosa	Polygonaceae	[105]
Senecio scandens	Asteraceae	[105]
Silene vulgaris	Caryophyllaceae	[98]
Solanum nigrum	Solanaceae	[70]
Telosma cordata	Apocynaceae	[105]

All these species (Table 6.1) have displayed different efficiencies for remediation of mercury from contaminated soils. It has been found that the biomass of several plant species decreases because of mercury toxicity [36, 156], while some species like *A. semibaccata, B. juncea, B. napus*, etc., can grow comfortably even in mercury-contaminated soils [80]. Hence, selecting a suitable plant species for phytoremediation technique should be a top priority. The plants, which are mercury tolerant, have high bioaccumulation and transfer factor; and such plants should be preferred. The other factors include: the bioavailability of mercury in soils, the soil properties (pH, cation-exchange capacity, electric conductivity, nutritive elements, organic matter, etc.), and the weather conditions [75].

6.6.2 CADMIUM

The average concentration of cadmium in soil ranges from 0.06 to 1.1 mg/kg [53]. It is regarded as an important HM because it is 2–20 times toxic compared to other metals [62]. In comparison to natural sources, anthropogenic sources add up 3–10 times more cadmium to the environment [57]. Cadmium is carcinogenic and mutagenic, and other effects of toxicity in humans include lung damage and endocrine disruption [43, 117]. Anthropogenic activities have released 2.2×10^7 kg of cadmium during the last five decades [31]. Hence, several approaches have been developed to remediate the cadmium in polluted soils and the phytoremediation process has been a preferred technology [53]. The process uses various hyperaccumulator plants to extract cadmium from the contaminated soils [86]. Some plant species with the potential to remediate cadmium are listed in Table 6.3.

TABLE 6.3 Plant Species Used for Remediation of Cadmium Contaminated Soils

Plant Species	Family	References
Amaranthus hybridus	Amaranthaceae	[165]
Arabidopsis halleri	Brassicaceae	[28]
Arabis gemmifera	Brassicaceae	[65]
Arabis gemmifera	Brassicaceae	[168]
Arabis paniculata	Brassicaceae	[134]
Arthrocnemum macrostachyum	Chenopodiaceae	[109]
Azolla pinnata	Azollaceae	[106]
Beta vulgaris var. cicla L.	Chenopodiaceae	[71]
Bidens pilosa L.	Asteraceae	[132]
Brassica napus L.	Brassicaceae	[112]

TABLE 6.3 *(Continued)*

Plant Species	Family	References
Brassica oleracea	Brassicaceae	[118]
Carthamus tinctorius L.	Asteraceae	[120]
Chara aculealata	Characeae	[125]
Chromolaena odorata	Asteraceae	[137]
Echinochloa polystachya	Poaceae	[124]
Eleocharis acicularis	Cyperaceae	[115]
Gynura pseudochina (L.) DC.	Asteraceae	[101]
Helianthus tuberosus L.	Asteraceae	[27]
Iris lactea var. chinensis	Iridaceae	[49]
Lonicera japonica Thunb.	Caprifoliaceae	[79]
Lycopersicon esculentum cv. Shenbaofen-2	Solanaceae	[51]
Mirabilis jalapa L.	Nyctaginaceae	[76]
Phytolacca americana	Phytolaccaceae	[97]
Picris divaricata	Asteraceae	[135]
Potentilla griffithii	Rosaceae	[54]
Prosopis laevigata	Fabaceae	[15]
Rorippa globosa	Brassicaceae	[152]
Sedum alfredii	Crassulaceae	[34]
Siegesbeckia orientalis L.	Asteraceae	[163]
Solanum nigrum L.	Solanaceae	[130]
Solanum photeinocarpum	Solanaceae	[164]
Tagetes patula	Asteraceae	[132]
Thlaspi caerulescens	Brassicaceae	[100]
Thlaspi praecox	Brassicaceae	[73]
Viola boashanensis	Violaceae	[153]

Extensive research has been carried out for the identification of plant species for remediation of cadmium contaminated soils, especially with respect to hyperaccumulators and hypertolerators [12]. Most cadmium hyperaccumulating species (such as *Thlaspi rotundifolium, Arabidopsis halleri, Thlaspi praecox, Sedum alfredii, Salsola kali, Viola baoshanensis*, etc.), are found in naturally occurring contaminated sites [150]; but *Solanum nigrum* [151] and *Rorippa golobosa* [152] with cadmium hyperaccumulation capability has been endorsed from clean soils. The cadmium hyperaccumulation in plants is credited to the large root-to-shoot translocation of cadmium instead of apoplastic-symplastic root uptake [69].

6.6.3 LEAD

Lead is released to the environment from diverse sources. The lead concentration ranges from 1 to 200 mg/kg with an average value of 13 mg/kg in the soil media [59]. It is frequently used in mining, chemical fertilizers and pesticides, smelting, burning of fossil fuels (that contain lead addictive), dumping of municipal waste, etc. Some commercial products, such as paints, television glass, medical equipment, ceramic glazes, ammunition, electrical equipment and batteries also contain lead [38].

It is a non-biodegradable metal, which makes it a significant pollutant owing to its toxicity even at low concentrations [59]. Lead toxicity results in memory loss, poor intelligence, coordination dilemma, developmental impairment and disabilities in learning and cardiovascular ailment in kids [154]. Numerous phytoremediation studies conducted on lead-contaminated soils have been found successful using several plant species [19, 41, 64]. Some of the plant species used for remediation of lead-contaminated soils are listed in Table 6.4.

TABLE 6.4 Plant Species Used for Remediation of Lead Contaminated Soils

Plant Species	Family	References
Alyssum lesbiacum	Brassicaceae	[7]
Alyssum murale	Brassicaceae	[7]
Ambrosia artemisiifolia	Asteraceae	[55]
Arabidopsis thaliana	Brassicaceae	[66]
Arabis paniculata	Brassicaceae	[134]
Astragalus bisulcatus	Fabaceae	[7]
Baccharis latifolia	Asteraceae	[11]
Bidens triplinervia	Asteraceae	[11]
Brassica juncea	Brassicaceae	[160]
Brassica oleracea	Brassicaceae	[7]
Buckwheat	Polygonaceae	[23]
Cynara cardunculus	Asteraceae	[39]
Euphorbia cheiradenia	Euphorbiaceae	[21]
Fagopyrum esculentum	Polygonaceae	[133]
Helianthus annuus	Helianthoideae	[23]

TABLE 6.4 *(Continued)*

Plant Species	Family	References
Hemidesmus indicus	Apocynaceae	[122]
Indian mustard	Brassicaceae	[23]
Jatropha curcas	Euphorbiaceae	[60]
Lepidium bipinnatifidum	Brassicaceae	[11]
Noea mucronata	Amaranthaceae	[22]
Onchus oleraceus	Asteraceae	[11]
Phaseolus vulgaris	Fabaceae	[82]
Pisum sativum	Fabaceae	[23]
Plantago orbignyana	Plantaginaceae	[11]
Poa pratensis	Poaceae	[51]
Populus deltoides	Salicaceae	[114]
Populus nigra	Salicaceae	[114]
Populus trichocarpa	Salicaceae	[114]
Raphanus sativus	Brassicaceae	[24]
Sedum alfredii	Crassulaceae	[46]
Sesuvium portulacastrum	Aizoaceae	[11]
Sonchus asper	Asteraceae	[158]
Sonchus oleraceus	Asteraceae	[155]
Tagetes minuta	Asteraceae	[116]
Thlaspi caerulescens	Brassicaceae	[7]
Thlaspi rotundifolium	Brassicaceae	[110]
Trifolium alexandrinum	Fabaceae	[2]
Zea mays	Poaceae	[55]
Zea mays	Poaceae	[87]

Removing lead becomes difficult once it is released into the soil matrix. The uptake of lead by plant species can be enhanced by increasing the pH, soil/water Eh, phosphate level, cation exchange capacity, and organic carbon content [141]. Among various plant species, *Agrostemma githago* displayed the highest lead accumulation and *Taraxacum officinale* and *Ambrosia artemisiifolia* have been reported as lead accumulator species [102]. The good lead accumulator species belong to Asteraceae, Brassicaceae, Euphorbiaceae, Lamiaceae, and Scrophulariaceae families [4]. *Brassica juncea* has a good ability to translocate lead from roots to shoots [74]. In China, *Vetivera zizanioides* has been a successful plant for lead phytoremediation [93].

All plant species discussed under the sections of remediation of mercury, cadmium, and lead-contaminated soils have displayed the potential to sequester the respective metals. Most of the plants are reported for hyperaccumulation potential. The amount of metal sequestration varies with the species and hence it is important to optimize the species for obtaining maximum metal extraction from the contaminated media. The phytoremediation is a recent and developing research area, and still, new species are screened for the pollutant removal capacity. Most researchers have reported phytoremediation technology as an eco-friendly, economical, and efficient process for remediation of contaminated soils.

6.7 SUMMARY

The noteworthy benefits of phytoremediation are: it does not allow the HMs to enter into the food chain; the plant remains could be used for biofuel production; and the recovered metals from the process could be handy for industrial use. Various phytoremediation techniques for HM removal from the soil are: phytoextraction, phytostabilization, phytovolatilization, phytodegradation, and rhizofiltration. Because of several advantages, the phytoremediation process has been a preferred technology at present for remediation of HMs like mercury, cadmium, and lead from the contaminated soils.

ACKNOWLEDGMENTS

The authors acknowledge the financial support (Grant No. BT/PR12899/NDB/39/506/2015 dt.20/06/2017) from the Department of Biotechnology, Government of India-New Delhi.

KEYWORDS

- **heavy metals**
- **hyperaccumulator plants**
- **phytodegradation**
- **phytoremediation**
- **phytovolatilization**
- **rhizofilitration**
- **soil contamination**

REFERENCES

1. Agency for Toxic Substances and Disease Registry (ATSDR), (2005). Toxicological profile for nickel. In: *Agency for Toxic Substances and Disease Registry Report* (p. 105). Atlanta, GA, USA. www.atsdr.cdc.gov/ToxProfiles (accessed on 1 August 2021).
2. Ali, H., Naseer, M., & Sajad, M. A., (2012). Phytoremediation of heavy metals by *Trifolium alexandrinum*. *International Journal of Environmental Science, 2*, 1459–1469.
3. Ali, H., Khan, E., & Sajad, M. A., (2013). Phytoremediation of heavy metals: Concepts and applications. *Chemosphere, 91*, 869–881.
4. Alkorta, I., & Hernández-Allica, J., (2004). Recent findings on the phytoremediation of soils contaminated with environmentally toxic heavy metals and metalloids such as zinc, cadmium, lead, and arsenic. *Reviews in Environmental Science and Biotechnology, 3*, 71–90.
5. Almeida, J. C., & Cardoso, C. E. D., (2019). Chromium removal from contaminated waters using nanomaterials: Review. *TrAC Trends in Analytical Chemistry, 118*, 277–291.
6. Awa, S. H., & Hadibarata, T., (2020). Removal of heavy metals in contaminated soil by phytoremediation mechanism: Review. *Water, Air, and Soil Pollution, 231*(2), 47–53.
7. Baker, A. J. M., Reeves, R. D., & McGrath, S. P., (1991). *In-situ* decontamination of heavy metal polluted soils using crops of metal-accumulating plants-a feasibility study. In: *In-situ Bioreclamation* (pp. 600–605). Butterworth-Heinemann, London.
8. Bamforth, S. M., & Singleton, I., (2005). Bioremediation of polycyclic aromatic hydrocarbons: Current knowledge and future directions. *Journal of Chemical Technology & Biotechnology: International Research in Process, Environmental & Clean Technology, 80*(7), 723–736.
9. Barceló, J., & Poschenrieder, C., (2003). Phytoremediation: Principles and perspectives. *Contributions to Science, 2003*, 333–344.
10. Basharat, Z., Novo, L. A. B., & Yasmin, A., (2018). Genome editing weds CRISPR: What is in it for phytoremediation? *Plants* (*Basel*), *7*(3), 51–55.
11. Bech, J., Duran, P., Roca, N., & Poma, W., (2012). Shoot accumulation of several trace elements in native plant species from contaminated soils in the Peruvian Andes. *Journal of Geochemical Exploration, 113*, 106–111.
12. Bhargava, A., Carmona, F. F., Bhargava, M., & Srivastava, S., (2012). Approaches for enhanced phytoextraction of heavy metals. *Journal of Environmental Management, 105*, 103–120.
13. Blaylock, M. J., & Salt, D. E., (1997). Enhanced accumulation of Pb in Indian mustard by soil-applied chelating agents. *Environmental Science and Technology, 31*, 860–865.
14. Broadley, M. R., & White, P. J., (2007). Zinc in plants. *New Phytologist, 173*(4), 677–702.
15. Buendia-Gonzalez, L., & Orozco-Villafuerte, J., (2010). *Prosopis laevigata* a potential chromium (VI) and cadmium (II) hyperaccumulator desert plant. *Bioresource Technology, 101*, 5862–5867.
16. Cassina, L., Tassi, E., Pedron, F., & Petruzzelli, G., (2012). Using a plant hormone and a thioligand to improve phytoremediation of Hg-contaminated soil from a petrochemical plant. *Journal of Hazardous Materials, 231*, 36–42.
17. Çelik, Ö., & Akdas, E. Y., (2019). Tissue-specific transcriptional regulation of seven heavy metal stress-responsive miRNAs and their putative targets in nickel indicator

castor bean (*R. communis* L.) plants. *Ecotoxicology and Environmental Safety, 170*, 682–690.

18. Chamba, I., Rosado, D., & Kalinhoff, C., (2017). *Erato polymnioides* - novel Hg hyperaccumulator plant in Ecuadorian rainforest acid soils with potential of microbe-associated phytoremediation. *Chemosphere, 188*, 633–641.
19. Chandrasekhar, C., & Ray, J. G., (2019). Lead accumulation, growth responses and biochemical changes of three plant species exposed to soil amended with different concentrations of lead nitrate. *Ecotoxicology and Environmental Safety, 171*, 26–36.
20. Chaney, R. L., Malikz, M., & Li, Y. M., (1997). Phytoremediation of soil metals. *Current Opinion in Biotechnology, 8*, 279–284.
21. Chehregani, A., & Malayeri, B. E., (2007). Removal of heavy metals by native accumulator plants. *International Journal of Agriculture and Biology, 9*, 462–465.
22. Chehregani, A., Noori, M., & Yazdi, H. L., (2009). Phytoremediation of heavy-metal-polluted soils: Screening for new accumulator plants in angouran mine (Iran) and evaluation of removal ability. *Ecotoxicology and Environmental Safety, 72*, 1349–1353.
23. Chen, Y., Li, X., & Shen, Z., (2004). Leaching and uptake of heavy metals by ten different species of plants during an EDTA-assisted phytoextraction process. *Chemosphere, 57*, 187–196.
24. Chen, Y. X., Lin, Q., Luo, Y. M., & He, Y. F., (2003). The role of citric acid on the phytoremediation of heavy metal contaminated soil. *Chemosphere, 50*(6), 807–811.
25. Chen, A. Y. Y., & Olsen, T., (2016). Chromated copper arsenate treated wood: A potential source of arsenic exposure and toxicity in dermatology. *International Journal of Women's Dermatology, 2*, 28–30.
26. Chen, J., Shiyab, S., Han, F. X., & Monts, D. L., (2009). Bioaccumulation and physiological effects of mercury in *Pteris vittata* and *Nephrolepis exaltata. Ecotoxicology, 18,* 110–121.
27. Chen, L., Long, X., Zheng, X., & Liu, Z., (2011). Effect on the photosynthetic characteristics of Cd uptake and translocation in seedlings of two *Helianthus tuberosus* varieties. *Acta Prataculturae Sinica, 20*(6), 60–67.
28. Chiang, H. C., Lo, J. C., & Yeh, K. C., (2006). Genes associated with heavy metal tolerance and accumulation in Zn/Cd hyperaccumulator *Arabidopsis halleri*: Genomic survey with cDNA microarray. *Environmental Science and Technology, 40*, 6792–6798.
29. Colpaert, J. V., Wevers, J. H., Krznaric, E., & Adriaensen, K., (2011). How metal-tolerant ecotypes of ectomycorrhizal fungi protect plants from heavy metal pollution. *Annals of Forest Science, 68*(1), 17–24.
30. Conte, F., Copat, C., & Longo, S., (2016). Polycyclic aromatic hydrocarbons in *Haliotis tuberculata* (Linnaeus, 1758) (Mollusca, Gastropoda): Considerations on food safety and source investigation. *Food and Chemical Toxicology, 94*, 57–63.
31. Costa, E. T. S., & Guilherme, L. R. G., (2012). Assessing the tolerance of castor bean to cd and pb for phytoremediation purposes. *Biological Trace Element Research, 145*, 93–100.
32. Cristaldi, A., Conti, G. O., & Jho, E. H., (2017). Phytoremediation of contaminated soils by heavy metals and PAHs: Brief review. *Environmental Technology and Innovation, 8*, 309–326.
33. Cullen, W. R., & Reimer, K. J., (1989). Arsenic speciation in the environment. *Chemical Reviews, 89*(4), 713–764.

34. Deng, D. M., Deng, J. C., Li, J. T., Zhang, J., Hu, M., Lin, Z., & Liao, B., (2008). Accumulation of zinc, cadmium, and lead in four populations of *Sedum alfredii* growing on lead/zinc mine spoils. *Journal of Integrative Plant Biology, 50*, 691–698.
35. Deng, Z., & Cao, L., (2017). Fungal endophytes and their interactions with plants in phytoremediation: Review. *Chemosphere, 168*, 1100–1106.
36. Ding, W., Zhang, J., Wu, S. C., Zhang, S., Christie, P., & Liang, P., (2019). Responses of the grass *Paspalum distichum* L. to Hg-stress: Proteomic study. *Ecotoxicology and Environmental Safety, 183*. Article ID: 109549.
37. Duval, J. F., (2013). Dynamics of metal uptake by charged biointerphases: Bioavailability and bulk depletion. *Physical Chemistry Chemical Physics, 15*, 7873–7888.
38. Environment Writer, (2000). *Lead (Pb) Chemical Backgrounder*. https://www.nsc.org/community-safety/safety-topics/other-poisons/lead-poisoning-prevention (accessed on 5 August 2021).
39. Epelde, L., Mijangos, I., Becerril, J. M., & Garbisu, C., (2009). soil microbial community as bioindicator of the recovery of soil functioning derived from metal phytoextraction with sorghum. *Soil Biology and Biochemistry, 41*, 1788–1794.
40. Etteieb, S., Magdouli, S., Zolfaghari, M., & Brar, S., (2020). Monitoring and analysis of selenium as an emerging contaminant in mining industry: Critical review. *Science of the Total Environment, 698*. Article ID: 134339.
41. Fanna, A. G., & Yadji, G., (2018). Phytoextraction of Pb, Cd, Cu and Zn by *Ricinus communis*. *Environment and Water Sciences, Public Health and Territorial Intelligence Journal, 2*, 56–62.
42. Gomez-Garrido, M., Mora, N. J., Murcia, N. F. J., & Faz, C. A., (2018). The chelating effect of citric acid, oxalic acid, amino acids, and *Pseudomonas fluorescens* bacteria on phytoremediation of Cu, Zn, and Cr from soil using *Suaeda vera*. *International Journal of Phytoremediation, 20*, 1033–1042.
43. Gratao, P. L., Monteiro, C. C., & Carvalho, R., (2012). Biochemical dissection of diageotropica and never ripe tomato mutants to cd-stressful conditions. *Plant Physiology and Biochemistry, 56*, 79–96.
44. Gulati, K., Banerjee, B., Lall, S. B., & Ray, A., (2010). Effects of diesel exhaust, heavy metals and pesticides on various organ systems: Possible mechanisms and strategies for prevention and treatment. *Indian Journal of Experimental Biology, 48*, 710–721.
45. Gupta, D. K., Huang, H. G., & Nicoloso, F. T., (2013). Effect of Hg, As, and Pb on biomass production, photosynthetic rate, nutrients uptake and phytochelatin induction in *Pfaffia glomerata*. *Ecotoxicology, 22,*1403–1412.
46. Gupta, D. K., Huang, H. G., Yang, X. E., Razafindrabe, B. H., & Inouhe, M., (2010). The detoxification of lead in *Sedum alfredii* H. is not related to phytochelatins but the glutathione. *Journal of Hazardous Materials, 177*, 437–444.
47. Hall, J. L., (2002). Cellular mechanisms for heavy metal detoxification and tolerance. *Journal of Experimental Botany, 53*(366), 1–11.
48. Hameed, A., Qadri, T. N., Mahmooduzzafar, T., Siddiqi, O., & Iqbal, M., (2011). Differential activation of the enzymatic antioxidant system of *Abelmoschus esculentus* L. under CdCl2 and HgCl2 exposure. *Brazilian Journal of Plant Physiology, 23*, 45–55.
49. Han, Y. L., Yuan, H. Y., Huang, S. Z., Guo, Z., Xia, B., & Gu, J. G., (2007). cadmium tolerance and accumulation by two species of *Iris*. *Ecotoxicology, 16*, 557–563.

50. Hashemi, S. A., & Tabibian, S., (2018). Application of *Mulberry nigra* to absorb heavy metal, mercury, from the environment of green space city. *Toxicology Reports, 5*, 644–646.
51. He, L. Y., Chen, Z. J., Ren, G. D., Zhang, Y. F., Qian, M., & Sheng, X. F., (2009). Increased cadmium and lead uptake of a cadmium hyperaccumulator tomato by cadmium-resistant bacteria. *Ecotoxicology and Environmental Safety, 72*, 1343–1348.
52. He, H., Dong, Z., Pang, J., Wu, G. L., Zheng, J., & Zhang, X., (2018). Phytoextraction of rhenium by lucerne (*Medicago sativa*) and erect milkvetch (*Astragalus adsurgens*) from alkaline soils amended with coal fly ash. *Science of the Total Environment, 630*, 570–577.
53. He, S., He, Z., Yang, X., Stoffella, P. J., & Baligar, V. C., (2015). Soil biogeochemistry, plant physiology, and phytoremediation of cadmium-contaminated soils. In: *Advances in Agronomy* (Vol. 134, pp. 135–225). Academic Press.
54. Hu, P., Qiu, R., & Senthilkumar, P., (2009). Tolerance, Accumulation and distribution of zinc and cadmium in hyperaccumulator *Potentilla griffithii. Environmental and Experimental Botany, 66*, 317–325.
55. Huang, J. W., & Cunningham, S. D., (1996). Lead phytoextraction: species variation in lead uptake and translocation. *New Phytologist, 13*, 75–84.
56. Huang, J. W., Chen, J., Berti, W. R., & Cunningham, S. D., (1997). Phytoremediation of lead-contaminated soils: Role of synthetic chelates in lead phytoextraction. *Environmental Science and Technology, 31*, 800–805.
57. Irwin, R., Van, M. M., Stevens, L., Seese, M. D., & Basham, W., (1997). *Environmental Contaminants Encyclopedia Arsenic Entry* (p. 75). Fort Collins: Colorado: Water resources division: National Park Service (NPS).
58. Jagetiya, B., & Sharma, A., (2013). Optimization of chelators to enhance uranium uptake from tailings for phytoremediation. *Chemosphere, 91*, 692–696.
59. Jagetiya, B., & Kumar, S., (2020). Phytoremediation of lead: Review. In: *Lead in Plants and the Environment* (pp. 171–202). Cham: Springer Nature.
60. Jamil, S., Abhilash, P. C., Singh, N., & Sharma, P. N., (2009). *Jatropha curcas*: Potential crop for phytoremediation of coal fly-ash. *Journal of Hazardous Materials, 172*, 269–275.
61. Jing, Y., He, Z., & Yang, X., (2007). Role of soil rhizobacteria in phytoremediation of heavy metal contaminated soils. *Journal of Zhejiang University Science B, 8*(3), 192–207.
62. Kabata-Pendias, A., (2001). *Trace Elements in Soils and Plants* (p. 432). Boca Raton, FL, USA: CRC Press. http://base.dnsgb.com.ua/files/book/Agriculture/Soil/Trace-Elements-in-Soils-and-Plants.pdf (accessed on 1 August 2021).
63. Kang, W., Bao, J., Zheng, J., Xu, F., & Wang, L., (2018). Phytoremediation of heavy metal contaminated soil potential by woody plants on tong Lushan ancient copper spoil heap in China. *International Journal of Phytoremediation, 2018*, 20, 1–7.
64. Kapourchal, S. A., Kapourchal, S. A., & Pazira, E., (2009). Assessing radish (*Raphanus sativus* L.) potential for phytoremediation of lead-polluted soils resulting from air pollution. *Plant Soil and Environment, 55*, 202–206.
65. Kashem, M. A., & Singh, B. R., (2007). Assessing the potential of *Arabidopsis halleri ssp. gemmifera* as a new cadmium hyperaccumulator grown in hydroponics. *Canadian Journal of Plant Science, 87*, 499–502.

66. Lasat, M. M., (2000). Phytoextraction of metals from contaminated soil: Review of plant/ soil/ metal interaction and assessment of pertinent agronomic issues. *Journal of Hazardous Substance, 2*(5), 1–25.
67. Lee, M., & Yang, M., (2010). Rhizofiltration using sunflower (*Helianthus annuus* L.) and bean (*Phaseolus vulgaris* L. var. *vulgaris*) to remediate uranium-contaminated groundwater. *Journal of Hazardous Materials, 173*, 589–596.
68. Leyssens, L., Vinck, B., Van, D. S. C., Wuyts, F., & Maes, L., (2017). Cobalt toxicity in humans: Review of the potential sources and systemic health effects. *Toxicology, 387*, 43–56.
69. Li, J. T., Liao, B., Zhu, R., Dai, Z. Y., Lan, C. Y., & Shu, W. S., (2011). Characteristics of Cd uptake, translocation and accumulation in a novel Cd-accumulating tree, star fruit (*Averrhoa carambola L., Oxalidaceae*). *Environmental and Experimental Botany, 71*, 352–358.
70. Li, X. X., Zhang, X., Wang, X. L., & Cui, Z. J., (2019). Phytoremediation of multi-metal contaminated mine tailings with *Solanum nigrum* L. and biochar/attapulgite amendments. *Ecotoxicology and Environmental Safety, 180*, 517–525.
71. Li, Y. S., Sun, L. N., Sun, T. H., & Wang, H., (2007). Cadmium hyperaccumulator *beta vulgaris var. cicla* L. and its accumulating characteristics. *Journal of Agriculture and Environmental Science (in Chinese)*, *26*, 1386–1389.
72. Li, Y. Y., Wang, Y. J., & Zhang, Q. J., (2019). Elemental sulfur amendment enhance methylmercury accumulation in rice (*Oryza sativa* L.) grown in Hg mining polluted soil. *Journal of Hazardous Materials, 379*. Article ID: 120701.
73. Likar, M., Pongrac, P., Vogel-Mikus, K., & Regva, M., (2010). Molecular diversity and metal accumulation of different *Thlaspi praecox* populations from Slovenia. *Plant Soil, 330*, 195–205.
74. Liu, D., Jiang, W., Liu, C., Xin, C., & Hou, W., (2000). Uptake and accumulation of lead by roots, hypocotyls, and shoots of Indian mustard (*Brassica juncea* L.). *Bioresource Technology, 71*, 273–277.
75. Liu, J., Chen, B., Wang, L., & Urbanovich, O., (2020). Review on phytoremediation of mercury-contaminated soils. *Journal of Hazardous Materials, 2020*, 8. doi: https://doi.org/10.1016/j.jhazmat.2020.123138.
76. Liu, J. N., Zhou, Q. X., Sun, T., Ma, L. Q., & Wang, S., (2008). Growth responses of three ornamental plants to Cd and CdePb stress and their metal accumulation characteristics. *Journal of Hazardous Materials, 151*, 261–267.
77. Liu, Z. C., Wang, L. A., Ding, S. M., & Xiao, H. Y., (2018). Enhancer assisted-phytoremediation of mercury-contaminated soils by *Oxalis corniculata* L., and rhizosphere microorganism distribution of *Oxalis corniculata* L. *Ecotoxicology and Environmental Safety, 160*, 171–177.
78. Liu, Z. C., Wang, L. A., Zeng, F. T., & Al-Hamadani, S. M. Z. F., (2015). The absorption and enrichment condition of mercury by three plant species. *Polish Journal of Environmental Studies, 24*(2), 887–891.
79. Liu, Z. L., He, X. Y., Chen, W., & Yuan, F. H., (2009). Accumulation and tolerance characteristics of cadmium in a potential hyperaccumulator - *Lonicera japonica* Thunb. *Journal of Hazardous Materials, 169*, 170–175.
80. Lomonte, C., Doronila, A. I., & Gregory, D., (2010). Phytotoxicity of biosolids and screening of selected plant species with potential for mercury phytoextraction. *Journal of Hazardous Materials, 173*, 494–501.

81. Lu, Q., Xu, X., Liang, L., Xu, Z., Shang, L., Guo, J., Xiao, D., & Qiu, G., (2019). Barium concentration, phytoavailability, and risk assessment in soil-rice systems from an active barium mining region. *Applied Geochemistry, 106*, 142–148.
82. Luo, C., Shen, Z., & Li, X., (2005). Enhanced phytoextraction of Cu, Pb, Zn and Cd with EDTA and EDDS. *Chemosphere, 59*, 1–11.
83. Mahbub, K. R., Krishnan, K., Naidu, R., Andrews, S., & Megharaj, M., (2017). Mercury toxicity to terrestrial biota. *Ecological Indicators, 74*, 451–462.
84. Marrugo-Negrete, J., & Durango-Hernández, J., (2016). Mercury uptake and effects on growth in *Jatropha curcas*. *Journal of Environmental Sciences, 48*, 120–125.
85. Mbanga, O., Ncube, S., Tutu, H., Chimuka, L., & Cukrowska, E., (2019). Mercury accumulation and transportation in wetland biota affected by gold mining. *Environmental Monitoring and Assessment, 191*, 186–189.
86. McGrath, S. P., & Lombi, E., (2006). Field evaluation of Cd and Zn phytoremediation potential by the hyperaccumulators *Thlaspi caerulescens* and *Arabidopsis halleri*. *Environmental Pollution, 141*, 115–125.
87. Meers, E., Slycken, S. V., & Adriaensen, K., (2010). Use of bioenergy crops (*Zea mays*) for phytoattenuation' of heavy metals on moderately contaminated soils: A field experiment. *Chemosphere, 78*, 35–41.
88. Mengel, K., & Kirkby, E. A., (1987). *Principles of Plant Nutrition* (p. 673). Dordrecht: Springer.
89. Miri, M., Derakhshan, Z., & Allahabadi, A., (2016). Mortality and morbidity due to exposure to outdoor air pollution in Mashhad metropolis, Iran: AirQ model approach. *Environmental Research, 151*, 451–457.
90. Mironyuk, I., Tatarchuk, T., Vasylyeva, H., Naushad, M., & Mykytyn, I., (2019). Adsorption of Sr (II) cations onto phosphated mesoporous titanium dioxide: Mechanism, isotherm and kinetics studies. *Journal of Environmental Chemical Engineering, 7*. Article ID: 103430.
91. Nakajima, H., Fujimoto, N., Yamamoto, Y., Amemiya, T., & Itoh, K., (2019). Response of secondary metabolites to Cu in the Cu-hyperaccumulator lichen *Stereocaulon japonicum*. *Environmental Science and Pollution Research, 26*, 905–912.
92. Nelson, M., Adams, T., Ojo, C., Carroll, M. A., & Catapane, E. J., (2018). Manganese toxicity is targeting an early step in the dopamine signal transduction pathway that controls lateral cilia activity in the bivalve mollusks *Crassostrea virginica*. *Comparative Biochemistry and Physiology - Part C, 213*, 1–6.
93. Oh, K., Cao, T., Li, T., & Cheng, H. Y., (2014). Study on application of phytoremediation technology in management and remediation of contaminated soils. *Journal of Clean Energy Technologies, 2*, 216–220.
94. Pacyna, E. G., Pacyna, J. M., Steenhuisen, F., & Wilson, S., (2006). Global anthropogenic mercury emission inventory for 2000. *Atmospheric Environment, 40*(22), 4048–4051.
95. Paz-Ferreiro, J., Gascó, G., Méndez, A., & Reichman, S. M., (2018). Soil pollution and remediation. *International Journal of Environmental Research and Public Health, 15*, 1657.
96. Peer, W. A., & Baxter, I. R., (2005). Phytoremediation and hyperaccumulator plants. *Topics in Current Genetics, 14*, 84–89.
97. Peng, K. J., Luo, C. L., You, W. X., Lian, C. L., Li, X. D., & Shen, Z. G., (2008). Manganese uptake and interactions with cadmium in the hyperaccumulator - *Phytolacca Americana* L. *Journal of Hazardous Materials, 154*, 674–681.

98. Perez-Palacios, P., Agostini, E., & Ibanez, S. G., (2017). Removal of copper from aqueous solutions by rhizofiltration using genetically modified hairy roots expressing a bacterial Cu-binding protein. *Environmental Technology, 38*, 2877–2888.
99. Pérez-Sanz, A., Millán, R., & Sierra, M. J., (2012). Mercury uptake by *Silene vulgaris* grown on contaminated spiked soils. *Journal of Environmental Management, 95*, S233-S237.
100. Perronnet, K., Schwartz, C., & Morel, J. L., (2003). Distribution of cadmium and zinc in the hyperaccumulator *Thlaspi caerulescens* grown on multicontaminated soil - distribution of metals in *Thlaspi caerulescens*. *Plant Soil, 249*, 19–25.
101. Phaenark, C., Pokethitiyook, P., Kruatrachue, M., & Ngernsansaruay, C., (2009). Cd and Zn accumulation in plants from the padaeng zinc mine area. *International Journal of Phytoremediation, 11*, 479–495.
102. Pichtel, J., Kuroiwa, K., & Sawyerr, H. T., (2000). Distribution of Pb, Cd and Ba in soils and plants of two contaminated soils. *Environmental Pollution, 110*, 171–178.
103. Pignattelli, S., Colzi, I., & Buccianti, A., (2012). Exploring element accumulation patterns of a metal excluder plant naturally colonizing a highly contaminated soil. *Journal of Hazardous Material, 227*, 362–369.
104. Pulford, I. D., & Watson, C., (2003). Phytoremediation of heavy metal-contaminated land by trees: Review. *Environment International, 29*, 529–540.
105. Qian, X. L., Wu, Y. G., Zhou, H. Y., & Xu, X. H., (2018). Total mercury and methylmercury accumulation in wild plants grown at wastelands composed of mine tailings: Insights into potential candidates for phytoremediation. *Environmental Pollution, 239*, 757–767.
106. Rai, P. K., (2008). Phytoremediation of Hg and Cd from industrial effluents using an aquatic free-floating macrophyte *Azolla pinnata*. *International Journal of Phytoremediation, 10*, 430–439.
107. Raskin, I., Smith, R. D., & Salt, D. E., (1997). Phytoremediation of metals: Using plants to remove pollutants from the environment. *Current Opinion in Biotechnology, 8*, 221–226.
108. Rather, M. Y., Tilwani, Y. M., & Dey, A., (2019). Assessment of heavy metal contamination in two edible fish species *Carassius carassius* and *Triplophysa kashmirensis* of Dal Lake, Srinagar, Kashmir, India. *Environmental Monitoring and Assessment, 191*(4), 242–246.
109. Redondo-Gomez, S., & Mateos-Naranjo, E., (2010). Accumulation and tolerance characteristics of cadmium in halophytic Cd- hyperaccumulator, *Arthrocnemum macrostachyum*. *Journal of Hazardous Materials, 184*, 299–307.
110. Reeves, R. D., & Brooks, R. R., (1983). Hyperaccumulation of lead and zinc by two metallophytes from mining areas of central Europe. *Environmental Pollution, 31*, 277–285.
111. Restrepo-Sánchez, N. E., & Acevedo-Betancourth, L., (2015). Remediation effect of compost on soluble mercury transfer in a crop of *Phaseolus vulgaris*. *Journal of Environmental Sciences, 31*, 61–67.
112. Rossi, G., Figliolia, A., Socciarelli, S., & Pennelli, B., (2002). Capability of *Brassica napus* to accumulate cadmium, zinc and copper from soil. *Biotechnolgia Acta, 22*, 133–140.
113. Rubio, C., Paz, S., & Tius, E., (2018). Metal contents in the most widely consumed commercial preparations of four different medicinal plants (Aloe, Senna, Ginseng, and Ginkgo) from Europe. *Biological Trace Element Research, 186*, 562–567.

114. Ruttens, A., Boulet, J., Weyens, N., & Smeets, K., (2011). Short rotation coppice culture of willows and poplars as energy crops on metal-contaminated agriculture soils. *International Journal of Phytoremediation, 13*, 194–207.
115. Sakakibara, M., Ohmori, Y., Ha, N. T. H., Sano, S., & Sera, K., (2011). Phytoremediation of heavy metal contaminated water and sediment by *Eleocharis acicularis*. *Clean – Soil, Air, Water, 39*, 735–741.
116. Salazar, M. J., & Pignata, M. L., (2014). Lead accumulation in plants grown in polluted soils: Screening of native species for phytoremediation. *Journal of Geochemical Exploration, 137*, 29–36.
117. Salem, H. M., Eweida, E. A., & Farag, A., (2000). Heavy metals in drinking water and their environmental impact on human health. In: *Proceedings of ICEHM 2000 Meeting* (pp. 542–556). Cairo University, Egypt.
118. Salt, D. E., & Rauser, W. E., (1995). Mg ATP-dependent transport of phytochelatins across the tonoplast of oat roots. *Plant Physiology, 107*, 1293–1301.
119. San-Miguel, A., Ravanel, P., & Raveton, M., (2013). Comparative study on the uptake and translocation of organochlorines by *Phragmites australis*. *Journal of Hazardous Materials, 244*, 60–69.
120. Sayyad, G., Afyuni, M., & Mousavi, S. F., (2010). Transport of Cd, Cu, Pb and Zn in a calcareous soil under wheat and safflower cultivation-a column study. *Chemosphere, 154*, 311–320.
121. Schaumlöffel, D., (2012). Nickel species: Analysis and toxic effects. *Journal of Trace Elements in Medicine and Biology, 26*, 1–6.
122. Sekhar, K. C., Kamala, C. T., Chary, N. S., Balaram, V., & Garcia, G., (2005). Potential of *Hemidesmus indicus* for phytoextraction of lead from industrially contaminated soils. *Chemosphere, 58*, 507–514.
123. Sharma, P., & Pandey, S., (2014). Status of phytoremediation in world scenario. *International Journal of Environmental Bioremediation and Biodegradation, 2*(4), 178–191.
124. Solís-Domínguez, F. A., & Gonzalez-Chavez, M. Z., (2007). Accumulation and localization of cadmium in *Echinochloa* polystachya grown within a hydroponic system. *Journal of Hazardous Materials, 141*, 630–636.
125. Sooksawat, N., & Meetam, M., (2003). Phytoremediation potential of charophytes: Bioaccumulation and toxicity studies of cadmium, lead and zinc. *Journal of Environmental Science, 25*, 596–604.
126. Sreelal, G., & Jayanthi, R., (2017). Review on phytoremediation technology for removal of soil contaminant. *Indian Journal of Scientific Research, 14*(1), 127–130.
127. Srivastava, N. K., & Majumder, C. B., (2008). Novel biofiltration methods for the treatment of heavy metals from industrial wastewater. *Journal of Hazardous Materials, 151*(1), 1–8.
128. Sterckeman, T., Gossiaux, L., Guimont, S., & Sirguey, C., (2019). How could phytoextraction reduce cd content in soils under annual crops? simulations in the French context. *Science of the Total Environment, 654*, 751–762.
129. Su, Y., Han, F. X. X., Chen, J., Sridhar, B. B. M., & Monts, D. L., (2008). Phytoextraction and accumulation of mercury in three plant species: Indian mustard (*Brassica Juncea*), beard grass (*Polypogon monospeliensis*), and Chinese brake fern (*Pteris vittata*). *International Journal of Phytoremediation, 10*, 547–560.

130. Sun, R. L., Zhou, Q. X., Sun, F. H., & Jin, C. X., (2007). Antioxidative defense and proline/phytochelatin accumulation in a newly discovered Cd-hyperaccumulator, *Solanum nigrum* L. *Environmental and Experimental Botany, 60*, 468–476.
131. Sun, Y., Xu, Y., Zhou, Q., Wang, L., Lin, D., & Liang, X., (2013). The potential of gibberellic acid 3(GA3) and tween-80 induced phytoremediation of co-contamination of Cd and Benzo[a]pyrene (B[a]P) using *Tagetes patula*. *Journal of Environmental Management, 114*, 202–208.
132. Sun, Y., Zhou, Q., Wang, L., & Liu, W., (2009). Cadmium tolerance and accumulation characteristics of *Bidens pilosa* L. as a potential Cd-hyperaccumulator. *Journal of Hazardous Materials, 161*, 808–814.
133. Tamura, H., Honda, M., Sato, T., & Kamachi, H., (2005). Pb hyperaccumulation and tolerance in common buckwheat (*Fagopyrum esculentum* Moench). *Journal of Plant Research, 118*(5), 355–359.
134. Tang, Y. T., Qiu, R. L., & Zeng, X. W., (2009). Lead, zinc, cadmium hyperaccumulation and growth stimulation in *Arabis paniculata* franch. *Environmental and Experimental Botany, 66*, 126–134.
135. Tang, Y. T., Qiu, R., Zeng, X., Fang, X., & Yu, F., (2009). Zn and Cd hyperaccumulating characteristics of *Picris divaricata* vant. *International Journal of Environment and Pollution, 38*, 26–38.
136. Tangahu, B. V., Sheikh, A. S. R., & Basri, H., (2011). Review on heavy metals (As, Pb, and Hg) uptake by plants through phytoremediation. *International Journal of Chemical Engineering, 2011*, 31–35.
137. Tanhan, P., Kruatrachue, M., Pokethitiyook, P., & Chaiyarat, R., (2007). Uptake and accumulation of cadmium, lead, and zinc by siam weed (*Chromolaena odorata* (L.) king and Robinson). *Chemosphere, 68*, 323–329.
138. Tchounwou, P. B., & Yedjou, C. G., (2014). Heavy *Metal Toxicity and the Environment* (p. 45). Washington, DC: NIH Public Access.
139. The United Nations Environment Programme (UN Environment)., (2019). *Global Mercury Assessment – 2018*. Online; https://www.unenvironment.org/resources/publication/global-mercury-assessment-2018 (accessed on 1 August 2021).
140. Tüzen, M., (2003). Determination of heavy metals in soil, mushroom and plant samples by atomic absorption spectrometry. *Microchemical Journal, 74*(3), 289–297.
141. USEPA, (1992). *Selection of Control Technologies for Remediation of Lead Battery Recycling Sites* (p. 89). EPA/540/S-92/011. US Environmental Protection Agency, Office of Emergency and Remedial Response, Washington, DC, USA.
142. Van-Oosten, M. J., & Maggio, A., (2014). Functional biology of halophytes in the phytoremediation of heavy metal contaminated soils. *Environmental and Experimental Botany, 111*, 135–146.
143. Van-Sprang, P., & Vangheluwe, M., (2008). Voluntary risk assessment of copper, copper (II) sulphate pentahydrate, copper (I) oxide, copper (II) Oxide, dicopper chloride trihydroxide. available at: European chemicals agency (ECHA). https://echa.europa.eu/it/copper-voluntary-risk-assessment-reports (accessed on 5 August 2021).
144. Vardhan, K. H., Kumar, P. S., & Panda, R. C., (2019). Review on heavy metal pollution, toxicity and remedial measures: Current trends and future perspectives. *Journal of Molecular Liquids, 290*. Article ID: 111197.
145. Wang, J. L., (2002). *Immobilization Techniques for Biocatalysts and Water Pollution Contamination*. Science Press, Beijing.

146. Wang, J., Feng, X., Anderson, C. W. N., Xing, Y., & Shang, L., (2012). Remediation of mercury-contaminated sites: A review. *Journal of Hazardous Materials, 221*, 1–18.
147. Wang, J. X., Feng, X. B., & Anderson, C. W. N., (2011). Ammonium thiosulphate enhanced phytoextraction from mercury-contaminated soil - results from a greenhouse study. *Journal of Hazardous Materials, 186*, 119–127.
148. Wani, K. A., Manzoor, J., Dar, A. A., Shuab, R., & Lone, R., (2020). Heavy metal intrusion and accumulation in aquatic ecosystems. In: Qadri, H., Bhat, R. A., Mehmood, M. A., & Dar, G. H., (eds.), *Fresh Water Pollution Dynamics and Remediation* (pp. 83–104). Springer, Singapore.
149. Watanabe, M. E., (1997). Phytoremediation on the brink of commercialization. *Environmental Science and Technology, 31*(4), 182–186.
150. Wei, S., Zhou, Q., & Koval, P. V., (2006). Flowering stage characteristics of cadmium hyperaccumulator *Solanum nigrum* L. and their significance to phytoremediation. *Science of the Total Environment, 369*, 441–446.
151. Wei, S., Zhou, Q., Wang, X., Zhang, K., Guo, G., & Ma, L., (2005). Newly-discovered Cd-hyperaccumulator *Solanum nigrum* L. *Chinese Science Bulletin, 50*, 33–38.
152. Wei, S., & Zhou, Q., (2006). Phytoremediation of cadmium-contaminated soils by *Rorippa globose* using two-phase planting. *Environmental Science and Pollution Research, 13*, 151–155.
153. Wu, C., Liao, B., Wang, S. L., Zhang, J., & Li, J. T., (2010). Pb and Zn accumulation in a Cd-hyperaccumulator (*Viola baoshanensis*). *International Journal of Phytoremediation, 12*, 574–585.
154. Wuana, R. A., & Okieimen, F. E., (2011). Heavy metals in contaminated soils: Review of sources, chemistry, risks and best available strategies for remediation. *ISRN Ecology, 2011*, 20. Article ID 402647. https://doi.org/10.5402/2011/402647.
155. Xiong, Z. T., (1997). Bioaccumulation and physiological effects of excess lead in a roadside pioneer species *Sonchus oleraceus* L. *Environmental Pollution, 97*, 275–279.
156. Xun, Y., Feng, L., Li, Y. D., & Dong, H. C., (2017). Mercury accumulation plant *Cyrtomium macrophyllum* and its potential for phytoremediation of mercury polluted sites. *Chemosphere, 189*, 161–170.
157. Yang, S. X., Liao, B., Yang, Z. H., Chai, L. Y., & Li, J. T., (2016). Revegetation of extremely acid mine soils based on aided phytostabilization: A case study from Southern China. *Science of the Total Environment, 562*, 427–434.
158. Yanqun, Z., Yuan, L., Jianjun, C., Haiyan, C., Li, Q., & Schvartz, C., (2005). Hyperaccumulation of Pb, Zn and Cd in herbaceous grown on lead-zinc mining area in Yunnan, China. *Environment International, 31*(5), 755–762.
159. Yin, H., Tan, Q., Chen, Y., Lv, G., He, D., & Hou, X., (2011). Polycyclic aromatic hydrocarbons (PAHs) pollution recorded in annual rings of gingko (*Gingko biloba* L,): Translocation, radial diffusion, degradation and modeling. *Microchemical Journal, 97*, 131–137.
160. Zaier, H., Ghnaya, T., & Lakhdar, A., (2010). Comparative study of Pb - phytoextraction potential in *Sesuvium portulacastrum* and *Brassica juncea*: Tolerance and accumulation. *Journal of Hazardous Materials, 183*, 609–615.
161. Zalewska, T., & Danowska, B., (2017). Marine environment status assessment based on macrophytobenthic plants as bioindicators of heavy metals pollution. *Marine Pollution Bulletin, 118*, 281–288.

162. Zhang, H., Zheng, L. C., & Yi, X. Y., (2009). Remediation of soil co-contaminated with pyrene and cadmium by growing maize (*Zea mays* L). *International Journal of Environmental Science and Technology, 6*, 249–258.
163. Zhang, S., Lin, H., Deng, L., & Gong, G., (2013). Cadmium tolerance and accumulation characteristics of *Siegesbeckia orientalis* L. *Ecological Engineering, 51*, 133–139.
164. Zhang, X., Xia, H., Li, Z., Zhuang, P., & Gao, B., (2011). Identification of a new potential Cd-hyperaccumulator *Solanum photeinocarpum* by soil seed bank-metal concentration gradient method. *Journal of Hazardous Materials, 189*, 414–419.
165. Zhang, X. C., Zhang, S. R., Xu, X. X., & Li, T., (2010). Tolerance and accumulation characteristics of cadmium in *Amaranthus hybridus* L. *Journal of Hazardous Materials, 180*, 303–308.
166. Zhang, Y. Y., Liu, J. H., & Zhou, Y. M., (2013). Enhanced phytoremediation of mixed heavy metal (Mercury) - organic pollutants (Trichloroethylene) with transgenic alfalfa co-expressing glutathione s-transferase and human P450-2E1. *Journal of Hazardous Materials, 260*, 1100–1107.
167. Zhao, A. Q., Gao, L. Y., Chen, B. Q., & Feng, L., (2019). Phytoremediation potential of *Miscanthus sinensis* for mercury-polluted sites and its impacts on soil microbial community. *Environmental Science and Pollution Research, 26*, 34818–34829.
168. Zhao, F. J., & Jiang, R. F., (2006). Cadmium uptake, translocation, and tolerance in the hyperaccumulator *Arabidopsis halleri*. *New Phytologist, 172*, 646–654.
169. Zhao, L., Li, T., Yu, H., Zhang, X., & Zheng, Z., (2016). Effects of [S,S]-ethylene diaminedi succinic acid and nitrilotriacetic acid on the efficiency of Pb phytostabilization by *Athyrium wardii* (Hook.) grown in Pb-contaminated soils. *Journal of Environmental Management, 182*, 94–100.
170. Zulfiqar, U., Farooq, M., & Hussain, S., (2019). Lead toxicity in plants: Impacts and remediation. *Journal of Environmental Management, 250*. Article ID: 109557.

CHAPTER 7

MICROBIAL AND PHYTOREMEDIATION-BASED REMOVAL OF POLYCYCLIC AROMATIC HYDROCARBONS (PAHS) IN SOIL ENVIRONMENTS

PODDUTURI SINDHURA, PODDUTURI VANAMALA, TANAJI VASAVILATHA, and MIR ZAHOOR GUL

ABSTRACT

The green technologies using microorganisms (bioremediation) and plants (phytoremediation) are economic and efficient alternative approach than other remediation methods to decontaminate polluted soils. Different strategies and protocols for bioremediation and phytoremediation have been designed, optimized, and adapted to restore contaminated terrestrial environments with PAHs. Various bacterial and fungal species and plants have the ability of bio-transforming these pollutants into less hazardous ones. Nonetheless, there are constraints for effective implementation of these bioremediation technologies, such as insufficient monitoring tools and flawed risk management. The use of advanced and contemporary molecular techniques to understand PAH-contaminated sites is revealing an increasing number of yet uncultured microorganisms, whose degrading ability remains unexplored.

7.1 SOIL: A VITAL RESOURCE

Soil is a natural platform that supplies the vitals for the sustenance of every living organism, including humans. Soil performs several services which can be categorized into four general categories: productivity and sustainability, biodiversity, environmental quality, and human well-being. Some of these functions are storing and cycling nutrients, decomposition of organic matter, storage of carbon, providing anchorage for roots, holding water and nutrients that are essential for optimal plant growth for providing food and biomass production.

Soil consortium have an important role in soil formation, biogeochemical cycling, organic matter decomposition, and bioremediation. Soils are fundamental in the preservation of environmental quality at a worldwide level. Sustainable management of soils and land supports agricultural productivity, food security, climate change mitigation and a range of ecosystem services. Healthy soils are key to our food security and sustainable future.

The present chapter presents a review of the existing remediation methods and strategies for the elimination of different PAHs from the contaminated soils.

7.2 SOIL DEGRADATION

Soil pollution refers to the presence of chemicals or substances out of place and/or present at a higher-than-normal concentrations that have adverse effects on any targeted organism [37]. Owing to increased human activities, improper disposal of wastes, and demand for agricultural land soil contamination is becoming a serious global environmental crisis since it harbors many health hazards. Soil pollution results in undesirable changes in the physical, biological, and chemical characteristics, which results in adverse effects on plant growth and animal health. Soil contamination affects the ability of soil to perform some of its vital functions in the ecosystem. Soil pollution is often accompanied by decreased soil fertility and agricultural output decreases availability of nutrients to plants and cause an imbalance of the ecosystem in the soil.

Plants grown in polluted soil accumulate high concentrations of soil pollutants through bioaccumulation. Crops and fodder grown on polluted soil can pass the contaminants on to consumers and the food chain and can manifest as diseases in humans and the aquatic life. Soil pollution can occur because of human activities, natural processes, dumping of industrial wastes, a large number of organic chemical compounds (such as pesticides), polycyclic aromatic hydrocarbons (PAHs), petroleum hydrocarbons (PHs), phthalate esters and nitroaromatic compounds [68]. Biological contaminants, such as pathogens (*Clostridium tetani*) and parasites (hookworms) are also responsible for soil pollution, causing negative impacts on human health.

7.3 TYPES OF SOIL CONTAMINANTS

Several chemicals may contaminate soils and can be categorized into inorganic contaminants and organic contaminants. Inorganic contaminants include: heavy metals (Cd, Cr, Hg, Mn, Pb) metalloids, non-metals (Se), actinides, and halogens-(F and I) [59]. Organic contaminants include: PAHs, dioxins, polychlorinated biphenyls (PCB), solvents (benzene), DDT (dichloro diphenyl trichloroethane), BTEX compounds.

7.3.1 AROMATIC COMPOUNDS IN SOIL

The contamination of soil with aromatic compounds has become one of the most serious ecological problems during recent years due to rapid industrialization and urbanization throughout the world. These compounds pose serious risk to human health because of the associated toxigenic, mutagenic, and carcinogenic properties. The most common types of aromatic contaminants in soil include:

- Monocyclic aromatic compounds (BTEX: benzene, toluene, ethylbenzene, and xylene);
- Organophosphorus and carbamate insecticides (pesticides);
- Per and polyfluorinated alkyl substances;
- Pharmaceuticals;
- Polybrominated biphenyls;
- Polychlorinated biphenyls (PCBs);
- Polychlorinated dibenzofurans;
- Polycyclic aromatic hydrocarbons (PAHs);
- Solvents.

7.3.1.1 MONO AROMATIC COMPOUNDS

BTEX group of contaminants is among the most abundantly produced chemicals with a worldwide annual production of 8–10 million tons of benzene, 5–10 million tons of toluene, ethylbenzene, and xylene [35]. These non-oxygenated mono aromatic hydrocarbons have been recognized as environmental priority pollutants and are of concern because of their toxic, carcinogenic, mutagenic, and genotoxic properties [61]. Chemicals of petroleum products, such as gasoline, diesel fuel, heating, and lubricating oils from leaking tanks are the most important sources of BTEX contamination of soil and groundwater.

7.3.1.2 POLYCYCLIC AROMATIC HYDROCARBONS (PAHS)

Polycyclic aromatic hydrocarbons (PAHs) represent a broad group of persistent organic pollutants (POPs) generated primarily by incomplete combustion of coal gas oil and wood. PAHs are ubiquitous environmental pollutants that consist of two or more fused aromatic rings in a linear, cluster, or angular arrangement. Soil contamination by PAHs is of serious concern as it possesses detrimental biological effects, mutagenicity, toxigenicity, and carcinogenicity. PAHs are lipophilic and readily soluble in organic solvents. The sources of PAH can be either anthropogenic or natural.

Natural sources of PAH formation include volcanic eruptions, burning of vegetation in forest and bush fires, erosion of sedimentary rocks containing PHs. On the contrary, anthropogenic sources are considered as significant contributors to PAH formation and result from a wide range of human activities such as smoke from wood, burning stoves, automotive emission and spillage of petroleum products, high-temperature industrial processes and usage of fertilizers in agricultural production.

Three major sources of PAHs are: petrogenic, pyrogenic, and biogenic. Petro genic PAHs are from petroleum and petroleum-derived products and are often marked as in an abundance of alkyl-substituted PAHs (such as alkyl naphthalene's, alkyl phenanthrenes and alkyl dibenzothiophenes (DBT)). Pyrogenic PAHs are produced from combustion processes and are comprised of predominantly unsubstituted PAHs. Biogenic aromatic compounds, including aromatic amino acids, lignin compounds, and their derivatives are of biotransformation origin [82].

PAHs can be classified into two types: low molecular weight (LMW) PAHs and high molecular weight (HMW) PAHs based on the presence of several aromatic benzene rings. LMW PAHs (anthracene, phenanthrene, naphthalene) comprises of 2 or 3 benzene rings and HMW PAHs (chrysene, benzo(a)pyrene, coronene, etc.), consists of more than three benzene rings [79]. PAHs accumulate in agricultural soils and have been identified to pose a serious threat to the health and well-being of humans because of their toxicological relevance. Frequent exposure to sub-lethal doses of these compounds can cause several physiological impairments, such as liver damage, hemolytic anemia, weight loss, gastrointestinal deterioration, impaired immune system and reduced productivity [57].

The most frequent PAHs are: anthracene, naphthalene, pyrene, benzopyrene, and phenanthrene. Several remediation technologies (ReTec) have been developed for the removal of PAHs from environments like biodegradation, photo-oxidation, chemical oxidation, leaching, and bioaccumulation [2]. Among them, bioremediation is the most promising relatively efficient, safe, and cost-effective option for remediation of soils adulterated with aromatic compounds. PAH can be degraded by employing naturally occurring soil microorganisms that are aerobic as well as anaerobic organisms belonging to the genera *Pseudomonas, Alcaligenes, Ralstonia, Arthrobacter, Enterobacter, Corynebacterium, Mycobacteria, Micrococcus,* etc. Fungal species mainly *Pleurotus ostreatus, Irpex lacteus, Aspergillus, Phanerochaete chrysporium* were used for remediation of soils contaminated with PAHs [79].

7.4 BIOREMEDIATION APPROACHES

The increasing awareness of the adverse impact of contaminated lands has led to international efforts for the bioremediation of aromatic compounds to remove the contaminants in soil.

Bioremediation offers degradation of contaminants to non-hazardous or less hazardous substances, which are mediated by the microorganism. Bioremediation is the process where organic wastes are biologically degraded under controlled conditions to level below concentration limits established by regulatory authorities. Bioremediation is effective only when environmental conditions permit microbial growth and activity, but to enhance the rate of bioremediation often it involves the manipulation of environmental parameters. When the microorganisms are imported to the site of degradation at a faster level, the process can be considered as Bioaugmentation.

Various microorganisms (bacteria, fungi, algae) and plants have excellent potential for bioremediation. The actual process of biodegradation is very slow. Only a few species of bacteria and fungi reported their ability to degrade the complex pollutants significantly. Many factors influence the microbial metabolic activities to use pollutants as substrates to metabolize them. Therefore, understanding metabolic pathways, mechanisms, and the enzymes responsible, is an essential aspect for the efficient removal of environmental pollutants.

7.4.1 TYPES OF BIOREMEDIATIONS

7.4.1.1 BIOSTIMULATION

In this approach, specific nutrients are injected at the site to stimulate or to enhance the activities of indigenous microorganisms. In this process, firstly, nutrient factors or fertilizers are supplied at the site of treatment, then the controlled environment with optimal pH, temperature, and oxygen are provided to increase the metabolic activities of microorganisms [28].

7.4.1.2 BIOAUGMENTATION

This is a strategic approach that includes the addition of pollutant degrading microbes (natural/engineered) to increase the potential of biodegradation of indigenous microflora. For this process, the microbes with biodegradation capacity are collected from the remediation site, separately cultured, genetically modified and returned to the site [28, 58].

7.4.1.3 BIOATTENUATION

This method is a natural attenuation (NA) model which involves the eradication of pollutants from the surroundings. It is a biological process that can be carried out as an aerobic or anaerobic degradation mechanism. This process even includes physical phenomena like advection, dispersion, dilution, diffusion, volatilization, sorption [72].

7.4.1.4 BIOVENTING

In this approach, enough oxygen is provided at low airflow rates to sustain microbial activity. Oxygen is usually supplied through direct air injection into the contaminated soil through wells. Effective bioremediation of petroleum-contaminated soil has been reported in various studies [3, 60].

7.4.2 FACTORS AFFECTING BIOREMEDIATION EFFICIENCY

Nature has many potential microbes that can degrade any type of pollutant, but the reason for an increased spread in pollutants in soil is that the microbes and pollutants do not co-exists. Bioremediation can be achieved successfully by making these pollutants available to the microbes. Apart from this, many other factors play an important role in the bioremediation of soil contaminated with aromatic compounds. They include biotic (microbial activity) and abiotic (types of soil, temperature, pH, moisture, oxygen availability, presence or absence of electron donors and acceptors, etc.), factors. Any manipulations in the factors may cause a delay in the bioremediation of soil.

7.4.2.1 ABIOTIC FACTORS

7.4.2.1.1 Soil Types

When different types of soils are compared for the improvement in its quality by bioremediation after getting contaminated with aromatic compounds, it was found that sandy soil showed the highest recovery than any other types of soil [48]. This is because sandy soil is highly porous, which provides enough space for the growth of microbes [4]. Least recovery was shown by clay soil contaminated with TPH. This is because clay soil has low permeability and low oxygen availability. Also, it has high amounts of humic acid that protects the soil organic material and makes it unavailable to microbes [86]. Hence, bioremediation of clay soil is a challenging task [14]. But the recovery rate of clay soil also can be improved up to 57% within a month by improving the oxygen availability in the soil. Therefore, more the porosity of soil, more will be the availability of oxygen and more will be the space for the microbes to accommodate and hence more is the chance of recovery.

7.4.2.1.2 Temperature

Among all the factors, temperature is the most important factor as it greatly influences the microbial activity and the structural integrity of the pollutant [7]. For a microbe to carry out its metabolic activities, many enzymes are required whose catalytic activity depends on the temperature. At low temperatures, these enzymes can be inactive and make the microbe inactive. On the other hand, at very high temperatures, microbial enzymes will be completely denatured, killing the organism. Sub-zero temperatures freeze the cytoplasm of bacteria and make the bacteria completely inactive [65, 99]. Hence, the optimum temperature is to be maintained for effective uptake of the aromatic contaminant. A slow increase in temperature increases the microbial activity and reaches a maximum at optimum temperature. Later upon the further increase, microbial activity reduces and may be completely arrested due to the death of the organism.

7.4.2.1.3 pH

pH also plays a major role in the growth of microbes that are involved in biodegradation. Optimum pH is extremely important as low or high pH than the optimum would affect the metabolism to a wide extent [98]. Uptake of pollutants as nutrients is dependent on the pH of the surroundings, which also influences the net negative charge on the bacterial cell surface. To degrade an aromatic pollutant, the pollutant must be solubilized as it is hydrophobic. The pH of the surrounding environment plays an important role in the solubility of such hydrophobic compounds [76]. Optimum pH for the makes the aromatic compound available to the bacteria and thus speed up the process of biodegradation.

7.4.2.1.4 Soil Water Content

Water content in the soil depends on the type of soil because each type has its holding capacity [48]. One must remember that to carry out optimum microbial growth, water content should be maximum, but it should not cause the blockage of oxygen diffusion in soil [81]. However, much detail about the exact amount of moisture required for biodegradation is yet to be investigated. For the pollutant to be absorbed by the microbe, it must be first solubilized in the water so that it becomes easily available to the organism. The optimum moisture content required for maximum absorption is 10–20% [53].

7.4.2.1.5 Oxygen Availability

For aerobic organisms, oxygen acts as a terminal electron acceptor. Petroleum contamination is mostly recovered by aerobic organisms and least by anaerobic organisms. Anaerobic organism's terminal electron acceptor is other than oxygen, such as nitrate, sulfate, iron, etc. [63, 65].

7.4.2.1.6 Organic Matter

Use of organic matter (such as buffalo manure (BM), food and kitchen waste (FKW), vegetable and fruits waste (VFW) and activated sewage sludge (SS)) is more promising in improving the growth of microbes that help in biodegradation of soil contaminated with aromatic compounds [64]. This is because organic wastes are rich in organic matter that is easily available nutrients for microbes than any other chemicals. The usage of organic matter for bioremediation is cost-effective and eco-friendlier [73]. Moreover, such organic matter also improves soil structure [94, 101].

Composting of soil with cattle manure and mixed vegetable waste completely removed the contamination of creosote within 19 months [12]. Bioremediation of such heavy hydrocarbons can be obtained using either synthetic organic fertilizer or natural organic fertilizers. Synthetic organic fertilizers include nitrogen, phosphorus, and potassium (NPK), whereas natural organic fertilizers include cow dung and poultry litter. Use of these two nutrients did not show much significant difference; but above NPK, natural organic fertilizers are preferred as they are more economical and eco-friendlier [73]. Moreover, organic waste is highly heterogeneous that can greatly affect the process of biodegradation [39]. If the organic matter used is rich in some soluble fractions, the rate of biodegradation can be enhanced. This is because the soluble fraction acts as a direct source of energy for the microbes [45, 47].

Hence it was found that organic matter with high amounts of soluble fraction favored the growth of PAH degrading microorganisms and hence more favorable for bioremediation of soli with organic pollutants. Such soluble fractions were found to be high in BM and SS than in FKW and VFW. With this one can understand that BW and SS can be used for effective biodegradation of PAH. But it was found that BM also consists of more amounts of insoluble lignin than in SS. Lignin cannot be used by bacteria but can be used by fungi, which is a slow metabolizing organism. Hence comparatively SS is more favorable than BM. FKW consisted of a high content of acid detergent soluble (ADS), which includes more complex compounds such as hemicelluloses and large proteins [49, 103]. Hence it is also not easily degraded by bacteria. Similarly, VFW also has high amounts of hemicelluloses and very low amounts of soluble fractions and hence is also not easily degraded by bacteria.

7.4.2.1.7 Availability of Nutrients

Providing a balanced condition is very important to ensure the maximum growth of microorganisms. The addition of nutrients balances the requirements of microbes. Particularly, carbon, nitrogen, and phosphorus addition hasten the metabolic activity of the organism. These nutrients do exist naturally in the soil but low concentrations [65]. Hence, they must be provided in appropriate concentration to increase the rate of metabolism and thus the process of degradation Usually, the ratio of C:N:P should be 100:10:1 to ensure successful biodegradation by microbial activity [29, 74].

7.4.2.2 Biotic Factors

The main aim of biodegradation is to make use of some naturally available organisms that do not in turn harm the ecological system. This biodegradation is mainly achieved by the metabolic activity of bacteria, fungi, and plants. The focus is on the use of microbes for bioremediation of aromatically contaminated soil because they are the most reliable organisms for this purpose [38, 54]. The microbes that can use these organic contaminants as a source of carbon and as a source of energy can only be employed for this purpose. But under natural conditions, there exists tough competition between the organisms for the carbon source available. Also, biodegradation is affected by predator organisms like protozoa and bacteriophages. The major biological factors that immensely influence the process of biodegradation are mutations, the interaction between organisms, enzymatic activity, population size, horizontal gene transfers [20, 78].

Aerobic and anaerobic bacteria, fungi can be used in biodegradation based on the zone of clearance. Fungi, especially *Vascular Arbascular Mychorhizae* (VAM) depend on other organisms for shelter [52] and plant growth-promoting rhizobia [13]. With this association, fungi protect itself from harsh chemicals and other toxic substances within the surroundings [11, 85]. It was found that pure cultures are less efficient in degrading the contaminant than the mixed population [76]. When microbes exist as mixed cultures, they carry out gene transfers and confer resistance, tolerance, and chemical degrading ability among each other. Such transformations are not possible when pure cultures are used. Hence, mixed populations ensure successful biodegradation than pure cultures.

7.5 STRATEGIES FOR BIOREMEDIATION OF POLYCYCLIC AROMATIC HYDROCARBONS (PAHS)

The microbial remediation and phytoremediation are safe and inexpensive approaches that use microorganisms and plants to eradicate the various contaminants including PAHs from the soil environments (Figure 7.1). These approaches provide useful and efficient alternative strategies to physical and chemical remediation process as they are economically and ecologically viable options. This chapter deals with the existing remediation methods and strategies to remove various PAHs from the soils.

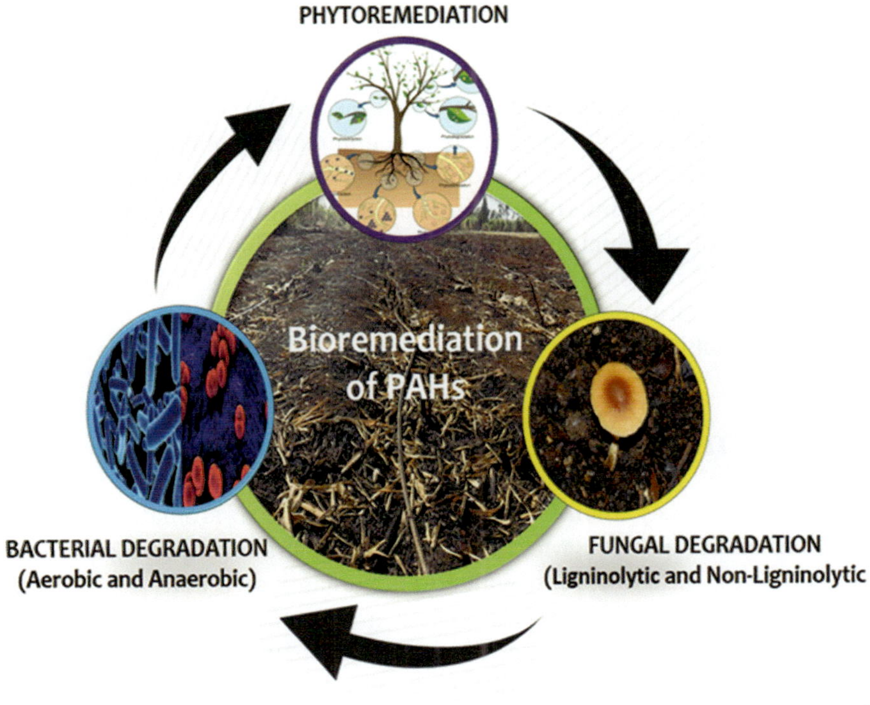

FIGURE 7.1 Bioremediation approaches to facilitate the eradication of polycyclic aromatic hydrocarbons (PAHs) from the soil environment.

7.5.1 MICROBIAL BIOREMEDIATION

Bioremediation is an effective solution for the transformation of hazardous fuel hydrocarbons into simple non-toxic forms or less toxic forms. Aromatic compounds are organic molecules containing one or more aromatic rings, specifically benzene rings. There are three major categories of aromatic compounds-PAHs, heterocyclics, and substituted aromatics. Specifically, the HMW PAHs are paid attention as they are recalcitrant in soils and less easy to degrade than many other organic compounds. They are difficult to remove from contaminated soil using the treatments that have been used successfully to clean soils contaminated with more degradable or volatile organic compounds (VOCs) such as alkanes.

Although PAH may undergo adsorption, volatilization, photolysis, and chemical degradation, microbial degradation is the major and eco-friendly degradation process. The rate of microbial degradation depends on pH, temperature, oxygen, microbial population, accessibility of nutrients, the chemical structure of a compound, bioavailability of the compound and chemical partitioning in the growth medium.

Another important factor in the bioremediation of contaminated soils is the availability of nitrogen and phosphorus, which allows the necessary increase in the size of the hydrocarbon-degrading microbial populations. It is a fact that each contaminated site can respond in a different way to distinct parameters that affect microbial degradation [28]. A better understanding of the diversity of the microbial communities inhabiting PAH-contaminated soils and their response to different biostimulation or bioaugmentation strategies could provide clues about the type of bacteria that can adapt to and exploit such habitats.

7.5.2 BACTERIAL BIODEGRADATION

Bacteria are actively involved in the biodegradation of aromatic petroleum hydrocarbon compounds. Most of the bacterial strains capable of biodegradation found in association with PAHs contaminated soil or sediments. Bacterial biodegradation involves various groups of bacteria and in biodegradation, few factors play a major role. The moisture content and aeration were determined to be the key factors associated with PAH bioremediation, and it is a very significant process in the natural decontamination of the environment. Enhanced biodegradation phenomenon is the basis for bioremediation technologies [6].

Several bacterial species are known to degrade PAHs, and most of them are isolated from contaminated soil or sediments. *Pseudomonas aeruginosa, Pseudomonas fluorescence, Mycobacterium* sps., *Rhodococcus* sps., *Haemophilus* sps, *Paenibacillus* spp. are found mostly in association with contaminated soils and play a significant role as PAH-degrading bacteria because of their quick adaptability to the environment. By developing several strategies, bacteria obtain the energy virtually from most of these aromatic compounds, hence they are considered as nature's ultimate scavengers. There are two major strategies associated with biodegradation of PAHs depending on the presence or absence of oxygen, and the rate can be enhanced by physical or chemical pretreatment of contaminated soil.

In the aerobic biodegradation, the oxygen is not only the final electron acceptor but also a co-substrate for the hydroxylation and oxygenolytic ring cleavage. In contrast, the anaerobic metabolism of aromatic compounds uses a completely different strategy, primarily based on reductive reactions [22, 40]. The anaerobic degradation pathways are most recently discovered compared to aerobic degradation. However, anoxic conditions are most prevalent in contaminated soils, where the anaerobic bacteria use alternative final electron acceptors such as nitrate, sulfate, or ferric ions [22, 40]. To enhance the potential of bacterial degradation of petroleum contaminants, the addition of biosurfactant-producing bacteria and light oils has a great impact. It increases the bioavailability of PAHs.

7.5.2.1 AEROBIC BIODEGRADATION

Biodegradation by aerobic bacteria usually involves the enzymatic action by dioxygenases, which are a multi-component system; for instance, naphthalene dioxygenase consists of a ferredoxin reductase, a ferredoxin, and a Fe-S protein [46]. Bacteria metabolize the PAHs via an oxygenase-mediated catabolic pathway. They also degrade PAHs by the cytochrome mediated 450 pathways, with the production of *trans*-dihydrodiols [71, 87]. The various aromatic intermediates formed in this catabolic process may serve as substrates for various other pathways and metabolize eventually by microbial enzyme activities.

Among all PAHs, Benzo(a)pyrene is the most toxic pollutant in soil. As per certain reviews, bacterial metabolic pathways can degrade these aromatic compounds; on incubation with *Sphingomonas paucimobilis* strain EPA 505, in liquid cultures. Few other bacterial strains, including *Pseudomonas,*

Agrobacterium, Bacillus, Burkholderia, and Sphingomonas spp, are found in association with contaminated soil and grow by using industrial wastes and other effluents from refineries. *Rhodococcus sp., Mycobacterium, and Flavobacterium spp.*, also been reported to be capable of Benzo(a)pyrene degradation [92, 97].

Pseudomonas fluorescence and *Haemophilus sps.*, the rod-shaped bacteria were found to be active in metabolizing the complex aromatic soil contaminants including acenaphthene, fluorine, phenanthrene, anthracene, pyrene by 70–100% in 40 days of initial treatment [100]. Recent studies described significant enhancement of the biodegradation of HMW PAHs and alkylated derivatives of these compounds from Casablanca crude oil caused by a microbial consortium in the presence of biosurfactant produced by *Pseudomonas aeruginosa* strain AT10 [1].

PAHs were degraded in a significant ratio and the biodegradation phenomenon found excessively when nutrients were not added, especially for benzo(a) anthracene and chrysene. All the treatment results showed that in first-order degradation rate during the first 45 days is higher, which was gradually reduced after 90 days. At the early stages of biodegradation, the proteobacteria group (genera *Sphingomonas* and *Azospirillum*) was the dominant group in all treatments. At later stages, there is a variation in microbial communities involved, especially the proteobacteria group (genus *Xanthomonas, Sphingomonas*) and the cytophaga-Flexibacter-Bacteroides group (Bacteroidetes) were the dominant groups in the non-nutrient treatment, while the proteobacteria group (genus *Xanthomonas*), and the proteobacteria group (genus *Sphingomonas*) were the dominant group in nutrient treatment [96].

Two other bacterial strains (*Mycobacterium flavescenes* and *Rhodococcus spp.*,) isolated from two different locations, are capable of PAH degradation and pyrene mineralization. Both these strains actively metabolize fluoranthene to produce 9-fluorenone-1-carboxylic acid [30]. Bacteria that are capable of degrading complex aromatic compounds, phenanthrene, are reported to be in association with mangrove sediments and the bacterial growth is directly proportional to the phenanthrene concentration [89]. Also, few bacterial strains produce monooxygenases that metabolize PAHs to trans-dihydrodiols.

7.5.2.2 ANAEROBIC BIODEGRADATION

Anaerobic bioremediation offers an attractive technology to reduce the soil contaminants, as these compounds are often present in the anoxic zones of the environment. The anaerobic biotransformation of aromatic hydrocarbons plays an important role when oxygen demand exceeds the supply in nature. Under such environmental conditions, anoxic or anaerobic degradation mediated by denitrifying or sulfate-reducing bacteria becomes significant for the cleanup of contaminated soils.

Most of the PAHs have been investigated with respect to anaerobic degradation. Benzene is persistent and most recalcitrant under anaerobic conditions. Anaerobic degradation of benzene has been observed in some sediments, microcosm, and microbial enrichments. In the mechanism of anaerobic degradation, the common strategy involves the reduction of the aromatic ring and converts most aromatic substrates into benzoyl-CoA which acts as a central intermediate. Most of the facultative anaerobic bacteria reduce benzoyl-CoA by an ATP-driven benzoyl-CoA reductase enzyme. By contrast, strictly anaerobic bacteria use a different class of benzoyl-CoA reductase that is ATP independent. The low redox potential that is required for benzoyl-CoA reduction is achieved by electron bifurcation. The mechanism involved is probably electrons from two reduced ferredoxins, split by a flavin cofactor in to low- and high-redox-potential electrons. Low-redox-potential electrons may reduce the aromatic ring, whereas high-redox-potential electrons may reduce NAD^+ [42].

Several anaerobic bacteria have been isolated, with degradation capacities. For instance, the bacteria belong to *Azoarcus tolulyticus* To14 [41], *Thauera aromatica* T1 [36], F, and *Magnetospirillum* strains, were described that can degrade toluene by utilizing nitrate as the electron acceptor [83]. The toluene oxidizing nitrate-reducing *Thauera* and *Azoarcus* species are facultative anaerobes. Most of these bacteria were isolated from anaerobic sludge or sediments [10].

7.5.3 ROLE OF FUNGI IN BIOREMEDIATION

Some HMW PAHs are mutagenic to indigenous bacteria and teratogenic or carcinogenic to the mammals found in the contaminated sites. Mycoremediation offers a unique and much useful alternative for reducing the soil pollution contaminated with complex aromatic compounds. The extensive research on fungi associated with the degradation of aromatic compounds has been reviewed in the last several years. There are following advantages of using fungi for biodegradation of aromatic hydrocarbons:

- Fungal species can absorb organic contaminants specifically benzo(a) pyrene into their cell walls and exhibit significant degradation [18, 90].
- Fungi are resistant to a high concentration of recalcitrant compounds, high temperatures and acidic conditions [9, 32].
- They can metabolize PAHs even under microaerophilic conditions. Hence, they are the potential organisms capable of degrading PAHs in saturated aquifers, waterlogged soils, paddy rice swamps and other anaerobic zones [31].
- They can secrete extracellular enzymes such as lignin peroxidase, versatile peroxidase, manganese dependent peroxidase, phenoloxidases (laccase, tyrosinase), H_2O_2 producing enzymes, cytochrome P 450 monooxygenase, dioxygenase, lipase, proteases, and epoxide hydrolase that can effectively oxidize PAHs [32, 77].
- They can spread themselves in a wide range of areas with their branched mycelium [18, 84].

Most fungi cannot grow by using PAHs as sole carbon and energy source, but they can co-metabolize them into a wide variety of oxidized products. Such metabolism is found with ligninolytic fungi (a group of fungi that produce enzymes to degrade lignin) and non-ligninolytic fungi (fungi that do not produce these enzymes).

7.5.3.1 LIGNINOLYTIC FUNGI

Lignin is a natural aromatic polymer that is structurally similar to PAHs. Ligninolytic fungi are the one which can produce enzymes that can degrade lignin. White rot fungi (WRF) and brown rot fungi are the ligninolytic fungi that are extensively used in bioremediation of soil contaminated with PAHs. The enzyme secreted by ligninolytic fungi has very low specificity for its substrate and hence can be employed in the degradation of many structurally related PAHs [56]. WRF are ligninolytic fungi that can not only degrade a carcinogenic PAH, i.e., benzo[a]pyrene but also detoxifies the soil and sediments with PAHs. Therefore, a wide variety of white-rot basidiomycetes such as *Phanerochaete chrysoporium, Trametes versicolor, Cirnipellis stipitaria, Bjerkandera adusta and Pleurotus ostreatus* can degrade PAHs.

LMW PAHs (naphthalene, phenanthrene, anthracene) can be degraded by *Aspergillus* sps., *Trichocladium canadense* and *Fusarium oxysporum* whereas HMW PAHs (chrysene, fluoranthene, pyrene) can be degraded by *Trichocladium canadense, Acremonium* sps., *Curvularia* sps., *Pythium* sps., *Aspergillus* sps., *Verticillium* sps., and *Achremonium* sps. [5, 84]. Ligninolytic fungi produce some extracellular ligninolytic enzymes also including lignin peroxidases, manganese peroxidases (MnP), and laccases that are involved in the degradation of aromatic compounds. These fungi make use of both peroxidases (LiP (lignin peroxidases) and MnP) and laccases [26, 51] to oxidize PAHs except *Pleurotus ostreatus* which metabolizes PAH using cytochrome P450 monooxygenase [16, 66, 91].

Three cytochrome P450 genes (CYP63A1, A2, A3) in *Phanerochaete chrysosporium* encode for the enzymes involved in the oxidation of both low as well as HMW PAHs [33, 34]. The expression of the genes to encode those enzymes usually depends on the availability of nutrients, cultural conditions, and the substrates used as inducers. The ligninolytic enzymes oxidize PAHs with the loss of an electron and cause the release of hydroxyl free radical [26, 50]. Therefore, instead of dihydrodiol PAH-quinone and acids will be formed. For example, in biodegradation of phenanthrene by *P. ostreatus,* phenanthrene forms 9,10-phenanthrenequinone and then 2,2-diphenic acid and CO_2, but the formation of phenanthrene 9,10-oxide and trans 9,10-phenanthrenequinone dihydrodiol is skipped [15]. With this, one can conclude that the rate of biodegradation varies with variation in the enzyme and the organism used. Ligninolytic enzymes cannot be induced into naive fungi [95]. When compared to bacterial enzymes, ligninolytic enzymes have a wide range of substrate specificity and hence are more beneficial than bacteria [25, 91].

7.5.3.2 NON-LIGNINOLYTIC FUNGI

The fungi that cannot produce any enzyme responsible for the degradation of lignin is non-ligninolytic fungi. Major groups of non-ligninolytic fungi; include the members of Zygomycetes, Ascomycetes, Hyphomycetes. Several species of *Aspergillus, Penicillium*, and *Cunninghamella*, involved in the initial transformation of PAHs catalyzed by a cytochrome P450 monooxygenase and oxygen to produce an arene oxide and water. The arene oxide is an unstable intermediate which is either hydrated to form a trans-dihydrodiols or rearranged non-enzymatically to form phenol [25].

PAH trans-dihydrodiols and phenols transform into sulfates, glucosides, glucuronides, or xylosides [21, 24]. Several reviews state clearly that much of the carcinogenesis imposed by coal tar is due to benzo[a]pyrene [75]. This compound is found to be even more carcinogenic than dibenzo [an,h] anthracene which is associated with causing skin tumors in mice. Other studies reported that the ultimate mammalian carcinogen is derived from benzo[a]pyrene, a stereoisomer of benzo[a]pyrene-7,8-diol-9,10-epoxide, which is a product formed by several enzymatic reactions involving monooxygenases and epoxide hydrolases [70].

Cunninghamella elegans produce cytochrome 450, while this fungus metabolizes PAHs with two to five aromatic rings. The metabolic products from PAHs produced by non-ligninolytic fungi are generally similar to the products formed by mammalian phase I and phase II reactions, but the stereoselectivity of the enzymes involved are different [23]. For instance, in the metabolism of phenanthrene, the major trans-9, 10-Dihydrodiol formed by *C. elegans* has an R, R configuration, but the S, S enantiomer is predominant in mammalian microsomal preparations [88]. Some other fungal species carry out various metabolic activities to degrade PAHs, which results in cleavage of the rings and production of carbon dioxide (CO_2).

7.6 REMOVAL OF POLYCYCLIC AROMATIC HYDROCARBONS (PAHS) VIA PHYTOREMEDIATION

Phytoremediation is an effective treatment technology that uses plants to bioremediate contaminants from the soil environment. Owing to its cost-effective and eco-friendly nature, phytoremediation is increasingly becoming an important application in environmental and ecological research. Plants have the potential to uptake chemicals from the soil. This property has been successfully used in plant-based bioremediation of metal contaminated soils.

Plants that cover a significantly large root surface area and that can adapt to the conditions of the soil are preferably chosen for phytoremediation. Studies have focused on Gramineae family since these species have very fibrous root systems that can penetrate deeper into the soil and can extend over a large surface area [67, 80]. There are several mechanisms by which plants clean up contaminated sites, and the efficiency of phytoremediation depends on the type of contaminant, bioavailability, and soil properties.

7.6.1 PHYTOEXTRACTION

This technique uses plants to absorb, concentrate, and precipitate toxic metals from contaminated soil by plant roots into the above groundmass (shoots, leaves, etc.). Metals, such as, nickel, zinc, and copper are the best candidates for removal by phytoextraction [17, 44].

7.6.2 PHYTOVOLATILIZATION

This involves the use of plants to take up contaminants from soil, transforming them into volatile forms and transpiring the volatile degradation products into the atmosphere [62, 93].

7.6.3 PHYTOSTIMULATION

This is also referred to as rhizodegradation that refers to the breakdown of contaminants within the plant root zone or rhizosphere. This process employs the use of several microorganisms prevalent in the rhizosphere to transform contaminants into nontoxic products. This technique is primarily useful in treating soils contaminated with PHs, PAHs, PCBs, pesticides, Benzene, toluene, ethylbenzene, and xylenes [93]. It is considered as plant assisted bioremediation, wherein plants release exudates or enzymes that stimulate bacterial and fungal growth and enhances microbial and fungal degradation [102].

7.6.4 PHYTODEGRADATION

This mechanism is also referred to as phytotransformation, which is the process of degradation of complex organic molecules to simpler molecules. Phytodegradation has been observed to remediate contaminants, such as herbicides, chlorinated solvents in the soil environment [93].

7.6.5 PHYTOSTABILIZATION

This process uses certain plant species to immobilize contaminants in the soil through absorption and accumulation by roots, adsorption onto roots, or precipitation within the rhizosphere [19].

7.6.6 RHIZOFILTRATION

Rhizofiltration is used for the remediation of groundwater. It is the adsorption or precipitation onto plant roots or absorption of contaminants in the solution that surrounds the root zone. Rhizofiltration can be used for the removal of metals or other inorganic compounds [93].

7.6.7 PHYTOREMEDIATION OF PAHS

Phytoremediation is a promising approach to remediate soils polluted with PAHs. PAH compounds have been detected in roots and shoots of alfalfa plants (*Medicago sativa* L.) [8]. During phytoremediation, plant enzymes aid in the degradation of contaminants. Since plant roots supply readily available carbon sources for microorganisms, they strengthen the microbial community by increasing microbial numbers, improve physicochemical conditions of soil, and accelerate the adsorption of pollutants on to rhizosphere.

Different plants have varied effects on microorganisms; thence the rate of degradation is dependent on the individual components of plant exudates [69]. Arbuscular mycorrhiza has been evaluated as a suited resistant fungus that helps the remediation of soils contaminated by PAHs, such as anthracene, chrysene, pyrene, anthracene, dibenzo(a,h)anthracene and benzopyrene [27, 55]. Arbuscular mycorrhizal phytoremediation (AMPR) was successfully employed to remediate sites polluted by phenanthrene and pyrene. The plant used was alfalfa (*Medicago sativa* L.) and after 70 days, studies showed that 98.1% of phenanthrene and 88.1% of pyrene were removed [43, 73].

7.7 SUMMARY

PAHs are biotransformed into less complex metabolites through mineralization into the organic minerals, H_2O, CO_2 (aerobic) or CH_4 (anaerobic); and the rate of biodegradation depends on different parameters such as presence or absence of oxygen, temperature, pH, microbial load, degree of acclimation, accessibility of nutrients, chemical structures of the compounds, cellular transport properties and the chemical portioning in the growth medium. PAH-contaminated soils can be remediated either individually or in combination using microorganisms: algae, bacteria, fungi, and plants. Mycoremediation offers a unique and much useful alternative for reducing the soil pollution contaminated with complex aromatic compounds. The main group fungi responsible for PAH-degradation are white-rot fungi, which have a range of enzymes, such as oxygenase, dehydrogenase, and ligninolytic enzymes. A holistic approach to physical, chemical, and biological degradation should be followed in an ecologically favorable manner to treat/remediate the polluted sites. In this context, future research endeavors should concentrate on a broader technological and economical assessment of all soil ReTec currently available not only for PAHs but also for pesticides and chlorinated organic pollutants.

ACKNOWLEDGMENT

Author (MZG) acknowledges Council of Scientific and Industrial Research, Government of India, New Delhi, India for the financial support vide award letter No. 09/132(0883)/2019-EMR-I). The author MZG is also thankful to Prof. Beedu Sashidhar Rao and Dr. Karuna Rupula, Department of Biochemistry, Osmania University, for their encouragement and support.

KEYWORDS

- **bioremediation**
- **contaminants**
- **ligninolytic enzymes**
- **microbial activity**
- **phytoremediation**
- **polycyclic aromatic hydrocarbons**
- **soil degradation**

REFERENCES

1. Abalos, A., Viñas, M., Sabaté, J., Manresa, M. A., & Solanas, A. M., (2004). Enhanced biodegradation of Casablanca crude oil by a microbial consortium in presence of a rhamnolipid produced by *Pseudomonas aeruginosa* AT10. *Biodegradation, 15*(4), 249–260.
2. Abdel-Shafy, H. I., & Mansour, M. S., (2016). Review on polycyclic aromatic hydrocarbons: Source, environmental impact, effect on human health and remediation. *Egyptian Journal of Petroleum, 25*(1), 107–123.

3. Agarry, S., & Latinwo, G. K., (2015). Biodegradation of diesel oil in soil and its enhancement by application of bioventing and amendment with brewery waste effluents as biostimulation-bioaugmentation agents. *Journal of Ecological Engineering, 16*, 82–91.
4. Akbari, A., & Ghoshal, S., (2015). Bioaccessible porosity in soil aggregates and implications for biodegradation of high molecular weight petroleum compounds. *Environmental Science and Technology, 49*(24), 14368–14375.
5. Akhtar, S., Mahmood-ul-Hassan, M., Ahmad, R., Suthor, V., & Yasin, M., (2013). Metal tolerance potential of filamentous fungi isolated from soils irrigated with untreated municipal effluent. *Soil and Environment, 32*(1), 55–62.
6. Alexander, M., (1999). *Biodegradation and Bioremediation* (2nd edn., p. 453). San Diego-USA: Academy Press, Elsevier.
7. Alkorta, I., Epelde, L., & Garbisu, C., (2017). Environmental parameters altered by climate change affect the activity of soil microorganisms involved in bioremediation. *FEMS Microbiology Letters, 364*(19), 18. Online, 10.1093/femsle/fnx200.
8. Alves, W. S., Manoel, E. A., & Santos, N. S., (2017). Detection of polycyclic aromatic hydrocarbons (PAHs) in *Medicago sativa* L. by fluorescence microscopy. *Micron, 95*, 23–30.
9. Anastasi, A., Tigini, V., & Varese, G. C., (2013). The bioremediation potential of different ecophysiological groups of fungi. In: Goltapeh, E., Danesh, Y., & Varma, A., (eds.), *Fungi as Bioremediators: Soil Biology* (Vol. 32, pp. 29–49). Heidelberg - Germany: Springer Verlag.
10. Anders, H. J., Kaetzke, A., Kampfer, P., Ludwig, W., & Fuchs, G., (**1995**). Taxonomic position of aromatic-degrading denitrifying pseudomonad strains K 172 and KB 740 and their description as new members of the genera *Thauera*, as *Thauera Aromatica Sp.* Nov., and *Azoarcus*, as *Azoarcus evansii* Sp. Nov., respectively, members of the beta subclass of the proteobacteria. International Journal of Systematic and Evolutionary Microbiology, *45*(2), 327–333.
11. Arriagada, C. A., Herrera, M. A., Borie, F., & Ocampo, J. A., (2007). Contribution of arbuscular mycorrhizal and saprobe fungi to the aluminum resistance of *Eucalyptus Globulus*. *Water, Air, and Soil Pollution, 182*(1–4), 383–394.
12. Atagana, H. I., Haynes, R. J., & Wallis, F. M., (2003). Co-composting of soil heavily contaminated with creosote with cattle manure and vegetable waste for the bioremediation of creosote-contaminated soil. *Soil and Sediment Contamination: An International Journal, 12*(6), 885–899.
13. Bais, H. P., Weir, T. L., Perry, L. G., Gilroy, S., & Vivanco, J. M., (2006). The role of root exudates in rhizosphere interactions with plants and other organisms. *Annual Review of Plant Biology, 57*(1), 233–266.
14. Balba, M. T., Al-Awadhi, N., & Al-Daher, R., (1998). Bioremediation of oil-contaminated soil: Microbiological methods for feasibility assessment and field evaluation. *Journal of Microbiology Methods, 32*(2), 155–164.
15. Bezalel, L., Hadar, Y., & Cerniglia, C. E., (1997). Enzymatic mechanisms involved in phenanthrene degradation by the white-rot fungus *Pleurotus ostreatus*. *Applied and Environmental Microbiology, 63*(7), 2495–2501.
16. Bezalel, L., Hadar, Y., Fu, P. P., Freeman, J. P., & Cerniglia, C. E., (1996). Initial oxidation products in the metabolism of pyrene, anthracene, fluorene, and dibenzothiophene by

the white-rot fungus *Pleurotus ostreatus*. *Applied and Environmental Microbiology, 62*(7), 2554–2559.
17. Bhargava, A., Carmona, F. F., Bhargava, M., & Srivastava, S., (2012). Approaches for enhanced phytoextraction of heavy metals. *Journal of Environmental Management, 105,* 103–120.
18. Bhattacharya, S., Das, A., & Prashanthi, K., (2014). Mycoremediation of benzo[a] pyrene by *Pleurotus ostreatus* in the presence of heavy metals and mediators. *3 Biotech, 4*(2), 205–211.
19. Bolan, N. S., Park, J. H., Robinson, B., Naidu, R., & Huh, K. Y., (2011). Phytostabilization: Green approach to contaminant. *Advances in Agronomy, 112,* 145–204.
20. Boopathy, R., (2000). Factors limiting bioremediation technologies. *Bioresource Technology, 74*(1), 63–67.
21. Capotorti, G., Cesti, P., Lombardi, A., & Guglielmetti, G., (2005). Formation of sulfate conjugates metabolites in the degradation of phenanthrene, anthracene, pyrene, and benzo[a]pyrene by the ascomycete *Aspergillus terreus*. *Polycyclic Aromatic Compounds, 25*(3), 197–213.
22. Carmona, M., Zamarro, M. T., & Blázquez, B., (2009). Anaerobic catabolism of aromatic compounds: A genetic and genomic view. *Microbiology and Molecular Biology Reviews, 73*(1), 71–133.
23. Casillas, R. P., Crow, S. A., Heinze, T. M., Deck, J., & Cerniglia, C. E., (1996). Initial oxidative and subsequent conjugative metabolites produced during the metabolism of phenanthrene by fungi. Journal of Industrial Microbiology and Biotechnology, *16*(4), 205–215.
24. Cerniglia, C. E., (1997). Fungal metabolism of polycyclic aromatic hydrocarbons: Past, present and future applications in bioremediation. *Journal of Industrial Microbiology and Biotechnology, 19,* 324–333.
25. Cerniglia, C. E., & Sutherland, J. B., (2009). Bioremediation of polycyclic aromatic hydrocarbons by ligninolytic and non-ligninolytic fungi. In: Gadd, G. M., (ed.), *Fungi in Bioremediation (British Mycological Society Symposia)* (pp. 136–187). Cambridge: Cambridge University Press.
26. Cerniglia, C. E., & Sutherland, J. B., (2010). Degradation of polycyclic aromatic hydrocarbons by fungi. In: Timmis, K. N., (ed.), *Handbook of Hydrocarbon and Lipid Microbiology* (pp. 2079–2110). Heidelberg Germany: Springer Verlag.
27. Cheema, S. A., Imran, K. M., Shen, C., & Tang, X., (2010). Degradation of phenanthrene and pyrene in spiked soils by single and combined plants cultivation. *Journal of Hazardous Materials, 177*(1–3), 384–389.
28. Chen, M., Xu, P., Zeng, G., Yang, C., Huang, D., & Zhang, J., (2015). Bioremediation of soils contaminated with polycyclic aromatic hydrocarbons, petroleum, pesticides, chlorophenols and heavy metals by composting: Applications, microbes and future research needs. *Biotechnology Advances, 33*(1), 745–755.
29. Couto, N., Fritt-Rasmussen, J., Jensen, P. E., & Højrup, M., (2014). Suitability of oil bioremediation in an artic soil using surplus heating from an incineration facility. *Environmental Science and Pollution Research, 21*(9), 6221–6227.
30. Dean-Ross, D., Moody, J., & Cerniglia, C. E., (2002). Utilization of mixtures of polycyclic aromatic hydrocarbons by bacteria isolated from contaminated sediment. *FEMS Microbiology Ecology, 41*(1), 1–7.

31. Deshmukh, R., Khardenavis, A. A., & Purohit, H. J., (2016). Diverse metabolic capacities of fungi for bioremediation. *Indian Journal of Microbiology, 56,* 247–264.
32. Ding, J., Chen, B. L., & Zhu, L. Z., (2013). Biosorption and biodegradation of polycyclic aromatic hydrocarbons by *Phanerochaete chrysosporium* in aqueous solution. *Chinese Science Bulletin, 58*(6), 613–621.
33. Doddapaneni, H., Subramanian, V., & Yadav, J. S., (2005). Physiological. regulation, xenobiotic induction, and heterologous expression of P450 monooxygenase gene PC-3(CYP63A3): New member of the CYP63 gene cluster in the white-rot fungus *Phanerochaete chrysosporium. Current Microbiology, 50*(6), 292–298.
34. Doddapaneni, H., & Yadav, J. S., (2004). Differential regulation and xenobiotic induction of tandem P450 monooxygenase genes PC-1(CYP63A1) and PC-2(CYP63A2) in the white-rot fungus *Phanerochaete chrysosporium. Applied Microbiology and Biotechnology, 65*(5), 559–565.
35. Ekere, N., (2015). Benzene, toluene, ethylbenzene and xylenes (Btex) contamination of soils and water bodies from alkyd resin and lubricants industrial production plant. *Journal of Chemical Society of Nigeria, 40,* 51–56.
36. Evans, P. J., Mang, D. T., Kim, K. S., & Young, L. Y., (1991). Anaerobic degradation of toluene by a denitrifying bacterium. *Applied and Environmental Microbiology, 57*(4), 1139–1145.
37. FAO (Food and Agriculture Organization), (2015). Status of the world's soil resources. *Intergovernmental Technical Panel on Soils*, 125.
38. Fennell, D. E., Du, S., Liu, F., Liu, H., & Häggblom, M. M., (2011). Dehalogenation of polychlorinated dibenzo-p-dioxins and dibenzofurans, polychlorinated biphenyls, and brominated flame retardants, and potential as a bioremediation strategy. In: Moo-Young, M., (ed.), *Comprehensive Biotechnology* (2nd edn., pp. 135–149). Amsterdam – Netherlands: Academic Press, Elsevier.
39. Fernández-Luqueño, F., & Marsch, R., (2008). Remediation of PAHs in a saline-alkaline soil amended with wastewater sludge and the effect on dynamics of C and N. *Science of the Total Environment, 402*(1), 18–28.
40. Foght, J., (2008). Anaerobic biodegradation of aromatic hydrocarbons: Pathways and prospects. *Journal of Molecular Microbiology and Biotechnology, 15*, 93–120.
41. Fries, M. R., Zhou, J., Chee-Sanford, J., & Tiedje, J. M., (1994). Isolation, characterization, and distribution of denitrifying toluene degraders from a variety of habitats. *Applied and Environmental Microbiology, 60*(8), 2802–2810.
42. Fuchs, G., Boll, M., & Heider, J., (2011). Microbial degradation of aromatic compounds-from one strategy to four. *Nature Reviews Microbiology, 9*(11), 803–816.
43. Gao, Y., Li, Q., Ling, W., & Zhu, X., (2011). Arbuscular mycorrhizal phytoremediation of soils contaminated with phenanthrene and pyrene. *Journal of Hazardous Materials, 185*(2–3), 703–709.
44. Garbisu, C., & Alkorta, I., (2001). Phytoextraction: Cost-effective plant-based technology for the removal of metals from the environment. *Bioresource Technology, 77*(3), 229–236.
45. Global Food Security Program, (2013). The importance of soil for ensuring food security. *Insight, 1*, 1–8.
46. Habe, H., & Omori, T., (2003). Genetics of polycyclic aromatic hydrocarbon metabolism in diverse aerobic bacteria. *Bioscience, Biotechnology and Biochemistry, 67*(2), 225–243.

47. Haderlein, A., Legros, R., & Ramsay, B. A., (2006). Pyrene mineralization capacity increases with compost maturity. *Biodegradation, 17*(4), 293–302.
48. Haghollahi, A., Fazaelipoor, M. H., & Schaffie, M., (2016). The effect of soil type on the bioremediation of petroleum contaminated soils. *Journal of Environmental Management, 180*, 197–201.
49. Hamby, D. M., (1996). Site remediation techniques supporting environmental restoration activities: A review. *Science of the Total Environment, 191*(3), 203–224.
50. Hammel, K. E., (1995). Mechanisms for polycyclic aromatic hydrocarbon degradation by ligninolytic fungi. *Environmental Health Perspectives, 103*(Suppl. 5), 41–43.
51. Hammel, K. E., (1989). Organopollutant degradation by ligninolytic fungi. *Enzyme and Microbial Technology, 11*(11), 776–777.
52. Harrison, M. J., (2005). Signaling in the arbuscular mycorrhizal symbiosis. *Annual Review of Microbiology, 59*(1), 19–42.
53. Hinchee, R. E., & Arthur, M., (1991). Bench-scale studies of the soil aeration process for bioremediation of petroleum hydrocarbons. *Biotechnology and Applied Biochemistry, 28*(1), 901–906.
54. Hiraishi, A., (2008). Biodiversity of de-halorespiring bacteria with special emphasis on polychlorinated biphenyl/dioxin dechlorinators. *Microbes and Environments, 23,* 1–12.
55. Joner, E. J., Leyval, C., & Colpaert, J. V., (2006). Ectomycorrhizas impede phytoremediation of polycyclic aromatic hydrocarbons (PAHs) both within and beyond the bhizosphere. *Environmental Pollution, 142*(1), 34–38.
56. Juckpech, K., Pinyakong, O., & Rerngsamran, P., (2012). Degradation of polycyclic aromatic hydrocarbons by newly isolated *Curvularia sp.* F18, *Lentinus sp.* S5, and *Phanerochaete sp.* T20. *Science Asia, 38*(2), 147–156.
57. Koshlaf, E., & Ball, A. S., (2017). Soil bioremediation approaches for petroleum hydrocarbon polluted environments. *AIMS Microbiology, 3(1)*, 25–49.
58. Kulshreshtha, S., (2013). Genetically engineered microorganisms: Problem solving approach for bioremediation. *Journal of Bioremediation and Biodegradation, 4*(4), 214–229.
59. Lamy, I., (2011). Trace elements in soils. In: Hooda, P. S., (ed.), *Analytical and Bioanalytical Chemistry* (Vol. 399, No. 6, pp. 2219, 2220). Heidelberg – Germany: Springer Verlag.
60. Lee, T. H., Byun, I. G., Kim, Y. O., Hwang, I. S., & Park, T. J., (2006). Monitoring biodegradation of diesel fuel in bioventing processes using *in situ* respiration rate. *Water Science and Technology, 53*(4, 5), 263–272.
61. Lee, Y., Lee, Y., & Jeon, C. O., (2019). Biodegradation of naphthalene, BTEX, and aliphatic hydrocarbons by *Paraburkholderia aromaticivorans* BN5 Isolated from petroleum-contaminated soil. *Scientific Reports, 9*(1), 860.
62. Limmer, M., & Burken, J., (2016). Phytovolatilization of organic contaminants. *Environmental Science and Technology, 50*, 6632–6643.
63. Lovley, D. R., (2003). Cleaning-up with genomics: applying molecular biology to bioremediation. *Nature Reviews Microbiology, 1*(1), 35–44.
64. Lukić, B., Huguenot, D., Panico, A., & Fabbricino, M., (2016). Importance of organic amendment characteristics on bioremediation of PAH-contaminated soil. *environmental Science and Pollution Research, 23*(15), 15041–15052.

65. Macaulay, B. M., (2015). Understanding the behavior of oil-degrading microorganisms to enhance the microbial remediation of spilled petroleum. *Applied Ecology and Environmental Research, 13*(1), 247–262.
66. Masaphy, S., Levanon, D., Henis, Y., Venkateswarlu, K., & Kelly, S. L., (1996). Evidence for cytochrome P-450 and P-450-mediated benzo[a]pyrene hydroxylation in the white-rot fungus *Phanerochaete chrysosporium. FEMS Microbiology Letters, 135*(1), 51–55.
67. McCutcheon, S. C., & Jørgensen, S. E., (2018). Phytoremediation. In: Fath, B., (ed.), *Encyclopedia of Ecology* (2nd edn., pp. 568–582). Amsterdam -Netherlands: Academic Press, Elsevier.
68. Megharaj, M., Ramakrishnan, B., Venkateswarlu, K., Sethunathan, N., & Naidu, R., (2011). Bioremediation approaches for organic pollutants: A critical perspective. *Environment International, 37*, 1362–1375.
69. Merkl, N., Schultze-Kraft, R., & Infante, C., (2005). Phytoremediation in the tropics - influence of heavy crude oil on root morphological characteristics of graminoids. *Environmental Pollution, 138*(1), 86–91.
70. Miller, K. P., & Ramos, K. S., (2001). Impact of cellular metabolism on the biological effects of benzo[a]pyrene and related hydrocarbons. *Drug Metabolism Reviews, 33*(1), 1–35.
71. Moody, J. D., Freeman, J. P., Fu, P. P., & Cerniglia, C. E., (2004). Degradation of benzo[a]pyrene by *Mycobacterium Vanbaalenii* PYR-1. *Applied and Environmental Microbiology, 70*(1), 340–345.
72. Mulligan, C. N., & Yong, R. N., (2004). Natural attenuation of contaminated soils. *Environment International, 30*(4), 587–601.
73. Nduka, J. K., Umeh, L. N., & Okerulu, I. O., (2012). Utilization of different microbes in bioremediation of hydrocarbon contaminated soils stimulated with inorganic and organic fertilizers. *Journal of Petroleum & Environmental Biotechnology, 3*(2).
74. Neina, D., (2019). The role of soil pH in plant nutrition and soil remediation. *Applied and Environmental Soil Science, 2019*, pp. 9.
75. Phillips, D. H., (1983). Fifty years of benzo[a]pyrene. *Nature, 303*, 468–472.
76. Pieper, D. H., & Reineke, W., (2000). Engineering bacteria for bioremediation. *Current Opinion in Biotechnology, 11*(3), 262–270.
77. Pozdnyakova, N. N., (2012). Involvement of the ligninolytic system of white-rot and litter-decomposing fungi in the degradation of polycyclic aromatic hydrocarbons. *Biotechnology Research International, 2012*, 1–20.
78. Qixing, Z., & Tao, H., (2004). Bioremediation: A review of applications and problems to be resolved. *Progress in Natural Science, 14*(11), 937 944.
79. Sakshi, Singh, S. K., & Haritash, A. K., (2019). Polycyclic aromatic hydrocarbons: Soil pollution and remediation. *International Journal of Environmental Science and Technology, 16*, 6489–6512.
80. Sarwar, N., Imran, M., & Shaheen, M. R., (2017). Phytoremediation strategies for soils contaminated with heavy metals: Modifications and future perspectives. *Chemosphere, 171*, 710–721.
81. Schjønning, P., Thomsen, I. K., Petersen, S. O., Kristensen, K., & Christensen, B. T., (2011). Relating soil microbial activity to water content and tillage-induced differences in soil structure. *Geoderma, 163*(3, 4), 256–264.

82. Seo, J. S., Keum, Y. S., & Li, Q. X., (2009). Bacterial degradation of aromatic compounds. *International Journal of Environmental Research and Public Health, 6*(1), 278–309.
83. Shinoda, Y., Akagi, J., Uchihashi, Y., & Hiraishi, A., (2005). Anaerobic degradation of aromatic compounds by *Magneto spirillum* strains: Isolation and degradation genes. *Bioscience, Biotechnology and Biochemistry, 69*(8), 1483–1491.
84. Silva, I. S., Grossman, M., & Durrant, L. R., (2009). Degradation of polycyclic aromatic hydrocarbons (2-7 rings) under microaerobic and very low oxygen conditions by soil fungi. *International Biodeterioration and Biodegradation, 63*(2), 224–229.
85. Srivastava, J., Shukla, D., Chand, V., Naraian, R., Chandra, H., & Nautiyal, A. R., (2010). Mycorrhizal colonization affects the survival of *Vetiveria zizanioides* (L.) Nash grown in water containing As(III). *Clean - Soil, Air, Water, 38*(8), 771–774.
86. Stott, D. E., & Martin, J. P., (2015). Synthesis and degradation of natural and synthetic humic material in soils. In: MacCarthy, P., Clapp, C. E., Malcolm, R. L., & Bloom, P. R., (eds.), *Humic Substances in Soil and Crop Sciences: Selected Readings* (pp. 37–63). Wisconsin - USA: American Society of Agronomy, Inc.
87. Sutherland, J. B., Freeman, J. P., Selby, A. L., Fu, P. P., Miller, D. W., & Cerniglia, C. E., (1990). Stereoselective formation of a K-region dihydrodiol from phenanthrene by *Streptomyces flavovirens*. Archives *of* Microbiology, *154*(3), 260–266.
88. Sutherland, J. B., Fu, P. P., Yang, S. K., & Von, T. L. S., (1993). Enantiomeric composition of the trans-dihydrodiols produced from phenanthrene by fungi. *Applied and Environmental Microbiology, 59*(7), 2145–2149.
89. Tam, N. F. Y., Guo, C. L., Yau, W. Y., & Wong, Y. S., (2002). Preliminary study on biodegradation of phenanthrene by bacteria isolated from mangrove sediments in hong kong. *Marine Pollution Bulletin, 45*(1–2), 316–324.
90. Tekere, M., Read, J. S., & Mattiasson, B., (2005). Polycyclic aromatic hydrocarbon biodegradation in extracellular fluids and static batch cultures of selected sub-tropical white-rot fungi. *Journal of Biotechnology, 115*(4), 367–377.
91. Tortella, G. R., Diez, M. C., & Durán, N., (2005). Fungal diversity and use in decomposition of environmental pollutants. *Critical Reviews in Microbiology, 31*(4), 197–212.
92. Trzesicka-Mlynarz, D., & Ward, O. P., (1995). Degradation of polycyclic aromatic hydrocarbons (PAHs) by a mixed culture and its component pure cultures, obtained from PAH-contaminated soil. *Canadian Journal of Microbiology, 41*(6), 470–476.
93. United States Environmental Protection Agency. (2015). *2012 Citizen's Guide to Bioremediation.* https://www.epa.gov/sites/production/files/2015-04/documents/a_citizens_guide_to_bioremediation.pdf (accessed on 2 August 2021).
94. Van, G. K., Mergaert, J., Swings, J., Coosemans, J., & Ryckeboer, J., (2003). Bioremediation of diesel oil-contaminated soil by composting with biowaste. *Environmental Pollution, 125*(3), 361–368.
95. Verdin, A., Sahraoui, A. L. H., & Durand, R., (2004). Degradation of benzo[a]pyrene by mitosporic fungi and extracellular oxidative enzymes. *International Biodeterioration and Biodegradation, 53*(2), 65–70.
96. Viñas, M., Sabaté, J., Espuny, M. J., & Solanas, A. M., (2005). Bacterial community dynamics and polycyclic aromatic hydrocarbon degradation during bioremediation of heavily creosote-contaminated soil. *Applied and Environmental Microbiology, 71*(11), 7008–7018.

97. Walter, U., Beyer, M., Klein, J., & Rehm, H. J., (1991). Degradation of pyrene by *Rhodococcus* sp. UW1. *Applied Microbiology and Biotechnology, 34*(5), 671–676.
98. Wang, Q., Zhang, S., Li, Y., & Klassen, W., (2011). Potential approaches to improving biodegradation of hydrocarbons for bioremediation of crude oil pollution. *Journal of Environmental Protection* (Irvine, Calif), *2*(1), 47–55.
99. Yang, S. Z., Jin, H. J., Wei, Z., He, R. X., Ji, Y. J., Li, X. M., & Mei, Y. S., (2009). Bioremediation of oil spills in cold environment: Review. *Pedosphere, 19*(3), 371–381.
100. Yuan, S. Y., Shiung, L. C., & Chang, B. V., (2002). Biodegradation of polycyclic aromatic hydrocarbons by inoculated microorganisms in soil. *Bulletin of* Environmental *Contamination and Toxicology, 69*(1), 66–73.
101. Zhang, Q., Fu, X., Ji, L., Qi, Z., & Wang, J., (2017). Bioremediation of oil sludge contaminated soil by eco-pile. *Chinese Journal of Chemical Engineering, 11*(11), 6142–6146.
102. Zhuang, P., Ye, Z. H., Lan, C. Y., Xie, Z. W., & Shu, W. S., (2005). Chemically assisted phytoextraction of heavy metal contaminated soils using three plant species. *Plant Soil, 276*(1, 2), 153–162.
103. Zuloaga, O., Navarro, P., & Bizkarguenaga, E., (2012). Extraction, clean-up and detection techniques for the determination of organic pollutants in sewage sludge: Review. *Analytica Chimica Acta, 736*, 7–29.

CHAPTER 8

MICROBIAL BIOREMEDIATION AND AMELIORATION OF PESTICIDE-CONTAMINATED SOILS

M. SRINIVASULU, M. SUBHOSH CHANDRA, and
NAGA RAJU MADDELA

ABSTRACT

Indiscriminate and excessive use of pesticides (such as organophosphates, carbamates, organochlorines, and chlorophenoxy acids) has led to the accumulation of pesticide residues in the soil environments. Pesticides mainly affect the non-target organisms and cause various complications in soil microorganisms and their biological activities, including humans. Among several soil remediation technologies (ReTec) available today for decontamination and detoxification of pesticide-contaminated soils, bioremediation is environmentally safe, eco-friendly, and cost-effective method. The present chapter is an assessment of the current approaches in the process of bioremediation of the soils adulterated with pesticides.

8.1 INTRODUCTION

Application of pesticides, either concurrently or one after another, to protect crops from a variety of pests often leads to accumulation of pesticide residues in the soils. The excessive use of pesticides has evoked possible side effects on the soil microorganisms. Because of great concern over the environmental hazards caused by these compounds, a fundamental understanding of microbial degradation potential is vital in maximizing the capacity of microbial groups to combat pollution problems.

Survey of literature has revealed that microorganisms are the most efficient bioremedial agents [12, 18, 45, 81, 85]. Although many physico-chemical and environmental factors can affect pesticide degradation, yet biodegradation has proved to be more advantageous. Among the soil micro-organisms, bacteria are major means in detoxification of many pesticides in the environment, which utilize them as a source of energy or nutrient or by co-metabolism [55, 113, 120].

Due to population explosion, the production of crops also needs to be enhanced [22, 38], which may be accomplished by decreasing damage to the crops by pests [101]. The pests that attack the agricultural crops include insects, weeds, rodents, nematodes, and microorganisms (bacteria, algae, and fungi) [17, 39]. The pesticides are classified based on their target, which proved beneficial for agricultural crop yields by decreasing loss of crop yield by pests [4]. Indiscriminate use of pesticides at an unmanageable rate has led to the accumulation of pesticide residues in soils and agricultural runoffs, which are biomagnified in water bodies, thereby enhancing the toxicity level at every stage in the food web [90, 106, 118]. Besides, pesticides also affect the function of human organs and cause damage to DNA at a molecular level leading to neurological disorders and cancer [42, 111, 121].

Pesticides also accumulate in the food materials from fields [7, 63, 99]. Several reports have revealed the presence of pesticides in beverages and other foodstuffs, such as milk, fruit juice, and seaweeds [13, 50, 63]. Therefore, it is obligatory to minimize the use of synthetic pesticides and organic farming should be encouraged [116]. Elimination and degradation of pesticides are carried out by conventional techniques of bioremediation [58, 79].

The bioremediation process essentially involves environmental cleansing of pesticides by microbiological processes whether *in-situ* or *ex-situ* [85]. *In-situ* bioremediation involves clean up with no removal of the soil from the site, while *ex-situ* treatment includes unearthing of soil at the site [86], and both are outstanding and eco-friendly alternatives for degrading the pollutants in the environment [81]. Through advancement in scientific research methodologies, gene editing and systemic biology apparatus are applied in bioremediation of heavy metals (HMs), persistent organic pollutants (POPs), acid drainage, petroleum, and xenobiotics [12, 18, 31, 45, 71]. The bioremediation includes the association of chemical components which exhibit a tendency to degrade and tangled microbial metabolic web in the contaminated state [82]. The systems biology approach provides information about the microbial systems [49], which respond differently under diverse subsets of conditions [78].

Interactions between the communities of microorganisms were observed with systems biology approach [61]. In addition, this approach is extremely useful in prospective survival of microorganisms under diverse environmental conditions of extreme temperature and pressure [16]. The omics such as transcriptomics, genomics, metabolomics, and proteomics aid the systems biology studies of microorganisms in studying the genetic level regulation for the bioremediation processes [32]. The improvement in the sequencing with high throughput sequencing (HTS) and next generation sequencing resolve the novel genes involved in the biodegradation pathways of a variety of persistent pollutants [114, 123].

Owing to the popularization of mechanized agricultural production, the pesticides are mostly applied via spraying by machines and even drones, which could inevitably pollute the soil, water, and atmosphere. The agricultural runoff also contributes to the pollution of adjacent water bodies. Thus, the pesticide contamination in the environment is mostly caused by [2, 41, 52]:

- Condensation of volatile pesticides;
- Surface runoff from agricultural and earlier contaminated soil;
- Inappropriate spraying of pesticides;
- Unacceptable transportation and storage of pesticides;
- Removal of inadequately treated or non-treated wastewater from pesticide industries.

The entry of pesticides into the environment is linked to various environmental and health hazards like toxicity of pesticide itself inhibiting growth of plants and animals. For example, organochlorine pesticides (OCPs) can affect the development of animals by interfering with their nervous system [18]. A general overview of the sources of pesticides, benefits, and damage caused by pesticides, trophic level transfer, and routes of human exposure to the pesticides is presented in Figure 8.1.

The pesticides enter into the food chain and their slow bioaccumulation occurs in higher trophic level organisms or yet human body by biomagnification process. Once exposed to pesticides, a sequence of symptoms like itchy skin, inflammation of the nose and throat, headache, rash, and blisters on the skin, nausea, vomiting, tingling of the eyes and skin, diarrhea, dizziness, blindness, blurred vision and even few deaths are possible usually within

24 h. However, in most of the cases, the severe effects of pesticides are not serious and do not require medical attention every time [18].

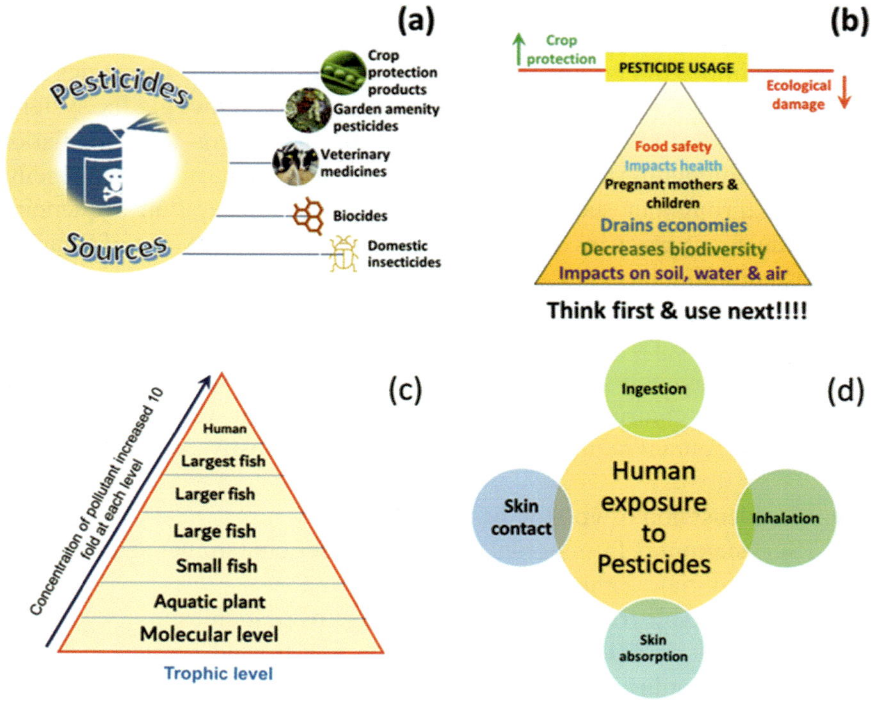

FIGURE 8.1 General view of pesticides: Sources of pesticides (a); balance between benefits and damage due to pesticide usage (b); trophic transfer of pesticides (c); and routes of human exposure to pesticides (d).

According to a few reports, atrazine, simazine, glyphosate, mecoprop, imidacloprid, and isoproturon are frequently used pesticides in agriculture [14, 107]. Particularly, OCPs are among the original pesticides and have been extensively used globally [132]. Owing to the introduction of DDT in polar organisms, the problem of pesticide contamination has attracted interest from researchers and the public [134, 140]. OCPs, classified as POPs, have been banned in many countries because of their great environmental persistence, considerable biotoxicity and meager biodegradability [8].

In this chapter, attention is focused on the interactions between pesticides and soil microorganisms and the recent advancements in microbial bioremediation that can be useful in remediation of the contaminated soils.

8.2 INTERACTION BETWEEN PESTICIDES AND SOIL MICROORGANISMS

The modern agriculture mainly relies on the widespread application of agro-chemicals, which include pesticides. The unsystematic, long term and excessive applications of the pesticides have severe effects on soil ecology that may lead to alterations in the beneficial or plant probiotic soil microorganisms. The weathered soils lose their capability to maintain improved production of crops or grains in similar soils. Although budding concern about the environmental contamination and the sustainable use of cropping land have emphasized inculcation of knowledge and the broader application of tools, yet methods and products that do not pollute the environment at all or have only meager ecological concerns [60]. The pesticides, which include fungicides, herbicides, and insecticides, are broadly used in agroecosystems. Since 1950s, intensive farming techniques have improved the range and effects of these agrochemicals [44].

Once a pesticide is released into soil matrix, its effect will depend both on its characteristics and of the soil [47]. The reactivity with soil biological compartment is one of the most important interactions between soil constituents and pesticides, which are well-known to be degraded by microorganisms or mineralized with telluric fungi [51, 97] and bacteria [37, 92]. In fact, microorganisms in soil have enzyme pools which can degrade natural and xenobiotic substrates. However, while microorganisms control the fate of pesticides in soils, pesticides can have a negative effect on their activities [122].

Though their use has enhanced crop yields and value, pesticides are known as a source of potential adverse environmental impacts, and their persistence in the surface and groundwater has grown significantly [28, 117]. Numerous bioremediation approaches have been proposed to decrease the existence of pesticides in the soil from which they can reach groundwater, such as remediation by enhancing the population of microorganisms capable of degrading the target compounds. This approach has been followed by the addition of organic matter from different sources [1, 33, 34].

It is hard to estimate the quantity of pesticides released into the environment through the production of pesticide, use, and also formulation. Only 0.1% of pesticide applied for crop production is utilized, while the remaining 99.9% of pesticide is released directly into the soil, ultimately into water biota and the food chain, thus leading to the pollution problem [89].

8.2.1 INFLUENCE OF PESTICIDES ON SOIL MICROFLORA

The agricultural usage of pesticides leaves residues in soil from fallout from aerial spraying, in rain or dust, or from plants or animals which become incorporated into soil [135]. Thus, soils serve as environmental reservoir and are becoming more polluted by pesticides because of wide and often indiscriminate use of pesticides in agricultural practice. As inhabitants of the soil environment, there are distinct opportunities for microflora to interact with them.

Soil microorganisms largely, *Azospirillum* sp., play an important role in nitrogen fixation in the soil. Nitrifiers are involved in the oxidation of reduced nitrogen to nitrite and then to nitrate. The bacteria and fungi are important to soil nutrition, because they play a key role in the decay of plant and organic matter in the soil. Anything that disrupts their populations could be expected to affect the nutritional quality of soils and may have serious consequences [113].

8.3 BIOREMEDIATION OF PESTICIDES

The soil, groundwater, and sediments are the definitive sinks for pesticides, where they may be broken down to simpler forms or remain persistent. Few pesticides are vulnerable to biomagnification and cause more hazards to the environment. The secure and inexpensive disposal of excess pesticide waste is a problem of significant magnitude. The pesticides are considered as xenobiotic compounds because they are the synthetic organic compounds produced chemically and have no apparent counterpart in the natural world.

The majority of pesticides are recalcitrant and resist their biodegradation. More than 1,000 pesticides have been sold for control of pests, which include herbicides, insecticides, bactericides, and fungicides, etc. Toxicity and bioaccumulation potential of chlorophenol increase with the degree of chlorination and lipophilicity [66].

Bioremediation and biodegradation are identical methods, and both are based on the conversion or metabolism of pesticides by microorganisms. The biodegradation is a natural process while bioremediation is a technology. In the case of bioremediation, microorganisms are used to degrade the pesticides *in situ*. The successful bioremediation method needs a potential bacterial strain able to degrade the largest contaminant to the lowest level. A sufficient rate of biodegradation is needed to achieve an acceptable level of pesticide residues or their metabolites at polluted site in a limited time framework. The extent of biodegradation in the soil mainly depends on four variables viz:

- Accessibility of pesticide or metabolite to the microorganisms;
- Physiological status of the microorganisms;
- Survival or propagation of pesticide degrading microorganisms at polluted site; and
- Sustainable inhabitants of microorganisms.

Thus, to reach the realistic bioremediation, it is desirable to design a unique alcove or microhabitat for preferred microorganisms, so that they are successfully exploited. At this point, only the complexity is limited knowledge on the population dynamics of pesticide degrading microorganisms in relation to other microorganisms present in a similar habitat. Temperature, water potential, pH, nutrients, and the quantity of pesticides/metabolites in soil environment can also act as limiting factors for the pesticide degrading microorganisms, which require further study in relation to the total population of microorganisms and their biochemical activities [36].

The degradation of xenobiotics by microorganisms is a process known as bioremediation, and is a cost-effective method of eliminating contaminants from the environment. The biodegradation involves the biological reactions that change the chemical structures of the compounds, which in turn involves a decrease in the toxicity of pollutants. Once pesticide degradation takes place, it typically involves more than one microorganism. Each microorganism contributes to the pesticide biodegradation reactions however there is no example of mineralization by a single strain has been illustrated. Several bacteria have been widely used for the bioremediation of pollutants [115].

Bacterial classes, gammaproteobacteria (*Pseudomonas, Aerobacter, Acinetobacter, Moraxella, Plesiomonas*), beta-proteobacteria (*Burkholderia, Neisseria*), alpha-proteobacteria (*Sphingomonas*), Actinobacteria (*Micrococcus*), and Flavobacteria (*Flavobacterium*) are considered to be active microbial degraders [46, 73, 75]. The bacterial genera capable of degrading endosulfan belong to the genera *Klebsiella, Acinetobacter, Alcaligenes, Flavobacterium, and Bacillus*. During the biodegradation, metabolites of endosulfan like endosulfan diol and endosulfan lactone, ether are also produced. However, they had low toxicity when compared with the original compound (i.e., endosulfan) [59].

More recently, *Bacterium raoultella sp.* has also found to degrading pesticides. Total biodegradation of the pesticide involves oxidation of the parent compound, resulting in the release of CO_2 and water, which provide energy to the microorganisms. The innate microorganisms present in the native soil may not be able to deal with pesticides. Hence, the external addition of pesticide-degrading microorganisms is recommended. The degradation/detoxification of the pesticides by microorganisms depends not only on the enzymatic system but also on certain environmental conditions like pH, temperature, and the availability of nutrients in soil. Few of the pesticides are easily degraded; though, some of them are recalcitrant due to the existence of anionic species in the compound.

In addition to the organophosphate compounds, the neonicotinoids are degraded by *Pseudomonas* species [55, 120]. Modification of the different bacterial samples using genetic mutation also increases the efficiency of applied microorganisms. Biodegradable removal of the pesticides has affirmative effects on the fertility of agricultural soil [55].

8.3.1 BIOREMEDIATION: AN EMERGING TECHNOLOGY FOR ENVIRONMENTAL CLEANUP

Manufacturing and the use of synthetic pesticides has been increasing extremely in India, and the generated waste from the pesticide industry has turn into a serious environmental problem due to the present insufficient and ineffective waste treatment technology involving physicochemical and biological treatments. As a result, recalcitrant pesticide residues remain in surface soil, leading to the toxicity of the soil-water environment. In view of the toxic effects of pesticides, it has become crucial to eliminate these chemo-contaminants from the soil environment. Bioremediation processes including phytoremediation and rhizoremediation are expected to be useful clean-up methods for soils contaminated with persistent organic pesticides including dieldrin and endrin [40].

The traditional processes of pesticide detoxification have been mainly relied on the landfills and incineration, which in turn generate secondary pollution problems because of leaching of pesticides into the surrounding soil and groundwater supplies and the production of potentially toxic by-product emissions. Furthermore, incineration, even though permitted by US EPA (United States Environmental Protection Agency), is an extremely costly process necessitates huge quantities of energy [24]. From this point of view, biodegradation is a much safer and economical process for detoxifying pesticides. Several methods of dispersal like landfill disposal, collection, removal, and incineration merely dilute or sequester the pollutants or transport them to a different environment.

In contrast, bioremediation may be viewed as the most effective and eco-friendly benign cleaning technology, because it results in the partial or complete bioconversion of the organic contaminants, such as pesticides, to microbial biomass and stable non-toxic end products [6]. It is a comparatively current addition to the battery of clean-up approaches presently employed to restore the polluted sites.

Bioremediation is defined as a biological reaction to the environmental exploitation. Therefore, it is the use of living organisms, mainly microorganisms, to degrade/detoxify the environmental pollutants into low toxic forms. It is a combination of two words-bio for biological, and remediation which means to remediate or eliminate. Depending on the type of organism used in the process, bioremediation may be of different types: Phytoremediation (the use of plants to clean-up the environment) and fungal remediation (the use of fungi for the clean-up of polluted environment). The use of fungal technology for the remediation of contaminated soil has assumed significance since 1985, while the white-rot fungus (WRF) *Phanerochaete chrysosporium* was capable to degrade a large number of significant environmental pollutants [19].

8.3.2 BIOREMEDIATION TECHNOLOGIES

There are numerous advantages of bioremediation technology: (i) It harnesses usual biogeochemical processes; (ii) It is cost-effective and optional; (iii) the toxic chemicals are damaged or removed from the environment and not simply separated; (iv) little capital intensive; (v) low energy requiring when compared with other technologies; (vi) Low manual supervision requirement.

The bioremediation technologies have been generally divided into two types based on whether biodegradation is enthused *in situ* or carried out *ex situ* in compost heaps or bioreactors. Organic pollutants can be bioremediated using either *in situ* or *ex situ* systems. *In situ* bioremediation methods involve improvement of the biodegradation rate of organic pollutants in the affected area (surface or groundwater environments, soil, and sediment). *In situ* is the clean-up technology that mainly involves the direct contact between microorganisms and the dissolved and sorbed pollutants for the biotransformation [3].

8.4 BIODEGRADATION PATHWAYS

The pyrethroid insecticides undergo several pathways once they reach the soil environment, including transformation or degradation, volatilization, sorption, desorption, uptake by plants, overflow into surface water, and transfer into groundwater. The modification or degradation is one of the most important methods that directs the environmental fate and transportation of pyrethroids, which also consist of different techniques including abiotic degradation (e.g., hydrolysis, photolysis, and oxidation) and biodegradation. Within these processes, pyrethroids have been transformed into degradation products or completely mineralized. But, the structure of a pyrethroid determines its innate biodegradation [130, 136, 139].

During the degradation of pyrethroid insecticides in the environment, ester cleavage is a main process that results in the release of 3-phenoxy benzyl alcohol, cyclopropane acid, 3-phenoxy benzoic acid or 3-phenoxy benzaldehyde (3-PBA). 3-Phenoxy benzyl alcohol is also the intermediate compound in the photocatabolism of pyrethroids, which undergo oxidation and results in the formation of corresponding carboxylic acid [124, 129]. Among the mentioned metabolites, 3-PBA frequently accumulates in the soil and further degradation of pyrethroids and 3-PBA may be partial or inhibited due to its strong antimicrobial activity [23, 127].

8.5 BACTERIAL REMEDIATION

A few organophosphate pesticides (OPs) are extremely toxic and are still extensively used for insect pest control. The extensive use of OPs causes serious concerns over food safety and environmental contamination. In recent years, a variety of bacterial strains were isolated for biodegradation of OPs. Serratia sp. SPL-2 can degrade/detoxify methidathion [68]. Pseudomonas aeruginosa Is-6 can degrade acephate, methyl parathion, methamidophos, malathion, and dimethoate [93].

Biodegradation of chlorpyrifos by soil bacterial communities comprising of seven different isolates of *Pseudomonas*, *Agrobacterium*, and *Bacillus* has been investigated [77]. *Bacillus subtilis*, *B. cereus*, *Brucella melitensis*, Klebsiella species, *Pseudomonas fluorescence*, *Serratia marcescens*, and *Pseudomonas aeruginosa* are able to degrade 46–72% of chlorpyrifos as sole source of carbon in an aqueous medium after 20 days of incubation [65].

Several factors that affect bacterial degradation of Ops include water holding capacity (WHC) of soil, pH, temperature, and even aging of the pesticides in soils earlier to inoculation. In the study of Enterobacter sp., aging of chlorpyrifos resulted in around 10% persistence in soil [110]. Diazinon-degrading Serratia marcescens is also capable of degrading chlorpyrifos, parathion, and fenitrothion [30]. Stenotrophomonas sp. SMSP-1 completely degrades methyl parathion, fenitrothion, and ethyl parathion, and degrades 37.8% of fenthion and 59.7% of phoxim. However, it cannot degrade chlorpyrifos at 20 mg L^{-1} [105]. Stenotrophomonas sp. PF32 effectively degrades methyl parathion, sumithion, phoxim, triazophos, fenthion, and chlorpyrifos. PF32 strain degraded around 99% of 100 mg/L fenitrothion in 18 h, 99% of 100 mg L^{-1} methyl parathion, fenthion in 24 h, 99% of 100 mg L^{-1} phoxim in 30 h, 97% of 100 mg L^{-1} triazophos in 42 h, and 97% of 100 mg L^{-1} chlorpyrifos in 48 h [131].

Stenotrophomonas maltophilia MHF ENV20 is able to degrade chlorpyrifos with a half-life of 96 h and its metabolite trichlorophenol in addition to diethyl thiophosphate salt [64]. Stenotrophomonas sp. YC-1 also utilizes chlorpyrifos as a sole source of carbon, but cannot further degrade its metabolite trichlorophenol. YC-1 can completely degrade 100 mg L^{-1} of chlorpyrifos in 24 h [133]. Stenotrophomonas sp. DSP-4 completely degrades 100 mg L^{-1} chlorpyrifos in 24 h [69].

Substrate specificity varies among the bacterial species. Methyl parathion hydrolase is an important organophosphorus hydrolase that is present in many species of bacteria. Until now, the hydrolase has been found in methyl parathion degrading *Pseudaminobacter salicylatoxidans* mp-1, *Ochrobactrum tritici* mp-3, *Achromobacter xylosoxidans* mp-2, *B. melitensis* mp-7, Plesiomonas sp. M6, *Pseudomonas stutzeri* HS-D36, *Sphingopyxis* sp. DLP-2 [29, 125, 137], chlorpyrifos degrading Cupriavidus sp. DT-1 [70] and phoxim-degrading *Delftia* sp. XSP-1 [104].

Dissimilar species of bacteria have different specificity of substrate utilization. Thus, methyl parathion is the most favorable substrate for methyl parathion hydrolase enzyme isolated and purified from Pseudomonas sp. WBC-3 [56] which can degrade chlorpyrifos, sumithion, and ethyl parathion [27]. Burkholderia sp. FDS-1 degrades methyl parathion, parathion, and sumithion but no triazophos, chlorpyrifos, phoxim, and methamidophos [138]. The enzyme methyl parathion hydrolase is intracellular in nature in a few strains [138]. Different types of the bacteria used for the bioremediation of pesticides are presented in Table 8.1.

TABLE 8.1 Bacteria Used for the Bioremediation of Pesticides in Soils

Bacteria	Pesticide	References
Serratia sp. SPL-2	Methidathion	[68]
Pseudomonas aeruginosa Is-6	Acephate, methyl parathion, methamidophos, malathion, and dimethoate	[93]
Bacillus subtilis, *Bacillus cereus*, *Brucella melitensis*, *Klebsiella* species, *Pseudomonas fluorescence*, *Serratia marcescens* and *Pseudomonas aeroginosa*	Chlorpyrifos	[65]
Serratia marcescens	Chlorpyrifos, parathion, and fenitrothion	[30]
Stenotrophomonas sp. SMSP-1	Methyl parathion, fenitrothion, ethyl parathion, fenthion and phoxim	[105]
Stenotrophomonas sp. PF32	Methyl parathion, sumithion, phoxim, triazophos, fenthion, and chlorpyrifos	[131]
Stenotrophomonas maltophilia MHF ENV20	Chlorpyrifos	[64]
Stenotrophomonas sp. YC-1		[133]
Stenotrophomonas sp. DSP-4		[69]
Pseudomonas sp. WBC-3	Chlorpyrifos, sumithion, and ethyl parathion	[27]
Burkholderia sp. FDS-1	Methyl parathion, parathion, and sumithion	[138]

8.6 FUNGAL REMEDIATION

Though microbial biodegradation of the pesticides has mostly been studied using bacteria, many fungal strains belonging to different genera including *Aspergillus* [102], *Penicillium* [87], *Trichoderma, Fusarium* [102]; *Phanerochaete* [26, 95, 100], *Rhizopus* [102], *Lentinus* [83] and *Trametes* [11], *Mortierella* [5] have been characterized for the biodegradation of various pesticides. In the year 1985, white-rot fungal species, *Phanerochaete chrysosporium*, has been reported to metabolize several important environmental contaminants [100]. This finding revealed the broad scope of the use of white-rot fungi (WRF) in cleaning up of environmental contaminants. Thus, different white-rot fungal strains such as *Trametes hirsutus*, *Pleurotus ostreatus P. sordida*, *Pleurotus* spp. (*P. florida*, *P. sajor-caju*, *P. eryngii*) *Lentinus subnudus* and *Cyathus bulleri were* reported to be capable of degrading recalcitrant chemical pesticides [54, 80, 83, 91, 98, 108, 109]. Similarly, different species of *Phlebia* (*P. acantho-cystis*, *P. brevispora* and *P. aurea*) were found capable of biodegrading different pesticides [128].

Nyakundi et al. [84] reported five WRF cultures and their mixtures in the degradation of various pesticides which contain different chemical groups. It was reported that the mixtures of various WRF and also the pure cultures were the efficient degraders, but the fungal consortium was more efficient when compared with pure cultures. Application of WRF in the bioremediation has many advantages over the use of other microorganisms in bioremediation as they spread their mycelial network. The WRF were also found to metabolically and cometabolically degrade insoluble and scarcely soluble chemicals [9].

The growth of mycelia facilitates rapid colonization of substrate, and hyphal extension allows penetration of soil reaching the contaminants [95]. Consequently, physical, chemical, and enzymatic contact with the surrounding environment enhances degradation [72]. Moreover, WRF can survive in a broad array of extreme environments such as pH, temperature, and as well as moisture variations [72]. Because the degradative process is provoked by nutrient deprivation, they do not necessitate preconditioning to a particular contaminant [9].

The WRF mainly use three main components of woody tissue, viz., cellulose, hemicellulose, and lignin for metabolism. Biodegradation of lignin resulted in the devastation of matrix, and consequently, the microorganisms are able to gain better access to the actual substrates, such as cellulose and hemicellulose [21, 43].

The energy required by these fungi is mostly gained by the degradation of hemicellulose and cellulose. The lignin-degrading capacity of WRF allows the fungi to target several organic pollutants resulting in their incomplete or total degradation [43]. Owing to its molecular size, it is not possible for lignin to be absorbed and degraded by the intracellular enzymes. Thus, ligninolytic enzymes are released extra-cellular by the degrading fungi which start the oxidation of substrates in extracellular environments. Therefore, WRF have been developed very non-specific mechanisms to degrade lignin extracellularly and in the process acquired the ability to target a variety of synthetic toxicants [9].

8.7 ALGAL REMEDIATION

The repeated use of pesticides causes serious environmental problems because of their high accumulation and persistent nature. In recent times, microalgae technology has received significant accomplishment in the efficient treatment of pesticides contamination [57]. Recently, application of the microalgae has shown promise as bioremediation technology [103, 136]. The application of microalgae to eliminate nutrients, pharmaceutical chemicals, pesticides, toxic elements, and oils from wastewater has been broadly reported [119].

Furthermore, microalgae (about 45,000 to 100,000 species of algae present on earth) have a broad range of applications, in the pharmaceutical industry, animal feed, environmental detection, biotechnology, renewable energy, cosmetic industry, food industry, etc. [25, 94, 112]. Only some species of the microalgae such as *Chlorella* and *Spirulina* were applied in commercial production and utilization purposes [96]. The microalgae are the primary producers in the food chain and exhibit strong adaptability to the environment. Hence, the microalgae have been effectively used in the field of environmental application and for the production of value-added products [15, 88]. They have gained meticulous attention from the researchers mostly because they can use wastewater organic matter as nutrient and carbon sources. Furthermore, microalgae are able to treat wastewater, eventually rendering the cultivation of the algae cost-effective [10].

Biodegradation is a very essential process in the microbial removal of pollutants in water [74]. Microalgal strains degrade the organic matter in water into the smaller molecules, which subsequently served as the nutrient sources for the growth of microalgae [88]. The efficacy of pesticide degradation by microalgae is comparatively higher than photodegradation through natural light [126].

Biodegradation of OPs by three filamental strains of algae (*N. muscorum, S. platensis* and *A. oryzae*) exhibited the mean removal efficiencies of 91%, 65% and 54%, respectively [53]. In another study, five green algae (i.e., *Scenedesmus sp.* MM1, *Chlamydomonas sp., Scenedesmus sp.* MM2, *Stichococcus sp., and Chlorella sp.*) were found to have high fenamiphos detoxification capability, particularly *Chlorella sp.,* which achieved above 99% removal efficacy [20].

Biodegradation of pesticides in the environment mostly depends on the metabolism of a variety of enzymes. Transferase, Esterase, and cytochrome P450 are considered the major enzymes involved in the processes of pesticide biotransformation and detoxification [35]. Additionally, hydrolase, phosphotriesterase, phosphatase oxidoreductases and oxygenase, are also involved in the degradation of pesticides [67].

In reality, the degradation of pesticides is a multistep process involving the enzyme metabolism as follows [48, 62, 67, 76]:

- The activation of pesticides in the absence of functional groups by cytochrome P450 by oxidation, reduction, and hydroxylation reactions to attain more hydrophilic, soluble, degradable low toxic compounds;
- Transfer of enzymes in cytosol to pesticides that are activated or contain functional groups, which form a conjugation with organic compounds such as glutathione (GSH), glucose, and malonate;
- Transport of these conjugates into the vacuoles via GSH transporters.

But the biodegradation mechanism of various pesticides requires further study as it really depends on the nature of pesticides and the microbial mechanism. The enzymes play an important role in the process of bioaccumulation and biodegradation, which can not only alleviate the damage by ROS to cells but also gradually reduce the toxicity of pesticides.

8.8 FUTURE DIRECTIONS

In these days, the use of pesticides has become an important part in agricultural production throughout the world. It has posed an extra burden on agriculture in order to produce more food and fiber to feed a growing population. Therefore, it is inevitable to use various types of pesticides to protect crops from the various pests, thereby increasing crop yields. Although the pesticides play a key role in improving crop yields in one way, on another mode, they turn into POPs that dramatically affect the beneficial soil microorganisms, including animals and humans. Therefore, many researchers have been continuously working on inventing technologies for the effective clean-up of contaminated sites. Although bioremediation is proved to be an effective technology for decontaminating the polluted soils by many researchers, in depth study is very much essential to achieve effective bioremediation of contaminated soils by using potential microbial strains. During the biremediation of contaminated soils, the following factors must be taken into consideration.

Several biotic and abiotic factors also significantly influence the bioremediation of pesticides by microorganisms in soils. Therefore, an understanding of the interactions between soil microorganisms and POPs during bioremediation of contaminated soil is an important and decisive factor to achieve an effective bioremediation of polluted environments. The biodegradation of a single pesticide may differ from the mixture of pesticides, because soil is the ultimate sink for many chemical contaminants. The soil conditions like pH, moisture, aeration, temperature, and the content of organic matter largely influence the rate of the microbial degradation due to their direct effect on the growth of soil microorganisms and their biological activity.

8.9 SUMMARY

The microorganisms play a tremendous role in decontaminating the contaminated soils throughout the world. Thus, many microbial communities have been used continuously to decontaminate the toxic compounds viz, POPs, OCPs, and OPs by many researchers. Increased attention has recently been paid to the isolation, identification, characterization, and biotechnological potential of microorganisms due to their unique biological features that tolerate them to survive under extreme environmental conditions like high temperature, alkaline conditions, and, water, and nutrient scarcity environments. Usually, physical and chemical cleanup technologies are expensive and sometimes not much effective. These non-biological methods may also produce other unwanted compounds that in turn contaminate the surrounding environment. However, the biological techniques such as bioremediation by using microorganisms (bacteria, fungi, and also algae) have been proven very effective, eco-friendly, and efficient. In this chapter, authors incorporated the pertinent data to understand the microbial interactions with pesticides in soils and discussed in detail about the biodegradation of pesticides by various microorganisms, which could be useful in determining the effective bioremediation strategies to decontaminate the polluted soils.

KEYWORDS

- **algae**
- **bacteria**
- **bioremediation**
- **fungi**
- **organochlorine pesticides**
- **persistent organic pollutants**
- **pesticides**

REFERENCES

1. Abdelhafid, R., Houot, S., & Barriuso, E., (2000). Dependence of atrazine degradation on c and n availability in adapted and non-adapted soils. *Soil Biology and Biochemistry, 32*, 389–401.
2. Agrawal, A., Pandey, R. S., & Sharma, B., (2010). Water pollution with special reference to pesticide contamination in India. *Journal of Water Resource & Protection, 2*(5), 432–448.
3. Alcalde, M., Ferrer, M., Plou, F. J., & Ballesteros, A., (2006). Environmental biocatalysis: From remediation with enzymes to novel green processes. *Trends in Biotechnology, 24*(6), 281–287.
4. Allmaras, R. R., Wilkins, D. E., Burnside, O. C., & Mulla, D. J., (2018). Agricultural technology and adoption of conservation practices. Chapter 6. In: Pierce, F. J., & Frye, W. W., (eds.), *Advances in Soil and Water Conservation* (pp. 99–158). New York: Routledge.
5. Badawi, N., Ronhede, S., & Olsson, S., (2009). Metabolites of the phenyl-urea herbicides chlorotoluron, diuron, isoproturon and linuron produced by the soil fungus *Mortierella* sp. *Environmental Pollution, 157*(10), 2806–2812.
6. Baker, K. H., & Herson, D. S., (1993). *Bioremediation*. (*Advanced Science & Technology*) (1st edn., p. 375). New York: McGraw Hill.
7. Bakirhan, N. K., Uslu, B., & Ozkan, S. A., (2018). The detection of pesticide in foods using electrochemical sensors. Chapter 5. In: Grumezescu, A., & Holban, A. M., (eds.), *Food Safety and Preservation* (pp. 91–141). Cambridge: Academic Press.
8. Barion, A., Balsaa, P., Werres, F., Neuhaus, U., & Schmidt, T. C., (2018). Stability of organochlorine pesticides during storage in water and loaded SPE disks containing sediment. *Chemosphere, 210*, 57–64.
9. Barr, D., & Aust, S., (1994). Mechanisms white rot fungi use to degrade pollutants. *Environmental Science and Technology, 28*(2), 78–87.
10. Barros, A. I., Goncalves, A. L., Simoes, M., & Pires, J. C. M., (2015). Harvesting techniques applied to microalgae: Review. *Renewable Sustainable Energy Reviews, 41*(C), 1489–1500.
11. Bastos, A. C., & Magan, N., (2009). *Trametes versicolor*: Potential for atrazine bioremediation in calcareous clay soil under low water availability conditions. *International Biodeterioration and Biodegradation, 63*(4), 389–394.
12. Basu, S., Rabara, R. C., Negi, S., & Shukla, P., (2018). Engineering PGPMOs through gene editing and systems biology: Solution for phytoremediation. *Trends Biotechnology, 36*, 499–510.
13. Bedi, J. S., Gill, J. P. S., Kaur, P., & Aulakh, R. S., (2018). Pesticide residues in milk and their relationship with pesticide contamination of feedstuffs supplied to dairy cattle in Punjab (India). *Journal of Animal and Feed Sciences, 27*, 18–25.
14. Benbrook, C. M., (2016). Trends in glyphosate herbicide use in the united states and globally. *Environmental Science Europe, 28*(3), 1–15.
15. Benemann, J. R., Woertz, I., & Lundquist, T., (2018). Autotrophic microalgae biomass production: From niche markets to commodities. *Industrial Biotechnology, 14*(1), 3–10.
16. Borja, A., (2018). Testing the efficiency of a bacterial community-based index (microgAMBI) to assess distinct impact sources in six locations around the world. *Ecological Indicators, 85*, 594–602.

17. Bottrell, D. G., & Schoenly, K. G., (2018). Integrated pest management for resource-limited farmers: Challenges for achieving ecological, social and economic sustainability. *Journal of Agriculture and Science, 156*, 1–19.
18. Boudh, S., & Singh, J. S., (2019). Pesticide contamination: Environmental problems and remediation strategies. Chapter 12. In: Ramnaresh, B., & Pankaj, C., (eds.), *Emerging and Eco-friendly Approaches for Waste Management* (pp. 245–270). Singapore: Springer Nature.
19. Bumpus, J. A., Tien, M., Wright, D., & Aust, S. D., (1985). Oxidation of persistent environmental pollutants by a white-rot fungus. *Science, 228*(4706), 1434–1436.
20. Caceres, T. P., Megharaj, M., & Naidu, R., (2008). Biodegradation of the pesticide fenamiphos by ten different species of green algae and cyanobacteria. *Current Microbiology, 57*(6), 643–646.
21. Canet, R., Birnstingl, J., Malcolm, D., Lopez-Real, J., & Beck, A., (2001). Biodegradation of polycyclic aromatic hydrocarbons (PAHs) by native microflora and combinations of white-rot fungi in a coal-tar contaminated soil. *Bioresource Technology, 76*, 113–117.
22. Cazalis, V., Loreau, M., & Henderson, K., (2018). Do we have to choose between feeding the human population and conserving nature?: Modeling the global dependence of people on ecosystem services. *Science of the Total Environment, 634*, 1463–1474.
23. Chen, S., Luo, J., Hu, M., Geng, P., & Zhang, Y., (2012). Microbial detoxification of bifenthrin by a novel yeast and its potential for contaminated soils treatment. *PLoS One*, 8. online, doi: 10.1371/journal.pone.0030862.
24. Chen, W., & Mulchandani, A., (1998). Use of live biocatalysts for pesticide detoxification. *Trends in Biotechnology, 16*(2), 71–76.
25. Chisti, Y., (2018). Society and microalgae: Understanding the past and present. Chapter 2. In: Levine, I. A., (ed.), *Microalgae in Health and Disease Prevention* (pp. 11–21). New York: Academic Press.
26. Chrinside, A. E., Ritter, W. F., & Radosevich, M., (2011). Biodegradation of aged residues of atrazine and alachlor in a mix-load site soil by fungal enzymes. *Applied and Environmental Soil Science,* 1–10.
27. Chu, X., Zhang, X., Chen, Y., Liu, H., & Song, D., (2003). Study on the properties of methyl parathion hydrolase from *Pseudomonas* sp.WBC-3. *Acta Microbiologica Sinica, 43*, 453–459.
28. Cox, L., Cecchi, A., Celis, R., Hermosin, M. C., Koskinen, W. C., & Cornejo, J., (2001). Effect of exogenous carbon on movement of simazine and 2,4-D in soils. *Soil Science Society of American Journal, 65*, 1688–1695.
29. Cui, Z. L., Li, S. P., & Fu, G. P., (2001). Isolation of methyl parathion-degrading strain m6 and cloning of the methyl parathion hydrolase gene. *Applied and Environmental Microbiology, 67*, 4922–4925.
30. Cycon, M., Mijowska, A. Z., Wojcik, M., & Piotrowska-Seget, Z., (2013). Biodegradation and bioremediation potential of diazinon-degrading *Serratia marcescens* to remove other organophosphorus pesticides from soils. *Journal of Environmental Management, 117*, 7–16.
31. Dai, Z., Zhang, S., Yang, Q., Zhang, W., Qian, X., Dong, W., Jiang, M., & Xin, F., (2018). Genetic tool development and systemic regulation in biosynthetic technology. *Biotechnology Biofuels, 11*, 152–158. Online, Doi: 10.1186/s13068-018-1153-5.
32. De Sousa, C. S., Hassan, S. S., & Pinto, A. C., (2018). Microbial omics: Applications in biotechnology. Chapter 1. In: Barh, D., & Azevedo, V., (eds.), *Omics Technologies and Bio-Engineering* (pp. 3–20). Cambridge, MA: Academic Press.

33. Delgado, M., & Pena, L. A., (2009). Compost and vermicompost of olive cake to bioremediate triazines - contaminated soil. *Science of the Total Environment, 407*, 1489–1495.
34. Delgado, M., & Pena, L. A., (2007). Organic amendments from olive-mill wastes as a strategy to modify degradation of sulfonylurea herbicides in soil. *Journal of Agriculture and Food Chemistry, 55*, 6213–6218.
35. Devonshire, A. L., & Field, L. M., (2011). Gene amplification and insecticide resistance. *Annual Reviews of Entomology, 67*(8), 1–23.
36. Dileep, K. S., (2008). Biodegradation and bioremediation of pesticide in soil: Concept, method and recent developments. *Indian Journal Microbiology, 48*, 35–40.
37. Dong, Y. J., Bartlam, M., & Sun, L., (2005). Crystal structure of methyl parathion hydrolase from *Pseudomonas* sp. WBC-3. *Journal of Molecular Biology, 353*, 655–663.
38. Drangert, J. O., Tonderski, K., & Conville, J. M., (2018). Extending the European union waste hierarchy to guide nutrient-effective urban sanitation toward global food security opportunities for phosphorus recovery. *Frontiers in Sustainable Food Systems, 2*, 3–10.
39. Duke, S. O., (2018). Interaction of chemical pesticides and their formulation ingredients with microbes associated with plants and plant pests. *Journal of Agricultural and Food Chemistry, 66*, 7553–7561.
40. Emiko, M., Kawanaka, Y., Yun, S. J., & Oyaizu, S., (2009). Bioremediation of the organochlorine pesticides, dieldrin and endrin, and their occurrence in the environment. *Applied Microbiology and Biotechnology, 84*(2), 205–16.
41. Eqani, A. M. A. S., Malik, R. N., & Cincinelli, A., (2013). Uptake of organochlorine pesticides (OCPs) and polychlorinated biphenyls (PCBs) by river water fish in river Chenab. *Science of the Total Environment, 450, 451*, 83–91.
42. Fatima, S. A., Hamid, A., Yaqub, G., Javed, A., & Akram, H., (2018). Detection of volatile organic compounds in the blood of farmers and their general health and safety profile. *Nature Environment and Pollution Technology, 17*, 657–660.
43. Field, J. A., De Jong, E., Costa, G. F., & Bont, D. J., (1992). Biodegradation of polycyclic aromatic hydrocarbons by new isolates of white-rot fungi. *Appl Environ Microbiol, 58*, 2219–2226.
44. Floch, C. C., Anne-Celine, J., Karine, C. Y., & Steven, C., (2011). Indicators of pesticide contamination: Soil enzyme compared to functional diversity of bacterial communities via biolog ecoplates. *European Journal of Soil Biology, 47*, 256–263.
45. Gaur, N., Narasimhulu, K., & Pydisetty, Y., (2018). Recent advances in the bioremediation of persistent organic pollutants and its effect on the environment. *Journal of Cleaner Production, 198*, 1602–1631.
46. Geetha, M., & Fulekar, M. H., (2008). Bioremediation of pesticides in surface soil treatment unit using microbial consortia. *African Journal of Environmental Science and Technology, 2*(2), 36–45.
47. Gevao, B., Semple, K. T., & Jones, K. C., (2000). Bound pesticide residues in soils: A review. *Environmental Pollution, 108*, 3–14.
48. Ghasemi, Y., Rasoul-Amini, S., & Fotooh-Abadi, E., (2011). The biotransformation, biodegradation, and bioremediation of organic compounds by microalgae. *Journal of Phycology, 47*(5), 969–980.
49. Hall, E., Bernhardt, E. S., Bier, R. L., & Bradford, M. A., (2018). Understanding how microbiomes influence the systems they inhabit. *Nature Microbiology, 3*, 977–982.

50. Hamedi, R., Bgaghaie, A., & Hadjmohammadi, M. R., (2018). Magnetic core micelles as a nanosorbent for the efficient removal and recovery of three organophosphorus pesticides from fruit juice and environmental water samples. *Journal of Separation Science, 41*, 2037–2045.
51. Hernandez-Rodriguez, D., & Sanchez, J. E., (2006). Degradation of endosulfan during substrate preparation and cultivation of *Pleurotus pulmonarius*. *World Journal of Microbiology and Biotechnology, 22*,753–760.
52. Huizhen, L., Zeng, E. Y., & Jing, Y., (2014). Mitigating pesticide pollution in china requires law enforcement, farmer training, and technological innovation. *Environmental Toxicology and Chemistry, 33*(5), 963–971.
53. Ibrahim, W. M., Karam, M. A., Elshahat, R. M., & Adway, A. A., (2014). Biodegradation and utilization of organophosphorus pesticide malathion by cyanobacteria. *Biomed Research International, 2*, 6.
54. Jauregui, J., Valderrama, B., Albores, A., & Vazquez-Duhalt, R., (2003). Microsomal transformation of organophosphorus pesticides by white rot fungi. *Biodegradation, 14*, 397–406.
55. Javaid, M. K., Ashiq, M., & Tahir, M., (2016). Potential of biological agents in decontamination of agricultural soil. *Scientifica*, (2016); pp. 9.
56. Jie, D. Y., Bartlam, M., Sun, L., & Zhou, Y. F., (2005). Crystal structure of methyl parathion hydrolase from *Pseudomonas* sp.WBC-3. *Journal of Molecular Biology, 353*, 655–663.
57. Jing, N., Yuqing, S., & Yaoyu, Z., (2020). Bioremediation of water containing pesticides by microalgae: Mechanisms, methods, and prospects for future research. *Science of the Total Environment, 707*, 1–8. Online, Doi. https://doi.org/10.1016/j.scitotenv.2019.136080.
58. John, E. M., Varghese, E. M., Krishnasree, N., & Jisha, M. S., (2018). *In situ* bioremediation of chlorpyrifos by *Klebsiella* sp. isolated from pesticide-contaminated agricultural soil. *International Journal of Current Microbiology and Applied Sciences, 7*, 1418–1429.
59. Kafilzadeh, F., Ebrahimnezhad, M., & Tahery, Y., (2015). Isolation and identification of endosulfan-degrading bacteria and evaluation of their bioremediation in Kor river, Iran. *Osong Public Health Research Perspectives, 6*(1), 39–46.
60. Kalia, A., & Gosal, S. K., (2011). Effect of pesticide application on soil microorganisms. *Archives of Agronomy and Soil Science, 57*(6), 569–596.
61. Kong, W., Meldgin, D. R., Collins, J. J., & Lu, T., (2018). Designing microbial consortia with defined social interactions. *Nature Chemistry Biology, 14*, 821–829.
62. Kumar, A., & Singh, J. S., (2017). Cyanoremediation: Green-clean tool for decontamination of synthetic pesticides from agro and aquatic ecosystems. Chapter 4. In: Singh, J. S., & Gamini, S., (eds.), *Agro-Environmental Sustainability* (pp. 59–84). New York: Springer International Publishing.
63. Kumar, D., Rai, D., Porwal, P., & Kumar, S., (2018). Compositional quality of milk and its contaminants on physical and chemical concern: Review. *International Journal of Current Microbiology and Applied Sciences, 7*, 1125–1132.
64. Kumari, K., & Fulekar, D. M. H., (2012). Chlorpyrifos bioremediation in Pennisetum rhizosphere by a novel potential degrader *Stenotrophomonas maltophilia* MHFENV 20. *World Journal of Microbiology and Biotechnology, 28*, 1715–1725.

65. Lakshmi, C. V., Kumar, M., & Khann, S., (2008). Biotransformation of chlorpyrifos and bioremediation of contaminated soil. *International Biodeterioration Biodegradation, 62*, 204–209.
66. Lakshmi, T., & Jitendra, S., (2012). Bioremediation of pesticides by microorganisms: General aspects and recent advances. Chapter 2. In: Maheshwari, D. K., & Dubey, R. C., (eds.), *Bioremediation of Pollutants* (pp. 24–48). New Delhi: I. K. International Publishing House.
67. Laura, M., Snchez-Salinas, E., Gonzlez, E. D., & Luisa, M., (2013). *Pesticide Biodegradation: Mechanisms, Genetics and Strategies to Enhance the Process* (Vol. 10, pp. 251–287). Intech open online.
68. Li, C., Lan, Y., Zhang, J., Chen, Z., Tang, D., & Xu, H., (2013). Biodegradation of methidathion by *Serratia* sp. in pure cultures using an orthogonal experiment design, and its application in detoxification of the insecticide on crops. *Annals of Microbiology, 63*, 451–459.
69. Li, X., Jiang, J., Gu, L., Ali, S., He, J., & Li, S., (2008). Diversity of chlorpyrifos-degrading bacteria isolated from chlorpyrifos-contaminated samples. *International Biodeterioration and Biodegradation, 62*, 331–335.
70. Lu, P., Li, Q., Liu, H., Feng, Z., Yan, X., Hong, Q., & Li, S., (2013). Biodegradation of chlorpyrifos and 3,5,6-trichloro-2-pyridinol by *Cupriavidus* sp. DT-1. *Bioresource Technology, 127*, 337–342.
71. Malla, M. A., Dubey, A., Yadav, S., Kumar, A., Hashem, A., & Abd, A. E. F., (2018). Understanding and designing the strategies for the microbe-mediated remediation of environmental contaminants using omics approaches. *Frontiers of Microbiology, 9*, 1132–1136. Online, doi: 10.3389/fmicb.2018.01132.
72. Maloney, S. E., (2001). Pesticide degradation. Chapter 8. In: Gadd, G. M., (ed.), *Fungi in Bioremediation* (Vol. 23, pp. 188–223). London: British Mycological Society - Cambridge University Press.
73. Mamta, K., Jaganadha, R. R., & Khursheed, A. W., (2015). Bioremediation of pesticides under the influence of bacteria and fungi. Chapter 3. In: Singh, S., (ed.), *Handbook of Research on Uncovering New Methods for Ecosystem Management through Bioremediation* (pp. 51–72). Philadelphia – PA: IGI Global Publisher.
74. Matamoros, V., Uggetti, E., Garca, A. J., & Bayona, J. M., (2016). Assessment of the mechanisms involved in the removal of emerging contaminants by microalgae from wastewater: A laboratory-scale study. *Journal of Hazardous Materials, 301*, 197–205.
75. Matsumoto, E., Kawanaka, Y., Yun, S. J., & Oyaizu, H., (2008). Isolation of dieldrin-and endrin-degrading bacteria using 1,2-epoxycyclohexane as a structural analog of both compounds. *Applied Microbiology and Biotechnology, 80*(6), 1095–1103.
76. Mau, A., Lavoie, M., Song, H., Jin, Y., Fu, Z., & Qian, H., (2017). Interaction of chiral herbicides with soil microorganisms, algae, and vascular plants. *Science of the Total Environment, 580*, 1287–1299.
77. Maya, K., Singh, R. S., Upadhyay, S. N., & Dubey, S. K., (2011). Kinetic analysis reveals bacterial efficacy for biodegradation of chlorpyrifos and its hydrolyzing metabolite TCP. *Process Biochem*istry*, 46*, 2130–2136.
78. Montiel-Rozas, M. M., & Dominguez, M. T., (2018). Long-term effects of organic amendments on bacterial and fungal communities in a degraded Mediterranean soil. *Geoderma, 332*, 20–28.

79. Moorman, T. B., (2018). Pesticide degradation by soil microorganisms: Environmental, ecological, and management effects. Chapter 5. In: Hatfield, J. L., (ed.), *Soil Biology* (pp. 127–172). Boca Raton, F. L: CRC Press.
80. Mougin, C., Pericaud, C., Malosse, C., Laugero, C., & Asther, M., (1996). Biotransformation of the insecticide lindane by the white-rot basidiomycete *Phanerochaete chrysosporium*. *Pesticide Science, 47*, 51–59.
81. Muangchinda, C., & Rungsihiranrut, A., (2018). The 16S metagenomic analysis reveals the adaptability of a mixed-PAH-degrading consortium isolated from crude oil-contaminated seawater to changing environmental conditions. *Journal of Hazardous Materials, 357*, 119–127.
82. Nikolopoulou, M., & Kalogerakis, N., (2010). Biostimulation strategies for enhanced bioremediation of marine oil spills, including chronic pollution. In: Kenneth, N. T., (ed.), *Handbook of Hydrocarbon and Lipid Microbiology* (pp. 2521–2529). Berlin: Springer.
83. Nwachukwu, E. O., & Osuji, J. O., (2007). Bioremedial degradation of some herbicides by indigenous white-rot fungus, *Lentinus subnudus*. *Journal of Plant Science, 2*, 619–624.
84. Nyakundi, W. O., Magoma, G., Ochora, J., & Nyende, A. B., (2012). Biodegradation of diazinon and methomyl pesticides by white rot fungi from selected horticultural farms in rift valley and central provinces, Kenya. *Scientific Conference Proceedings, 2012*, 639–654.
85. Ortiz-Hernandez, M. L., & Castrejon-Godinez, M. L., (2018). Strategies for biodegradation and bioremediation of pesticides in the environment. Chapter 6. In: Fuentes, M. S., & Colin, V. L., & Saez, J. M., (eds.), *Strategies for Bioremediation of Organic and Inorganic Pollutants* (pp. 95–115). Boca Raton, FL: CRC Press.
86. Parween, T., Bhandari, P., Sharma, R., Jan, S., Siddiqui, Z. H., & Patanjali, P. K., (2018). Bioremediation: Sustainable tool to prevent pesticide pollution. Chapter 12. In: Mohammad, O., & Mohammad, Z. K., & Iqbal, M. I. I., (eds.), *Modern Age Environmental Problems and their Remediation* (pp. 215–227). Cham: Springer.
87. Peng, C. L., Gu, P., Li, J., Chenq, Y., Feng, C. H., Luo, H. H., & Du, Y. J., (2012). Identification and field bioassay of the sex pheromone of *Trichophysetis cretacea* (Lepidoptera: Crambidae). *Journal of Economic Entomology, 105*, 1566–1572.
88. Perez-Legaspi, I. A., & Ortega-Clemente, L. A., (2016). Effect of the pesticide lindane on the biomass of the microalgae *Nannochloris oculata*. *Journal of Environmental Science & Health Part B, 51*(2), 103–106.
89. Pimentel, D., & Leviton, L., (1986). Pesticides: Amounts applied and amounts reaching pests. *Bioscience, 36*, 86–91.
90. Plattner, J., Kazner, C., Naidu, G., & Wintgens, T., (2018). Pesticide and microbial contaminants of groundwater and their removal methods: Review. *Journal Jaffna Science Association, 1*, 1–9.
91. Purnomo, A. S., Putra, S. R., Shimizu, K., & Kondo, R., (2014). Biodegradation of heptachlor and heptachlor epoxide-contaminated soils by white-rot fungal inocula. *Environmental Science and Pollution Research, 21*(19), 11305–11312.
92. Qiu, X. H., Bai, W. Q., Zhong, Q. Z., Li, M., He, F. Q., & Li, B. T., (2006). Isolation and characterization of a bacterial strain of the genus *Ochrobactrum* with methyl parathion mineralizing activity. *Journal of Applied Microbiology, 101*, 986–994.

93. Ramu, S., & Seetharaman, B., (2014). Biodegradation of acephate and methamidophos by a soil bacterium *Pseudomonas aeruginosa* strain IS-69. *Journal of Environmental Science B, 49*, 23–34.
94. Ravindran, B., Gupta, S. K., & Cho, W. M., (2016). Microalgae potential and multiple roles-current progress and future prospects: Overview. *Sustainability, 8*(12), 1215–1220.
95. Reddy, C. A., & Mathew, Z., (2001). Bioremediation of potential of white-rot fungi. Chapter 3. In: Gadd, G. M., (ed.), *Fungi in Bioremediation* (pp. 52–78). London: Cambridge University Press.
96. Rizwan, M., Mujtaba, G., Memon, S. A., Lee, K., & Rashid, N., (2018). Exploring the potential of microalgae for new biotechnology applications and beyond: Review. *Renewable Sustainable Energy Reviews, 92*, 394–404.
97. Ronhede, S., Sebastian, B., Sorensen, R., Jensen, B., & Aamand, J., (2007). Mineralization of hydroxyl atedisoproturon metabolites produced by fungi. *Soil Biology and Biochem*istry*, 39*, 1751–1758.
98. Sagar, V., & Singh, D. P., (2011). Biodegradation of lindane pesticide by non-white-rots soil fungus *Fusarium* sp. *World Journal of Microbiology and Biotechnology, 27*, 1747–1754.
99. Sarlio, S., (2018). Serving sustainable and healthy food to consumers and decision-makers: From commitments to action. Chapter 4. In: *Towards Healthy and Sustainable Diets* (pp. 63–82). Cham: Springer.
100. Sasek, V., (2003). Why myco-remediations have not yet come to practice. Chapter 24. In: Sasek, V., (ed.), *The Utilization of Bioremediation to Reduce Soil Contamination: Problems and Solutions* (Vol. 19, pp. 247–276). Dordrecht: Kluwer Academic Publishers.
101. Schmidt, J. R. A., & Nault, B. A., (2018). Crop spatiotemporal dominance is a better predictor of pest and predator abundance than traditional partial approaches. *Agriculture Ecosystem and Environment, 265*, 331–339.
102. Sene, L., Converti, A., Secchi, G. A., R, & Simao, R. C. G., (2010). New aspects on atrazine biodegradation. *Brazilian Archives of Biology and Technology, 53*(2), 487–496.
103. Shao, J., He, Y., Li, F., Zhang, H., Chen, A., Luo, S., & Gu, J. D., (2016). Growth inhibition and possible mechanism of oleamide against the toxin-producing cyanobacterium *Microcystis aeruginosa* NIES-843. *Ecotoxicology, 25*(1), 225–233.
104. Shen, Y. J., Hong, Y. F., Hong, Q., Jiang, X., & Li, S. P., (2007). Isolation, identification and characteristics of phoximd-degrading bacterium XSP-1. *Environmental Science, 28*, 2833–2837.
105. Shen, Y. Z., Lu, P., Mei, H., Yu, H. J., Hong, Q., & Li, S. P., (2010). Isolation of a methyl parathion - degrading strain *Stenotrophomonas* sp. SMSP-1 and cloning of Theophc2 gene. *Biodegradation, 21*, 785–792.
106. Silva-Barni, M. F., Gonzalez, M., Wania, F., Lei, Y. D., & Miglioranza, K. S. B., (2018). spatial and temporal distribution of pesticides and PCBs in the atmosphere using XAD-resin-based passive samplers: Case study in the Quequén Grande River watershed. *Argentina Atmospheric Pollution Research, 9*, 238–245.
107. Simondelso, N., Amaralrogers, V., & Belzunces, L. P., (2015). Systemic insecticides (neonicotinoids and fipronil): Trends, uses, mode of action and metabolites. *Environmental Science and Pollution Research, 22*(1), 5–34.
108. Singh, B. K., & Kuhad, R. C., (1999). Biodegradation of lindane (γ-hexachlorocyclohexane) by the white-rot fungus *Trametes hirsutus*. *Letters in Applied Microbiology, 28*(3), 238–241.

109. Singh, B. K., Kuhad, R. C., Singh, A., Tripathi, K. K., & Ghosh, P. K., (2000). Microbial degradation of the pesticide lindane (gamma-hexachlorocyclohexane). *Advances in Applied Microbiology, 47*, 269–298.
110. Singh, B. K., Walker, A., & Wright, W. J., (2006). Bioremedial potential of fenamiphos and chlorpyrifos degrading isolates: Influence of different environmental conditions. *Soil Biology Biochemistry, 38*, 2682–2693.
111. Singh, N. S., Sharma, R., Parween, T., & Patanjali, P. K., (2018). Pesticide contamination and human health risk factor. Chapter 3. In: Mohammad, O., & Mohammad, Z. K., & Iqbal, M. I. I., (eds.), *Modern Age Environmental Problems and their Remediation* (pp. 49–68). Singapore: Springer International Publisher.
112. Spolaore, P., Joannis-Cassan, C., Duran, E., & Isambert, A., (2006). Commercial applications of microalgae. *Journal of Bioscience and Bioengineering, 101*(2), 87–96.
113. Srinivasulu, M., & Rangaswamy, V., (2012). *Influence of Insecticides Alone and in Combination with Fungicides on Microflora of Groundnut (Arachis hypogaea L.) Soils* (p. 189). PhD Thesis; Sri Krishnadevaraya University, Anantapur, Andhra Pradesh, India.
114. Suenaga, H., (2012). Targeted metagenomics: A high-resolution metagenomics approach for specific gene clusters in complex microbial communities. *Environmental Microbiology, 14*, 13–22.
115. Sylvia, D. M., Fuhrmann, J. F., Hartel, P. G., & Zuberer, D. A., (2005). *Principles and Applications of Soil Microbiology* (2nd edn., p. 640). New Jersey, Pearson Education Inc.
116. Tal, A., (2018). Making conventional agriculture environmentally friendly: Moving beyond the glorification of organic agriculture and the demonization of conventional agriculture. *Sustainability, 10*, 1078.
117. Tejada, M., (2009). Evolution of soil biological properties after addition of glyphosate, diflufenican and glyphosate + diflufenican herbicides. *Chemosphere, 76*, 365–373.
118. Thomas, J. E., Ou, L. T., & Al-Agely, A., (2008). DDE remediation and degradation. Chapter 3. In: David, M. W., (ed.), *Reviews of Environmental Contamination and Toxicology* (pp. 55–69). New York, NY, Springer.
119. Ummalyma, S. B., Pandey, A., Sukumaran, R. K., & Sahoo, D., (2018). Bioremediation by microalgae: Current and emerging trends for effluents treatments for value addition of waste streams. Chapter 19. In: Sunita, J. V., Binod, P., Sunil, K., & Sunil, K. K., (eds.), *Biosynthetic Technology and Environmental Challenges* (pp. 355–375). Singapore: Springer Nature.
120. Uqab, B., Mudasir, S., & Nazir, R., (2016). Review on bioremediation of pesticides. *Journal of Bioremediation Biodegradation, 7*, 343–352. online, https://doi.org/10.4172/2155- 6199.1000343.
121. Vidart, D. E., B, N., & Bailey, H. D., (2018). Maternal residential pesticide use during pregnancy and risk of malignant childhood brain tumors: Pooled analysis of the ESCALE and ESTELLE studies (SFCE). *International Journal of Cancer, 142*, 489–497.
122. Vieuble-Gonod, L., Martin-Laurent, F., & Chenu, C., (2006). The 2,4-D impact on bacterial communities and the activity and genetic potential of 2,4-D degrading communities in soil. *FEMS Microbiology Ecology, 58*, 529–537.
123. Von, N. F., Granitsiotis, M. S., Szalay, A. R., & Lueders, T., (2018). Next-generation sequencing of functional marker genes for anaerobic degraders of petroleum

hydrocarbons in contaminated environments. Chapter 12. In: Boll, M., (ed.), *Handbook of Hydrocarbon and Lipid Microbiology* (pp. 257–276). Cham: Springer.
124. Wang, B., Ma, Y., Zhou, W., Zheng, J., Zhu, J., He, J., & Li, S. P., (2011). Biodegradation of synthetic pyrethroids by *Ochrobactrum tritici* strain PYD-1. *World Journal of Microbiology and Biotechnology, 27*, 2315–2324.
125. Wang, B., Xiong, L., & Zheng, Y., (2008). Cloning and expression of the mpd gene from a newly isolated methyl parathion - degrading strain of bacteria. *Acta Scientific Circum., 28*, 1969–1975.
126. Warren, N., Allan, I. J., Carter, J. E., House, W. A., & Parker, A., (2003). Pesticides and other microorganic contaminants in freshwater sedimentary environments: Review. *Applied Geochemistry, 18*(2), 159–194.
127. Xia, W. J., Zhou, J. M., Wang, H. Y., & Chen, X. Q., (2008). Effect of nitrogen on the degradation of cypermethrin and its metabolite 3-phenoxybenzoic acid in soil. *Pedosphere, 18*, 638–644.
128. Xiao, P., Mori, T., Kamei, I., Kiyota, H., Takagi, K., & Kondo, R., (2011). Novel metabolic pathways of organochlorine pesticides dieldrin and aldrin by the white-rot fungi of the genus *Phlebia*. *Chemosphere, 85*(2), 218–224.
129. Xiao, Y., Chen, S., Gao, Y., Hu, W., Hu, M., & Zhong, G., (2015). Isolation of a novel beta-cypermethrin degrading strain *Bacillus subtilis* BSF01 and its biodegradation pathway. *Applied Microbiology and Biotechnology, 99*, 2849–2859.
130. Xu, H., Li, W., & Schilmiller, A. L., (2019). Pyrethric acid of natural pyrethrin insecticide: Complete pathway elucidation and reconstitution in *Nicotiana benthamiana*. *New Phytologist, 223*, 751–765.
131. Xu, Y. X., Feng, Z. Z., Lu, P., Zhang, J., Li, S. P., & Huang, X., (2009). Isolation and characterization of capable of degrading parathion-methyl bacterium *Stenotrophomonas* sp. PF32. *Chinese Journal of Pesticide Science, 11*, 329–334.
132. Yadav, I. C., Devi, N. L., Syed, J. H., & Cheng, Z., (2015). Current status of persistent organic pesticides residues in air, water, and soil, and their possible effect on neighboring countries: Comprehensive review of India. *Science of the Total Environment, 511*, 123–137.
133. Yang, C., Liu, N., Guo, X., & Qiao, C., (2006). Cloning of MPD gene from a chlorpyrifos-degrading bacterium and use of this strain in bioremediation of contaminated soil. *FEMS Microbiology Letters, 265*, 118–125.
134. Yang, J., Hu, C. T., Zhu, X., Zhu, Q., Ward, M. D., & Kahr, B., (2017). DDT polymorphism and the lethality of crystal forms. *Angewandte Chem International Edition, 56*(34), 10165–10169.
135. Zabaloy, M. C., Garland, J. L., & Gomez, M. A., (2008). Integrated approach to evaluate the impacts of the herbicides glyphosate, 2,4-D and metsulfuron-methyl on soil microbial communities in the Pampas region, Argentina. *Applied Soil Ecology, 40*, 1–12.
136. Zhang, H., Zhang, Y., Hou, Z., Wang, X., & Wang, J., (2016). Biodegradation potential of deltamethrin by the *Bacillus cereus* strain Y1 in both culture and contaminated soil. *International Biodeterioration & Biodegradation, 106*, 53–59.
137. Zhang, R., Cui, Z., Jiang, J., He, J., Gu, X., & Li, S., (2005). Diversity of organophosphorus pesticide degrading bacteria in a polluted soil and conservation of their organophosphorus hydrolase genes. *Canadian Journal of Microbiology, 51*, 337–343.

138. Zhang, Z., Xu, J., Hong, Q., Zhang, X., & Li, S., (2006). Isolation and characterization of fenitrothion-degrading strain FDS-l (*Burkhohelria* sp.): Cloning and expression of mpd gene. *Biodegradation, 17*(3), 275–283.
139. Zhao, H., Geng, Y., Chen, L., Tao, K., & Hou, T., (2013). Biodegradation of cypermethrin by a novel *Catellibacterium* sp. strain CC-5 Isolated from contaminated soil. *Canadian Journal of Microbiology, 59*, 311–317.
140. Zhou, Y., Wu, S., & Zhou, H., (2018). Chiral pharmaceuticals: Environment sources, potential human health impacts, remediation technologies and future perspective. *Environment International, 121*(1), 523–537.

CHAPTER 9

BIOREMEDIATION OF PETROLEUM CONTAMINATED SOILS

VIKRANT B. BERDE, PALLAVAL V. BRAMHACHARI, and
CHANDA PARULEKAR BERDE

ABSTRACT

Petroleum is an important non-renewable energy source used for various purposes like transportation and in industries. Sometimes accidently or due to human interference, it can contaminate the soil. Cleaning of these polluted soils with mechanical or physicochemical treatments could be very expensive or difficult task. Bioremediation of these petroleum affected soils is an eco-friendly, economic, and efficient option. This chapter focuses on the use of bioremediation in treatment of petroleum-contaminated soils.

9.1 INTRODUCTION

Contamination of environment with petroleum hydrocarbons (PHs) occurs on account of accidental spillage, human activities, during refining and processing, transportation, leakage from pipelines and storage tankers and oil-well waxing and overhauling of refineries [20, 115, 120]. Soil is thus affected by petroleum components like hydrocarbons.

Petroleum contaminated soil affects the environment and is hazardous to human health. Treatment of the polluted soil can be done by physical, chemical, and biological methods. The microbiological method is being preferred, because it is efficient, economical, eco-friendly, and adaptable substitute to physicochemical procedures [42]. The indigenous flora of the soil being exposed to the hydrocarbons become adapted to the pollutants. These microorganisms synthesize enzymes for the degradation of the hydrocarbons, for utilizing these compounds for their carbon and energy requirements. Such organisms are useful in the removal of contaminants [59, 62, 84]. Improvements in the bacterial remediation technology have provided solutions for the removal of petroleum pollution from the environments [31, 32].

Bioremediation using bacteria cultures with hydrocarbon-degrading ability is a widely accepted technology. Many bacterial species have been characterized, studied, and applied in bioremediation of soil contaminated with hydrocarbons. However, numerous complications reducing the rate of biodeterioration have been encountered while practically applying the methods.

The present chapter is an overview of bioremediation techniques using various microorganisms in the treatment of petroleum contaminated soils. It describes major environmental issues created due to the petroleum industry and human use of petroleum in day-to-day life. It also highlights the role played by the microorganisms in the rejuvenation of the contaminated sites with suitable examples.

9.2 HAZARDOUS EFFECTS OF PETROLEUM POLLUTION ON ENVIRONMENT

harmful effects of petroleum pollution on the environment are well known [20, 92]. It affects not only humans but also animals as well as the microorganisms [30, 63, 71, 81] disturbing the ecosystems. Petroleum contaminated soil is hazardous to human health [35]. The contaminants from the soil leach out and enter the water bodies, and may even contaminate the water table, thus increasing the risk further. Petroleum contaminated soil has lower crop productivity due to the pollution [114].

Research studies have confirmed that there is a reduction in species richness and phylogenetic diversity of the affected areas following exposure to PHs due to the inhibitory effect of hydrocarbons and their metabolic intermediates, on the microflora [18, 19, 47, 56, 104]. The microorganisms that can tolerate and degrade the hydrocarbons survive in the conditions and are the dominant community in the polluted environments.

9.3 BIOREMEDIATION

Petroleum hydrocarbon pollutants can be removed by physical and chemical methods, which however are either expensive or not effective. Microbial bioremediation is the most promising solution to the total removal of petroleum pollution from any environment [1, 20, 46, 107]. Petroleum degrading microorganisms have enzymes for the breakdown of the complex petroleum components. PHs consist of four classes: the saturates, the aromatics, the asphaltenes and the resins [23]. Thus, hydrocarbon degradation is influenced by the nature of pollutant and its concentration. The more complex the nature of the contaminants, the more difficult and time-consuming is the bioremediation process [65], even though it is less costly as compared to other methods of remediation.

Bioremediation methodologies are largely classified as *in situ* (done at the site) or *ex-situ* (done at somewhere else) [3]. Organic pollutants are degraded to end-products carbon dioxide (CO_2) and water at the site of pollution. This method is called *in situ*, and it is less costly and eco-friendly, as compared to the *ex-situ* bioremediation technique that requires to excavate the contaminated samples and get them to the treatment site, which increases the cost.

Bioremediation process can be carried out using stimulating the growth of indigenous hydrocarbon-degrading bacteria or by adding grown cultures, i.e., by biostimulation or bioaugmentation. Thus, the degraders having the metabolic abilities to utilize the hydrocarbons survive, while the remaining microflora that is sensitive is inhibited. However, there is a need to pay more attention to the research that will add up efficient oil degraders and also standardization of methods for efficient and quick clean-up strategies.

9.4 ENVIRONMENTAL COMPONENTS REQUIRED FOR BIOREMEDIATION

9.4.1 ABIOTIC COMPONENTS

Bioremediation involves various mechanisms of degradation, eradication, immobilization, or detoxification of chemicals and hazardous materials that can pollute the environment. Bioremediation technology makes use of microbial cultures to get rid of toxic contaminants to harmless products (mainly CO_2, water, and other inorganic compounds) [73]. It is thus the application of microorganisms that degrade and transform the pollutants including hydrocarbons, oil, heavy metal (HM), pesticides, dyes, etc.

The rate of biodegradation depends on biotic and abiotic factors, such as pollutant concentration, nature of the pollutant, environmental conditions, physicochemical parameters, etc. [33]. Biodegradation reactions are highly influenced by temperature, nutrients, electron acceptors, and the pollutants, which play a key role in hydrocarbon bioremediation [108]. Hence, the results obtained in the laboratory and under field conditions will vary [45]. The bioremediation process also requires the availability of the substrate that the bacteria can degrade [90]. The rate at which the degradation of the hydrocarbons occur follows the order as linear [8, 58]:

alkanes > branched alkanes > low-molecular-weight alkyl aromatics > monoaromatics > cyclic alkanes > polyaromatics (such as asphaltenes)

The polycyclic aromatic compounds are degraded incompletely [6]. They are broken down into intermediate compounds; and for the complete degradation, more than one microorganism is involved. Other factors, such as soil type, presence of other contaminants, type, and amount of organic matter present, also play an important role in determining the bioremediation rate [12, 89].

Physical factors, such as, temperature affect the bioremediation process as the bacterial strains involved are temperature sensitive and also affect the hydrocarbons [2, 60, 119]. It was observed that PHs were degraded efficiently under laboratory conditions, whereas the same effects were not observed *in situ*. This was attributed to the on-site temperature variations, which could not be controlled [82, 83]. At low temperatures, petroleum products persist for a longer period. Low temperatures reduce the solubility, increase the viscosity of hydrocarbons, and delay the onset of degradation, and vice versa for elevated temperature. The increased degradation rate of hydrocarbons ensues at 30–40°C [4, 7, 11, 24, 38, 70]. However, at higher temperatures, again degradation rate is lowered as the elevated temperature will inhibit microbial growth [99].

Bacterial growth in the soil depends on nutrient availability, which includes nitrogen, phosphorus, carbon, hydrogen, etc. [21, 39]. Hydrocarbons are utilized by bacterial strains as carbon and energy sources. However, if other nutrients are deficient, the growth of the degraders is hampered, and consequently the degradation process is also affected. The addition of fertilizers is used as a stimulatory method in the bioremediation process [44, 69, 83, 84]. The use of hydrocarbon-degrading bacteria having the ability to fix nitrogen replaces the need for fertilizer supplementation [100].

The onset of hydrocarbon degradation is by incorporation of oxygen in the aromatic ring with the help of oxygenases. A step-by-step degradation process follows, whereby the enzymes act on the compound aerobically.

Thus, the presence of oxygen is the key requirement for the onset and rapid degradation of hydrocarbons [39]. In a study carried out under controlled aeration, 75% degradation was observed [41]. Aerated degradation processes however increase the cost associated with the process. Alternatively, methods to increase the soil permeability (such as the addition of a bulking agent) may be used. Another option is degradation under anaerobic conditions, which is less expensive as compared to the aerobic bioremediation method [15, 125].

9.4.2 BIOLOGICAL COMPONENTS: MICROORGANISMS AND THEIR ROLE IN BIOREMEDIATION

Bioremediation of petroleum hydrocarbon pollutants is an ecofriendly, less costly, and effective biological treatment. The microorganisms found in mesophilic conditions cannot function properly in harsh conditions and show reduced metabolism and hence can lower degradation abilities. Removal of pollutants from environments under extreme conditions is dependent on the microorganisms present in that niche, i.e., the extremophiles belonging to the archaea group, and they can grow in conditions unfavorable to eubacteria. Therefore, for the bioremediation of contaminated soils (such as estuarine mudflats, etc.), the focus is on members of archaea. *In situ* bioremediation can make use of a single or combination of microorganisms (i.e., consortia) [29]. This ensures complete breakdown of the pollutant.

9.4.2.1 BACTERIA

The primary degraders of petroleum during oil spills are bacteria [14, 78]. These bacterial cultures can utilize hydrocarbons as the carbon source [121], for metabolism and growth. There is an increase in their numbers as the breakdown occurs. Once the hydrocarbons are removed, the microbial count reduces with a simultaneous reduction in the physiological stress on the environment [43, 53, 124].

Several factors decide the fate of the contaminant in the environment and the degrader type and abundance will also depend on the type of petroleum product, degradation pathways/enzymes present in the organisms, physico-chemical parameters, etc. [40, 110].

The oil-degrading bacteria found in the ocean waters and sediments are subject to temperature and salinity variations. The oil-degrading organisms isolated are mostly from the domain bacteria [17] and few groups of

Archaea belonging to Marine Group II Archaea (namely: Euryarchaeota and Thaumarchaeota) [79]. Unculturable archaeal members were identified in marine sediments belonging to strains of Euryarchaeota [28, 48, 51].

These findings suggest that unculturable Archaea includes the haloarchaea and Methanogens, which are involved in the removal of oil contamination in most of the marine environments [126]. There are reports of an increase in the haloarchaeal species such as *Haloferax*, on the addition of oil droplets and nitrate [16, 116, 123].

According to Tremblay et al. [103], 79 genera having the hydrocarbon utilizing potential, have been studied, which includes [49, 62, 68, 87, 109, 111, 119]:

- *Achromobacter;*
- *Acinetobacter;*
- *Alkanindiges;*
- *Alteromonas;*
- *Arthrobacter;*
- *Burkholderia;*
- *Dietzia;*
- *Enterobacter;*
- *Kocuria;*
- *Marinobacter;*
- *Mycobacterium;*
- *Pandoraea;*
- *Pseudomonas;*
- *Rhodococcus;*
- *Staphylococcus;*
- *Streptobacillus;*
- *Streptococcus.*

Some investigators have reported the efficient bioremediation by consortia of cultures, such as OHCB (obligate hydrocarbon clastic bacteria). The bacterial members included in the consortia were: *Alkanindiges* sp., *Alcanivorax, Marinobacter, Thallassolituus, Cycloclasticus, Oleispira* [40, 121]. These reports also stated that almost negligible number of bacterial species were present during pre-exposure and their abundance was increased post-exposure in the presence of contaminants. A particular strain cannot use all the petroleum products or cannot breakdown the contaminants completely.

Wang et al. [118] studied the bacterial strain *Dietzia* sp. DQ12-45-1b that consumes n-alkanes (C6–C40) and related compounds utilizing them as the sole sources of carbon. Similarly, *Acinetobacter* sp. could metabolize n-alkanes having chain length C10–C40 [102]; while Ma et al. [61] reported *Achromobacter xylosoxidans* DN002 that could utilize a number of mono- and polyaromatic hydrocarbons efficiently. Some more bacterial genera, namely, *Gordonia*, *Brevibacterium*, *Aeromicrobium, Dietzia*, *Burkholderia,* and *Mycobacterium, Sphingomonas*, have also been reported with petroleum bioremediation activities [11, 26, 37].

Varjani et al. [113] worked with a consortium of halotolerant bacteria, named HUBC, which included *Ochrobactrum* sp., *Stenotrophomonas maltophilia*, and *Pseudomonas aeruginosa,* having very good degradation rates and similar work was reported by Tao et al. [98]; and Wang et al. [115]. These reports demonstrated that in the consortium, the bacterial strains had crude oil bioremediating potential in marine ecosystems. Szulc et al. [97] reported a hydrocarbon-degrading efficiency of 89% by consortia of bacterial cultures that included *Aeromonas hydrophila, Alcaligenes xylosoxidans, Gordonia* sp., *Pseudomonas putida, Pseudomonas fluorescens, Rhodococcus equi, S. maltophilia,* and *Xanthomonas* sp.

Therefore, bioremediation of petroleum waste can be improved by the application of consortia of bacteria having the ability to degrade these compounds and as a result, can degrade the pollutants completely.

9.4.2.2 FUNGI

Fungal cultures with petroleum bioremediation abilities are said to be better bioremediation agents as compared to bacteria [10, 11, 37, 101]. The fungal hyphae provide an advantage of penetration into the soil contaminated with hydrocarbons, thus reaching the deeper layers of soil. Fungal cultures can adapt to different variations in the environments and have mechanisms to cope with the hydrocarbon contaminants in the environment and bring about its degradation.

Fungal cultures (such as *Amorphoteca*, *Graphium, Neosartorya*, *Talaromyces, Candida, Yarrowia, and Pichia* sp.) were found to have hydrocarbon-degrading ability [19]. In another report, terrestrial fungi *Aspergillus*, *Cephalosporium,* and *Penicillium* were isolated and found to utilize crude oil efficiently [93].

Some yeast species for their petroleum degradation ability are: *Geotrichum* sp., *Candida lipolytica, Trichosporon mucoides* and *Rhodotorula mucilaginosa* that were isolated from contaminated waters [13]. Barnes et al. [10] has described studies on 10 fungal isolates that could degrade the crude oil. They were isolated from sediments from mangroves, the Arabian Sea and from tarballs. The fungal isolates were identified as *Aspergillus, Fusarium Penicillium*, and *Acremonium* species using its rDNA (recombinant deoxyribonucleic acid) sequencing.

Ezekoye et al. [36] have described the studies on petroleum hydrocarbon bioremediation using the fungal cultures, such as *Aspergillus fumigates*, *Aspergillus terrens*, *Aspergillus niger*, *Aspergillus nidulans*, *Aspergillus versicolor*, *Acremonium* sp., *Bipolaris* sp., *Candida tropicalis*, *Candida globrata*, *Cladosporium* sp., *Cladophialophora carrionii*, *Geotrichum candidum*, *Gliocladium* sp., *Phoma* sp., *Paecilomyces variotii*, *Scopulariopsis brevicaulis*, *Sepedonium* sp., *Trichophyton tonsurans*, *Trichophyton terrestre* and *Rhodotorula* sp. Additional reports describe the hydrocarbon degradation ability and their applications in bioremediation [22, 66, 75, 86, 122]. These isolates, along with bacterial cultures with good bioremediation efficiency can be utilized together for the clean-up of oil spills and other environmental contaminants, such as tarballs. Oil-Zapper is one such application used in the environmental cleanup (TERI, India).

Therefore, there is a need to conduct in depth study of the bioremediation process of both bacteria and fungi, the enzymes involved, efficiencies of different consortia, etc., to optimize the *in-situ* bioremediation of petroleum pollutants.

9.4.2.3 GENETICALLY MODIFIED ORGANISMS

As compared to natural isolates, the genetically engineered microorganisms (GEMs) have a higher capacity of environment cleanup. There are several reports on the biodegradation of pollutants by GEM according to Das and Chandran [25]. The GEMs include mostly strains of *Pseudomonas* species [34, 72, 95].

The GEMs have applications in monitoring the bioremediation process, monitoring the response of stress on the strains, toxicity analysis, etc. [25]. For process monitoring in PCB contaminated soils, *A. eutrophus* H850Lr is used [105]. Other strains used in process monitoring are: *P. putida* TVA8 and *P. fluorescens* HK44 for TCE, BTEX, and naphthalene, anthracene degradation, respectively [5, 88].

B. cepacia BRI6001L was used for the strain monitoring for 2,4-D degradation [64]. There is a report on the use of GEM strain *P. fluorescens* 10586s/pUCD607 to study response to stress during BTEX degradation

[94]. During the degradation of 2,4-dinitrophenol hydroquinone, Kelly et al. [52] has reported the application of the *Pseudomonas* strain Shk1 for the toxicity assessment studies. Layton et al. [57] described the application of *A. eutrophus* 2050 for endpoint analysis during the degradation of non-polar narcotics.

The only constraints involved are the environmental concerns and regulatory obligations. Before the application of the GEMs, a thorough study is needed on: the ecological consequences of the release of these strains, the biochemical pathways and enzymes involved, and the possibilities of strain reversion, etc.

9.5 PROS AND CONS OF BIOREMEDIATION FOR PETROLEUM CONTAMINATED SOILS

Most importantly, bioremediation is known for being safe to the environment and less costly [9, 74, 77]. It may be carried out in two ways: bioaugmentation (seeding, inoculation) and biostimulation [106], wherein either there is the addition of petroleum degraders to the site or the growth of indigenous flora is encouraged by supplementing with nutrients (mostly fertilizers) [21, 91, 112].

Degradation by indigenous microflora increases the microbial count during degradation. The pollutant is utilized by the microbes for their growth. Once the compound is degraded, their number start declining. CO_2 and water are the non-toxic end-products of the degradative process. Thus, the contaminants are removed completely, and not simply transferred to different environmental sites [91]. Bioremediation of pollutants by microorganisms is an eco-friendly natural way of treating the contaminants, and it is sustainable as the microbial cultures involved in the process remain in the site and can act on waste if added to the site again [27]. The sites can be used continuously. Hence the methods are non-intrusive, and there is ease of implementation [55].

The bioremediation process can be carried out *in situ*, and hence the efforts of carrying the contaminated soil can be avoided, and no transport charges are involved [67]. Physicochemical methods of bioremediation of petroleum contaminated soils are available [54, 117], but they are costly. These methods also lead to the generation of other waste and hence are not eco-friendly [85]. Successful examples are the disposal of contaminants by incineration, which leads to the production of toxic gases. More complications go with the combustion of petroleum contaminants.

Although it is known that biodegradation also leads to CO_2 formation, yet the quantity produced is negligible. Some workers have reported the ill-effects of landfilling with such contaminants [50, 80]. Over time, the contaminants leach out causing further harm to the water table [76, 96].

The bioremediation process has its drawbacks also. Firstly, it can be carried out for those compounds that can be broken down by microorganisms. Some compounds may not be completely degraded and the intermediates thus formed may be more toxic than the compound itself. These are called recalcitrant compounds. Sometimes 2 or more compounds may be involved in the degradation of a particular contaminant. Through *in-situ*, the situation is very different from laboratory conditions. The concentration of the contaminants is not known, the physicochemical parameters cannot be controlled, other biotic factors are involved, the contaminated site may have more than one compounds present, and also the state in which they are present, i.e., solids, liquids, and gases. Thus, the degree of success cannot be judged.

Application method and environmental conditions also matter such as in summer the temperature will be high while in monsoon, there are chances of wash off. The biggest drawback is the time factor, as these methods are very time-consuming. Being natural and dependent on the microbial activity, the degradation processes take long. The cultures to be used in the bioremediation process need to be cultured on a large scale, and hence there is a requirement to follow all set SOPs (standard operating procedure) for culture growth and application methods, along with the requirement of skilled labor [55].

9.6 ECOLOGICAL AND ECONOMIC IMPORTANCE OF BIOREMEDIATION

Using the endogenous microflora of the contaminated site by the biostimulation method of bioremediation, it does not affect the biotic components of the environment. After removal of the contaminant, the number of microflora start reducing and survive on the nutrients available. This is thus a natural process with no formation of harmful components or no side effects. As it is carried out on-site, the cost and dangers of transportation can be avoided. A lot of expenses are reduced due to the almost nil energy requirements, minimal equipment or instrumentation involved, and also low maintenance. Hence, bioremediation is economically feasible. Bioremediation has a higher acceptance from regulatory authorities as well as the public.

9.7 COMMERCIAL APPLICATIONS OF BIOREMEDIATION

Bioremediation can be used in several sites, where there is spillage of petroleum or its related products. It can be used in petroleum stations, where the spillage problem is well known. Another application is at the industrial sites and the sites used for landfills. Petroleum and petroleum-related compounds enter the environments through these portals. The products may leach out from the landfills and contaminate the water bodies. Thus, the bioremediation of these sites is necessary. Oils spills occurring at the sea, ports, oil rigs can affect marine life.

The removal of the spills requires the use of bioremediation. The successful implementation of bioremediation for the revival of hydrocarbon contaminated environments is the need of the day. Further, bioremediation technology should be well established with intense research and field trials with effective petroleum degrading microorganisms, including the much efficient genetically engineered organisms.

9.8 SUMMARY

The overuse of petroleum products leads to contamination of soils. Soil organisms play a crucial role in the hydrocarbon degradation in contaminated environments. The present chapter is based on the hazardous effects of petroleum contaminated soils and the remedial measures taken to treat these soils with microorganisms.

KEYWORDS

- **bioremediation**
- **genetically engineered microorganisms**
- **microorganisms**
- **obligate hydrocarbon clastic bacteria**
- **petroleum**
- **pollution**
- **toxicity**

REFERENCES

1. Abbasian, F., Lockington, R., Mallavarapu, M., & Naidu, R., (2015). A comprehensive review of aliphatic hydrocarbon biodegradation by bacteria. *Applied Biochemistry and Biotechnology, 176*(3), 670–699.
2. Abed, R. M. M., Al-Kharusi, S., & Al-Hinai, M., (2015). Effect of biostimulation, temperature and salinity on respiration activities and bacterial community composition in an oil-polluted desert soil. *International Biodeterioration and Biodegradation, 98*, 43–52.
3. Aggarwal, P. K., Means, J. L., & Hinchee, R. E., (1990). *Methods to Select Chemicals for in-situ Biodegradation of Fuel Hydrocarbons* (p. 14). Florida: Tyndall air force base, air force engineering and service center.
4. Aislabie, J., Saul, D., & Foght, J., (2006). Bioremediation of hydrocarbon-contaminated polar soils. *Extremophiles, 10,* 171–179.
5. Applegate, B. M., Kehrmeyer, S. R., & Sayler, G. S., (1998). A chromosomally based *Tod-Luxcdabe* whole-cell reporter for benzene, toluene, ethylbenzene, and xylene (BTEX) sensing. *Applied and Environmental Microbiology, 64*(7), 2730–2735.
6. Atlas, R., & Bragg, J., (2009). Bioremediation of marine oil spills: When and when not-the Exxon Valdez experience. *Microbial Biotechnology, 2*(2), 213–221.
7. Atlas, R. M., (1975). Effect of temperature and crude oil composition on petroleum biodegradation. *Journal Applied Microbiology, 30*(3), 396–403.
8. Atlas, R. M., (1981). Microbial degradation of petroleum hydrocarbons: An environmental perspective. *Microbiology Reviews, 45*(1), 180–209.
9. Atlas, R. M., & Bartha, R., (1998). *Microbial Ecology: Fundamentals and Applications* (4th edn., p. 259). Menlo Park, California: Benjamin/Cummings.
10. Barnes, N. M., Vishwas, B., Khodse, N. P., Lotlikar, R. M. M., & Damare, S. R., (2018). Bioremediation potential of hydrocarbon-utilizing fungi from select marine niches of India. *Biotech, 8*(1), 21–25.
11. Bartha, R., & Bossert, I., (1984). The treatment and disposal of petroleum wastes. In: Atlas, R. M., (ed.), *Petroleum Microbiology* (pp. 553–578). New York: Macmillan.
12. Bento, F. M., Camargo, F. A. O., Okeke, B. C., & Frankenberger, W. T., (2005). Comparative bioremediation of soils contaminated with diesel oil by natural attenuation, biostimulation and bioaugmentation. *Bioresource Technology, 96,* 1049–1055.
13. Bogusławska, E., & Dabrowski, W., (2001). The seasonal variability of yeasts and yeast-like organisms in water and bottom sediment of the Szczecin lagoon. *International Journal of Hygiene and Environmental Health, 203*(5–6), 451–458.
14. Brooijmans, R. J. W., Pastink, M. I., & Siezen, R. J., (2009). Hydrocarbon-degrading bacteria: The oil-spill clean-up crew. *Microbial Biotechnology, 2*(6), 587–594.
15. Brown, D. M., Okoro, S., & Van, G. J., (2017). Comparison of land farming amendments to improve bioremediation of petroleum hydrocarbons in Niger delta soils. *Science of the Total Environment, 596*, 284–292.
16. Campeão, M. E., Reis, L., & Leomil, L., (2017). The deep-sea microbial community from the Amazonian basin associated with oil degradation. *Frontiers in Microbiology, 8*. Article ID: 1019.
17. Catania, V., Cappello, S., & Di Giorgi, V., (2018). Microbial communities of polluted sub-surface marine sediments. *Marine Pollution Bulletin, 131*(A), 396–406.

18. Cerniglia, C. E., Freeman, J. P., Althaus, J. R., & Van, B. C., (1983). Metabolism and toxicity of 1-and 2-methylnaphthalene and their derivatives in cyanobacteria. *Archives of Microbiology, 136,* 177–183.
19. Chaillan, F., Le Fl`eche, A., & Bury, E., (2004). Identification and biodegradation potential of tropical aerobic hydrocarbon-degrading microorganisms. *Research in Microbiology, 155*(7), 587–595.
20. Chen, M., Xu, P., Zeng, G., Yang, C., Huang, D., & Zhang, J., (2015). Bioremediation of soils contaminated with polycyclic aromatic hydrocarbons, petroleum, pesticides, chlorophenols and heavy metals by composting: Applications, microbes and future research needs. *Biotechnology Advances, 33,* 745–755.
21. Choi, S. C., Kwon, K. K., Sohn, J. H., & Kim, S. J., (2002). Evaluation of fertilizer additions to stimulate oil biodegradation in sand seashore. *Journal of Microbiology and Biotechnology, 12,* 431–436.
22. Clemente, A. R., Anazawa, T. A., & Durrant, L. R., (2001). Biodegradation of polycyclic aromatic hydrocarbons by soil fungi. *Brazilian Journal of Microbiology, 32*, 255–261.
23. Colwell, R. R., Walker, J. D., & Cooney, J. J., (1977). Ecological aspects of microbial degradation of petroleum in the marine environment. *critical Reviews in Microbiology, 5*(4), 423–445.
24. Cooney, J. J., (1984). The fate of petroleum pollutants in freshwater ecosystems. In: Atlas, R. M., (ed.), *Petroleum Microbiology* (pp. 399–434). New York: Macmillan.
25. Das, N., & Chandran, P., (2011). Microbial degradation of petroleum hydrocarbon contaminants: An overview. *Biotechnology Research International, 2011*, 13. Article ID 941810, doi.org/10.4061/2011/941810.
26. Daugulis, A. J., & McCracken, C. M., (2003). Microbial degradation of high and low molecular weight polyaromatic hydrocarbons in a two-phase partitioning bioreactor by two strains of *Sphingomonas* sp. *Biotechnology Letters, 25*(17), 1441–1444.
27. Dell'Anno, A., Beolchini, F., Rocchetti, L., Luna, G. M., & Danovaro, R., (2012). High bacterial biodiversity increases the degradation performance of hydrocarbons during bioremediation of contaminated harbor marine sediments. *Environment Pollution, 167*, 85–92.
28. Dias, A. C. F., Dini-Andreote, F., & Taketani, R. G., (2011). Archaeal communities in the sediments of three contrasting mangroves. *Journal Soils and Sediments, 11*(8), 1466–1476.
29. Diaz, E., (2004). Bacterial degradation of aromatic pollutants: A paradigm of metabolic versatility. *International. Microbiology, 7,* 173–180.
30. Díez, S., Jover, E., Bayona, J. M., & Albaigés, J., (2007). Prestige oil spill, III: Fate of a heavy oil in the marine environment. *Environmental Science and Technology, 41*, 3075–3082.
31. Dombrowski, N., Donaho, J. A., Gutierrez, T., Seitz, K. W., Teske, A. P., & Baker, B. J., (2016). Reconstructing metabolic pathways of hydrocarbon-degrading bacteria from the deepwater horizon oil spill. *Nature Microbiology, 1.* Article ID: 16057.
32. Dvo˘rák, P., Nikel, P. I., Damborský, J., & De Lorenzo, V., (2017). Bioremediation version 3.0: Engineering pollutant - removing bacteria in times of systemic biology. *Biotechnology Advances, 35*, 845–866.
33. El-Fantroussi, S., & Agathos, S. N., (2005). Is bioaugmentation a feasible strategy for pollutant removal and site remediation? *Current Opinion in Microbiology, 8,* 268–275.

34. Erickson, B. D., & Mondello, F. J., (1993). Enhanced biodegradation of polychlorinated biphenyls after site-directed mutagenesis of a biphenyl dioxygenase gene. *Applied and Environmental Microbiology, 59*(11), 3858–3862.
35. Eugris. (2005). http://www.eugris.info/eugris-soil-water-encyclopedia.asp (accessed on 5 August 2021).
36. Ezekoye, C. C., Chikere, C. B., & Okpokwasili, G. C., (2018). Fungal diversity associated with crude oil impacted soil undergoing *in-situ* bioremediation. *Sustainable Chemistry and Pharmacy*, (2018) *10,* 148–152.
37. Floodgate, G., (1984). The fate of petroleum in marine ecosystems. In: Atlas, R. M., (ed.), *Petroleum Microbiology* (pp. 355–398). New York: Macmillan.
38. Foght, J. M., Westlake, D. W. S., Johnson, W. M., & Ridgway, H. M., (1996). Environmental gasoline utilizing isolates and clinical isolates of *Pseudomonas aeruginosa* are taxonomically indistinguishable by chemotaxonomic and molecular techniques. *Microbiology, 142*(9), 2333–2340.
39. Fritsche, W., & Hofritcher, M., (2000). Aerobic degradation by microorganisms. In: Klein, J., (ed.), *Environmental Processes-Soil Decontamination* (pp. 146–155). Weinheim, Germany: Wiley - VCH.
40. Fuentes, S., Barra, B., Caporaso, J. G., & Seeger, M., (2015). From rare to dominant: Fine-tuned soil bacterial bloom during petroleum hydrocarbon bioremediation. *Applied and Environmental Microbiology, 82*, 888–896.
41. Gogoi, B. K., Dutta, N. N., Goswami, P., & Mohan, T. K., (2003). Case study of bioremediation of petroleum-hydrocarbon contaminated soil at a crude oil spill site. *Advances in Environmental Research, 7,* 767–782.
42. Guerra, A. B., Oliveira, J. S., & Silva-Portela, R. C., (2018). Metagenome enrichment approach used for selection of oil-degrading bacteria consortia for drill cutting residue bioremediation. *Environment Pollution, 235*, 869–880.
43. Hazen, T. C., Dubinsky, E. A., DeSantis, T. Z., & Andersen, G. L., (2010). Deep-sea oil plume enriches indigenous oil-degrading bacteria. *Science, 330*(6001)*,* 204–208.
44. Hazen, T. C., Prince, R. C., & Mahmoudi, N., (2016). Marine oil biodegradation. *Environmental Science and Technology, 50,* 2121–2129.
45. Head, I. M., Jones, D. M., & Röling, W. F., (2006). Marine microorganisms make a meal of oil. *Nature Reviews Microbiology, 4,* 173–178.
46. Heider, J., Spormann, A. M., Beller, H. R., & Widdel, F., (1998). Anaerobic bacterial metabolism of hydrocarbons. FEMS. *Microbiology Reviews, 22*, 459–473.
47. Hou, N., Zhang, N., Jia, T., Sun, Y., Dai, Y., Wang, Q., Li, D., Luo, Z., & Li, C., (2018). Biodegradation of phenanthrene by biodemulsifier - producing strain *Achromobacter* sp. LH-1 and the study on its metabolisms and fermentation kinetics. *Ecotoxicology and Environmental Safety, 163*, 205–214.
48. Jeanbille, M., Gury, J., & Duran, R., (2016). Chronic polyaromatic hydrocarbon contamination is a marginal driver for community diversity and prokaryotic predicted functioning in coastal sediments. *Frontiers in Microbiology, 7,* 1303–1309.
49. Jin, H. M., Kim, J. M., Lee, H. J., Madsen, E. L., & Jeon, C. O., (2012). *Alteromonas* as key agent of polycyclic aromatic hydrocarbon biodegradation in crude oil-contaminated coastal sediment. *Environmental Science and Technology, 46,* 7731–7740.
50. Johnson, O. A., & Affam, A. C., (2019). Petroleum sludge treatment and disposal: A review. *Environmental Engineering Research, 24,* 191–201.

51. Jurelevicius, D., & De Almeida, C. C. R., (2014). Response of the archaeal community to simulated petroleum hydrocarbon contamination in marine and hypersaline ecosystems. *Water, Air and Soil Pollution, 225*(2). Article ID: 1871.
52. Kelly, C. J., Lajoie, C. A., Layton, A. C., & Sayler, G. S., (1999). Bioluminescent reporter bacterium for toxicity monitoring in biological wastewater treatment systems. *Water Environment Research, 71*(1), 31–35.
53. Kleindienst, S., Paul, J. H., & Joye, S. B., (2015). Using dispersants after oil spills: Impacts on the composition and activity of microbial communities. *Nature Reviews Microbiology, 13,* 388–396.
54. Kuiper, I., Lagendijk, E. L., Bloemberg, G. V., & Lugtenberg, J. J., (2004). Rhizoremediation: A beneficial plant-microbe interaction. *Molecular Plant-Microbe Interaction, 17*(1), 6–15.
55. Kumar, A., Bisht, B. S., & Joshi, V. D., (2011). Bioremediation of polluted environment: A management tool. *International Journal of Environmental Sciences, 1*(60), 1079–1093.
56. Labud, V., Garcia, C., & Hernandez, T., (2007). Effect of hydrocarbon pollution on the microbial properties of a sandy and clay soil. *Chemosphere, 66,* 1863–1871.
57. Layton, A. C., Gregory, B., Schultz, T. W., & Sayler, G. S., (1999). Validation of genetically engineered bioluminescent surfactant resistant bacteria as toxicity assessment tools. *Ecotoxicology and Environmental Safety, 43*(2), 222–228,
58. Leahy, J. H., & Colwell, R., (1990). Microbial degradation of hydrocarbons in the environment. *Microbiology Reviews, 54*(3), 305–315.
59. Lea-Smith, D. J., Biller, S. J., & Davey, M. P., (2015). Contribution of cyanobacterial alkane production to the ocean hydrocarbon cycle. *Proceedings National Academy of Science, 112,* 13591–13596.
60. Li, D., Xu, X., Zhai, Z., Yu, H., & Han, X., (2017). Isolation and identification of an N-hexadecane bacterial degrader from soil polluted by petroleum oil in momoge wetlands and its degradation characteristics. *Wetland Science, 15,* 85–91.
61. Ma, Y. L., Lu, W., Wan, L. L., & Luo, N., (2015). Elucidation of fluoranthene degradative characteristics in a newly isolated *Achromobacter xylosoxidans* DN002. *Applied Biochemistry and Biotechnology, 175,* 1294–1305.
62. Margesin, R., Labbé, D., Schinner, F., Greer, C. W., & Whyte, L. G., (2003). Characterization of hydrocarbon-degrading microbial populations in contaminated and pristine alpine soils. *Applied and Environmental Microbiology, 69*, 3085–3092.
63. Mason, O. U., Hazen, T. C., & Borglin, S., (2012). Metagenome, metatranscriptome and single-cell sequencing reveal microbial response to deepwater horizon oil spill. *ISME Journal, 6,* 1715–1727.
64. Masson, L., Tabashnik, B. E., & Mazza, A., (2002). Mutagenic analysis of a conserved region of domain iii in the cry1ac toxin of *Bacillus Thuringiensis*. *Applied and Environmental Microbiology, 68*(1), 194–200.
65. Medina-Bellver, J. I., & Marín, P., (2005). Evidence for *In situ* crude oil biodegradation after the prestige oil spill. *Environmental Microbiology, 7*(6), 773–779.
66. Mesyami, P., & Baheri, M., (2003). Pre-Screening of fungi and bulking agents for contaminated soil bioremediation. *Advances in Environmental Research, 7,* 881–887.
67. Montagnolli, R. N., Matos, L. P. R., & Bidoia, E. D., (2015). Assessing *Bacillus subtilis* biosurfactant effects on the biodegradation of petroleum products. *Environmental Monitoring and Assessment, 187*(4116), 1–17.

68. Nie, Y., Liang, J. L., Fang, H., Tang, Y. Q., & Wu, X. L., (2014). Characterization of a CYP153 alkane hydroxylase gene in a gram-positive *Dietzia* sp. DQ12- 45-1b and its "team role" with alkw1 in alkane degradation. *Applied Microbiology and Biotechnology, 98,* 163–173.
69. Nikolopoulou, M., & Kalogerakis, N., (2008). Enhanced bioremediation of crude oil utilizing lipophilic fertilizers combined with biosurfactant and molasses. *Marine Pollution Bulletin, 56,* 1855–1861.
70. Okoh, A. I., (2006). Biodegradation alternative in the cleanup of petroleum hydrocarbon pollutants. *Biotechnology and Molecular Biology Reviews, 1*(2), 38–50.
71. Overholt, W. A., & Marks, K. P., (2015). Hydrocarbon degrading bacteria exhibit a species specific response to dispersed oil while moderating ecotoxicity. *Applied and Environmental Microbiology, 82,* 518–527.
72. Panke, S., & Sanchez-Romero, J. M., & De Lorenzo, V., (1998). Engineering of quasi-natural *Pseudomonas putida* strains toluene metabolism through an ortho-cleavage degradation pathway. *Applied and Environmental Microbiology, 64*(2), 748–751.
73. Philip, J. C., Bamforth, S. M., Singleton, I., & Atlas, R. M., (2005). Environmental pollution and restoration: A role for bioremediation. In: Atlas, R. M., & Philip, J., (eds.), *Bioremediation: Applied Microbial Solutions for Real-World Environmental Cleanup* (pp. 1–48). Washington, DC: ASM Press.
74. Piskonen, R., & Itävaara, M., (2009). Evaluation of chemical pretreatment of contaminated soil for improved PAH bioremediation. *Applied Microbiology and Biotechnology, 65,* 627–634.
75. Prenafeta-Boldú, F. X., Andrea, K., & Luykx, D. M., (2011). Isolation and characterization of fungi growing on volatile aromatic hydrocarbons as their sole carbon and energy source. *Mycological Research, 105,* 477–484.
76. Radwan, S., (2008). Microbiology of oil-contaminated desert soils and coastal areas in the Arabian Gulf region. In: Dion, P., & Nautiyal, C. S., (eds.), *Microbiology of Extreme Soils-Soil Biology* (Vol. 13, pp. 275–298). Berlin: Springer.
77. Radwan, S. S., (2009). Phytoremediation for oily desert soils. In: Singh, A., & Kuhad, R. C., & Ward, O. P., (eds.), *Advances in Applied Bioremediation* (pp. 279–298). Berlin: Springer-Verlag.
78. Rahman, K. S. M., & Rahman, T. J., (2003). Enhanced bioremediation of N-alkane in petroleum sludge using bacterial consortium amended with rhamnolipid and micronutrients. *Bioresource Technology, 90*(2), 159–168.
79. Redmond, M. C., & Valentine, D. L., (2012). Natural gas and temperature structured a microbial community response to the deepwater horizon oil spill. *Proceedings of National Academy of Sciences-USA*, *109*(50), 20292–20297.
80. Renou, S., Givaudan, J. G., Poulain, S., Dirassouyan, F., & Moulin, P., (2008). Landfill leachate treatment: Review and opportunity. *Journal of Hazardous Materials, 150,* 468–493.
81. Rivers, A. R., Sharma, S., & Tringe, S. G., (2013). Transcriptional response of bathypelagic marine bacterioplankton to the deepwater horizon oil spill. *ISME Journal, 7.* Article ID: 2315.
82. Röling, W. F., Milner, M. G., & Jones, D. M., (2004). Bacterial community dynamics and hydrocarbon degradation during a field-scale evaluation of bioremediation on a mudflat beach contaminated with buried oil. *Applied and Environmental Microbiology, 70,* 2603–2613.

83. Röling, W. F., & Milner, M. G., (2002). Robust hydrocarbon degradation and dynamics of bacterial communities during nutrient-enhanced oil spill bioremediation. *Applied and Environmental Microbiology, 68,* 5537–5548.
84. Ron, E. Z., & Rosenberg, E., (2014). Enhanced bioremediation of oil spills in the sea. *Current Opinion in Biotechnology, 27,* 191–194.
85. Rosenberg, E., (1993). Exploiting microbial growth on hydrocarbons: New markets. *Trends in Biotechnology, 11,* 419–424.
86. Dos, S. E. O., & Da Rosa, C. C., (2008). Pre-screening of filamentous fungi isolated from a contaminated site in southern Brazil for bioaugmentation purposes. *African Journal of Biotechnology, 7*(9), 1314–1317.
87. Sarkar, P., Roy, A., & Pal, S., (2017). Enrichment and characterization of hydrocarbon-degrading bacteria from petroleum refinery waste as potent bioaugmentation agent for *in situ* bioremediation. *Bioresource Technology, 242,* 15–27.
88. Sayler, G. S., & Ripp, S., (2000). Field applications of genetically engineered microorganisms for bioremediation processes. *Current Opinion in Biotechnology, 11*(3), 286–289,
89. Sebate, J., Vinas, M., & Solanas, A. M., (2004). Laboratory-scale bioremediation experiments on hydrocarbon-contaminated soil. *International Biodeterioration and Biodegradation, 54*, 19–25.
90. Semple, K. T., Doick, K. J., Jones, K. C., Burauel, P., Craven, A., & Harms, H., (2004). Defining bioavailability and bioaccessibility of contaminated soil and sediment is complicated. *Environmental Science and Technology, 38,* 228A-231A.
91. Sharma, S., (2012). Bioremediation: Features, strategies and applications. *Asian Journal of Pharmacy and Life Science, 2*(2), 202–13.
92. Sikkema, J., De Bont, J. A., & Poolman, B., (1995). Mechanisms of membrane toxicity of hydrocarbons. *Microbiology Reviews, 59,* 201–222.
93. Singh, H., (2006). *Mycoremediation: Fungal Bioremediation* (p. 256). New York: Wiley- Interscience.
94. Sousa, C., De Lorenzo, V., & Cebolla, A., (1997). Modulation of gene expression through chromosomal positioning in *Escherichia coli. Microbiology, 143*(6), 2071–2078.
95. Suyama, A. R., & Iwakiri, N., (1996). Engineering hybrid *Pseudomonads* capable of utilizing a wide range of aromatic hydrocarbons and of efficient degradation of trichloroethylene. *Journal of Bacteriology, 178*(14), 4039–4046.
96. Sverdrup, L. E., Nielsen, T., & Krogh, P. H., (2002). Soil ecotoxicity of polycyclic aromatic hydrocarbons in relation to soil sorption, lipophilicity, and water solubility. *Environmental Science and Technology. Sci. Technol., 36*, 2429–2435.
97. Szulc, A., & Ambro˙zewicz, D., (2014). The influence of bioaugmentation and biosurfactant addition on bioremediation efficiency of diesel-oil contaminated soil: Feasibility during field studies. *Journal of Environmental Management, 132,* 121–128.
98. Tao, K., Liu, X., Chen, X., Hu, X., Cao, L., & Yuan, X., (2017). Biodegradation of crude oil by a defined co-culture of indigenous bacterial consortium and exogenous *Bacillus subtilis*. *Bioresource Technology, 224,* 327–332.
99. Tharmer, M., & Al-Kubaisi, A. R., (2013). Biodegradation of Kirkuk light crude oil by *Bacillus thuringiensis*, Northern of Iraq. *Nature Science, 5*(7), 865–873.
100. Thavasi, R., Jayalakshmi, S., Balasubramanian, T., & Banat, I. M., (2006). Biodegradation of crude oil by nitrogen-fixing marine bacteria *Azotobacter chroococcum. Research Journal of Microbiology, 1*, 401–408.

101. Thomas, S. A., (2000). Mushrooms: Higher macrofungi to clean up the environment. *Battelle Environmental Issues*, *Fall*, 1–23.
102. Throne-Holst, M., & Wentzel, A., (2007). Identification of novel genes involved in long-chain n-alkane degradation by *Acinetobacter* sp. strain DSM 17874. *Applied and Environmental Microbiology, 73*(10), 3327–3332.
103. Tremblay, J., & Yergeau, E., (2017). Chemical dispersants enhance the activity of oil and gas condensate - degrading marine bacteria. *ISME Journal, 11,* 2793–2808.
104. Van, D. J., Siciliano, S. D., Winsley, T., Snape, I., & Ferrari, B. C., (2014). Bacterial targets as potential indicators of diesel fuel toxicity in sub-antarctic soils. *Applied and Environmental Microbiology, 80,* 4021–4033.
105. Van, D. M. I., Lee, H., & Trevors, J. T., (1996). Survival of lux AB marked *Alcaligenes eutrophus* H850 in PCB-contaminated soil and sediment. *Journal of Chemical Technology and Biotechnology, 65*(2), 115–122.
106. Van, L. H., Top, E. M., & Verstraete, W., (1998). Bioaugmentation in activated sludge: Current features and future perspectives. *Applied Microbiology and Biotechnology, 50,* 16–23.
107. Varjani, S. J., (2017). Microbial degradation of petroleum hydrocarbons. *Bioresource Technology, 223,* 277–286.
108. Varjani, S. J., & Gnansounou, E., (2017). Microbial dynamics in petroleum oilfields and their relationship with physiological properties of petroleum oil reservoirs. *Bioresource Technology, 245*, 1258–1265.
109. Varjani, S. J., Rana, D. P., Jain, A. K., Bateja, S., & Upasani, V. N., (2015). Synergistic *Ex-situ* biodegradation of crude oil by halotolerant bacterial consortium of indigenous strains isolated from onshore sites of Gujarat - India. *International Biodeterioration and Biodegradation, 103,* 116–124.
110. Varjani, S. J., & Upasani, V. N., (2016). Biodegradation of petroleum hydrocarbons by oleophilic strain of *Pseudomonas aeruginosa* NCIM 5514. *Bioresource Technology, 222,* 195–201.
111. Varjani, S. J., & Upasani, V. N., (2012). Characterization of hydrocarbon utilizing *Pseudomonas* strains from crude oil contaminated samples. *International Journal Science and Computer, 6*(2), 120–127.
112. Varjani, S. J., & Upasani, V. N., (2017). Critical review on biosurfactant analysis, purification, and characterization using rhamnolipid as a model biosurfactant. *Bioresource Technology, 232*, 389–397.
113. Varjani, S. J., & Upasani, V. N., (2017). New look on factors affecting microbial degradation of petroleum hydrocarbon pollutants. *International Biodeterioration and Biodegradation, 120,* 71–83.
114. Wang, J., Zhang, Z. Z., Su, Y. M., He, W., He, F., & Song, H. G., (2008). Phytoremediation of petroleum polluted soil. *Petroleum Science, 5*(2), 167.
115. Wang, C., Liu, X., Guo, J., Lv, Y., & Li, Y., (2018). Biodegradation of marine oil spill residues using aboriginal bacterial consortium based on Penglai 19-3 oil spill accident, China. *Ecotoxicology and Environmental Safety, 159,* 20–27.
116. Wang, L., Huang, X., & Zheng, T. L., (2016). Responses of bacterial and archaeal communities to nitrate stimulation after oil pollution in mangrove sediment revealed by Illumina sequencing. *Marine Pollution Bulletin, 109*(1), 281–289.

117. Wang, Q., Zhang, S., Li, Y., & Klassen, W., (2011). Potential approaches to improving biodegradation of hydrocarbons for bioremediation of crude oil pollution. *Journal of Environmental Protection, 2,* 47–55.
118. Wang, X. B., Chi, C. Q., Nie, Y., Tang, Y. Q., Tan, Y., Wu, G., & Wu, X. L., (2011). Degradation of petroleum hydrocarbons (C6–C40) and crude oil by a novel *Dietzia* strain. *Bioresource Technology, 102,* 7755–7761.
119. Xu, X., Zhai, Z., Li, H., Wang, Q., Han, X., & Yu, H., (2017). Synergetic effect of bio-photocatalytic hybrid system: G-C3N4, and *Acinetobacter*, sp. JLS1 for enhanced degradation of C16 alkane. *Chemical Engineering Journal, 323,* 520–529.
120. Xue, J., Yu, Y., Bai, Y., Wang, L., & Wu, Y., (2015). Marine oil-degrading microorganisms and biodegradation process of petroleum hydrocarbon in marine environments: A review. *Current Microbiology.* (2015), *71*, 220–228.
121. Yakimov, M. M., Timmis, K. N., & Golyshin, P. N., (2007). Obligate oil-degrading marine bacteria. *Current Opinion in Biotechnology*, *18,* 257–266.
122. Yamada, K., Mukumoto, H., Katsuyama, Y., & Tani, Y., (2002). Degradation of long-chain alkanes by a polyethylene-degrading fungus, *Penicillium simplicissimum* YK. *Enzyme and Microbial Technology, 30,* 828–831.
123. Yan, L., Yu, D., Hui, N., & Naanuri, E., (2018). Distribution of archaeal communities along the coast of the gulf of Finland and their response to oil contamination. *Frontiers in Microbiology, 1*(9), 15–20.
124. Yang, Y., Wang, J., Liao, J., Xie, S., & Huang, Y., (2015). Abundance and diversity of soil petroleum hydrocarbon-degrading microbial communities in oil exploring areas. *Applied Microbiology and Biotechnology, 99,* 1935–1946.
125. Zedelius, J., Rabus, R., & Grundmann, O., (2011). Alkane degradation under anoxic conditions by a nitrate-reducing bacterium with possible involvement of the electron acceptor in substrate activation. *Environmental Microbiology Reports, 3,* 125–135.
126. Zhang, Z., & Lo, I. M., (2015). Biostimulation of petroleum hydrocarbon- contaminated marine sediment with co-substrate: Involved metabolic process and microbial community. *Applied Microbiology and Biotechnology, 99*(13), 5683–5696.

CHAPTER 10

PHYTOREMEDIATION OF BIOMEDICAL WASTE CONTAMINATED SITES

PIMPI SAHU, KUMAR KRITARTHA KAUSHIK, CAIYUN ZHENG, QINGYAN ZENG, ZHUANGZHUANG ZHAO, KAI DONG, and TINGLI LU

ABSTRACT

Health, social, ethical, and environmental aspects are related to biomedical waste management. Bioremediation of waste through natural plants (roots, shoots, and leaves) is called phytoremediation. Root length and plant biomass is an essential feature of the phytoremediation process. Through the phytoremediation method, heavy metal (HM) contaminants can be extricated from groundwater and soil. In this chapter, the types of biomedical wastes are mentioned, and the different methods for their remediation are discussed.

10.1 INTRODUCTION

In hospitals and clinics, various materials (cotton swabs, syringes, plastics, biopsy devices, intrauterine devices, and other things) are needed for assisting patients. Anatomical waste is also generated in medical or animal research. All this waste can be characterized as biomedical waste. Biomedical waste can be liquid or solid, including its containers. The instruments used during diagnosis are also considered as biomedical waste [55]. The waste that is obtained from hospitals includes: blood sample vials, chemicals such as alcohol, used needles, bandages, discarded plaster, medical equipment, placenta from delivery, operated body parts, personal protection equipment kit, etc. [67].

Soil pollution takes place when these biomedical wastes are dumped in landfills. Discarding biomedical waste can also lead to air and water pollution [4]. Dumping of biomedical waste without pre-treatment leads to air pollution, or some microorganisms may invade into the atmosphere. It is not a safe practice to burn biomedical waste openly. During toxic incineration gases that are released into the atmosphere may cause respiratory problems in humans [38]. The practice of dumping biomedical waste in soil without scientific procedures can pollute the groundwater due to leaching.

Heavy metals (HMs) when leached underground make the water unsuitable for its consumption [47]. Cadmium, zinc, mercury are some HMs, which are found in the waste. When these HMs are being absorbed by the plants in a chain wise manner, it accumulates in our food chain [6] and might be removed through the phytoremediation processes (such as Phytoextraction, phytostabilization, rhizofiltration, and phytovolatilization) [1, 3, 64]. Phytoremediation is an eco-friendly and beneficial method to clean up the contaminants. The pollutants get absorbed by the roots and are converted into less toxic forms [16, 21, 32].

The present chapter discusses the effect of biomedical wastes on the environment, the need for the removal and their bioremediation using plants and hyperaccumulators.

10.2 IMPACT OF BIOMEDICAL WASTE ON THE ENVIRONMENT

There are different types of biomedical wastes, which are disposed of from the hospitals and other health centers (Figure 10.1). Disposal and treatment of biomedical waste is usually done through various techniques. Anatomical waste is buried under land. Microbiological waste is treated by autoclaving. Chemical treatment can be done with syringes. Autoclaving can also be an alternative treatment process for syringes. Medicines and some highly toxic drugs, which are discarded, undergo landfilling [4, 7, 62]. Liquid biomedical waste, when improperly disposed of, becomes a serious threat to the human health [55].

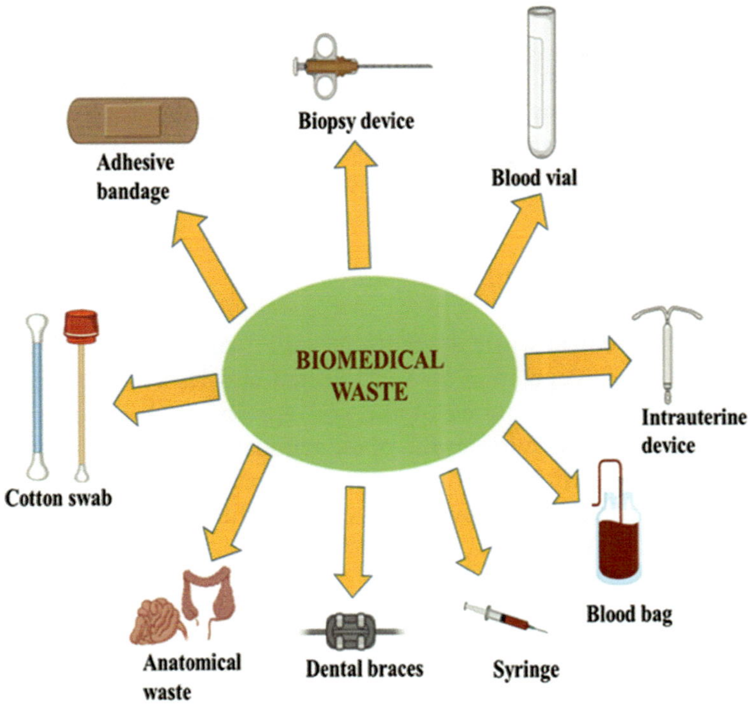

FIGURE 10.1 Biomedical waste.

10.2.1 IMPACT ON AIR

Emission of particulate matter and greenhouse gases are a peculiar challenge for the environment. Burning of biomedical waste is one of the significant issues regarding our health. Toxic pollutants are emitted during burning and can cause respiratory illnesses. Sulfate, ammonia, and black carbon are of 2.5 and 10 particulate matters. During summer these particulate matters are higher than the monsoon season in India. In a research, it has been found that the air is contaminated with organic compounds, such as cyclobutane, acetone diperoxide and carbon disulfide [29]. The amount of HMs present in the atmosphere varies from place to place. It has been found that Pb and Zn from hospital wastes are released in higher amounts than other HM [50]. However, cadmium, lead, and copper have also been noted as well.

10.2.2 IMPACT ON WATER

A higher concentration of polycyclic aromatic hydrocarbons (PAHs) is generated while burning the biomedical wastes, ultimately polluting the water bodies. Both surface and groundwater get polluted. Toxicity should be removed before the demolition into landfills [25, 59].

10.2.3 IMPACT ON SOIL

Dumping of biomedical waste degrades the soil quality. Due to the degradation of soil quality, vegetation also is affected and lessens the yields. The chemical and physical features of the soil are changed. Deterioration of the soil quality and reduction in the plant abundance is also comprehended due to the dumping of biomedical wastes (Figure 10.2). The pH and HM concentration of the soil is known to upsurge by the dumping of biomedical waste [2].

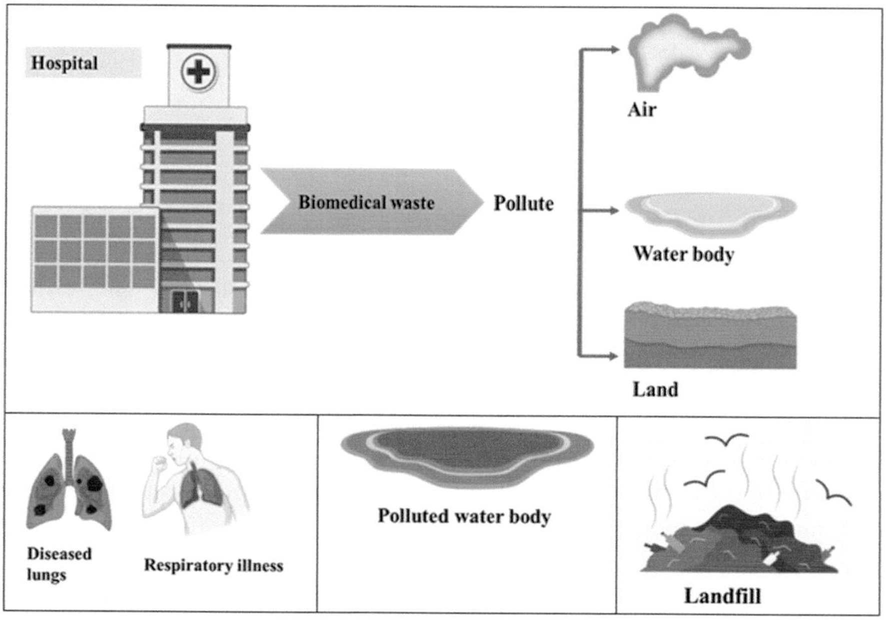

FIGURE 10.2 Biomedical waste generated from the hospital can pollute the air, water, and land quality. Polluted air may cause respiratory problems. Polluted water becomes unsuitable for human consumption and in the case of land, the soil quality gets deteriorated and hence lack of vegetation.

10.3 IMPORTANCE OF BIOMEDICAL WASTE MANAGEMENT

It is necessary to manage the waste generated from the hospitals to maintain proper hygiene [12]. Waste management is an essential part of managing a hospital. However, most of the people do not pay attention to manage biomedical waste. The target for maintaining biomedical waste is to prevent contamination and improve the natural environs and surroundings [28, 41].

10.3.1 HEALTH ASPECTS

If medical equipment (such as scalpels and knives) are not adequately treated after its use, it causes contamination when comes in human contact. There is a chance of nosocomial infection if the instruments are not sanitized. According to Ayurveda, Anupasargikarogas is a type of nosocomial infection [48]. Every biomedical waste should be properly disposed of to avoid contamination [39].

10.3.2 ETHICAL ASPECTS

A hospital should be a place for healing, not for getting infected. Every hospital worker should maintain hygiene [58]. Awareness programs should be organized in hospitals for the workers, which will make them aware regarding disposing of waste and its importance. Posters or through video screen, hygiene instructions can be shown, which will encourage the patients and their visitors to keep the hospital surroundings clean. Proper knowledge of waste handling is very crucial [31].

10.3.3 ENVIRONMENTAL REASONS

In a study, it has been found that contamination takes place by means of air, water, and soil. *M. tuberculosis* and *S. aureus* can enter into the host through contaminated air. *E. coli* have been reported to invade and contaminate water as it acts as a suitable medium for its propagation. *P. aeruginosa*, *Enterococci*, and *S. aureus* are known to contaminate soil [57].

10.4 PHYTOREMEDIATION OF BIOMEDICAL WASTES

Bioremediation and phytoremediation are eco-friendly tools based on the use of microorganisms and plants. It is comprehensive, advantageous, and environmentally safe. Certain plant species with related microbes can stabilize and clean the toxic substances from the air, water, and soil [7, 9, 52, 56]. Phytoremediation technology draws the attention of researchers due to the progress in the right evidences and also due to developing technologies. The public is becoming more concerned regarding the surroundings. By keeping the environment in a good state, one can also remain healthy. Consequently, the phytoremediation techniques and research are in progression [5, 11, 44, 63].

The physiology of plants, the soil chemistry and the associated microorganisms make phytoremediation a multidisciplinary approach [15]. Plants can translocate and extract toxic materials from the soil. They can also convert toxic substances into less toxic forms. In the root zone of plants, the toxic materials become confined, which inhibits leaching process [27]. The productivity of the soil remains intact. Plant litters can enhance in fastening the pollutant [40]. Tree plantation helps to prevent soil erosion and stride out pollutants into groundwater [13].

The increased number of medical facilities and the hospitals have consequently augmented the dimensions of biomedical waste generation. The biomedical wastes from the hospitals contain many toxic and hazardous chemical constituents, the pathogenic bacteria, mold, viruses, and radioactive substances, which pose a negative impact on the human health. However, the research and studies are much meager for the remediation of biomedical wastes [2–4, 8].

10.4.1 PHYTOREMEDIATION TECHNOLOGY

Phytoremediation is the process of cleaning-up of the environment from toxic materials (Figure 10.3). For the organic contamination, phytostabilization, rhizodegradation, rhizofiltration, phytodegradation, and phytovolatilization are involved [18, 42]. These mechanisms specific to organic contaminant removal are not being usually utilized. For the inorganic pollutants, mechanisms are phytostabilization, rhizofiltration, phytoaccumulation, and phytovolatilization. To reduce the availability of contaminants, the plant roots exudate to maintain the stability and shackle the pollutants in the matrix of the soil. This process is called phytostabilization [62].

The ability of plants to absorb and then volatilize the pollutant is called phytovolatilization. This technique is applicable to groundwater and soil treatment. Degradation of pollutants by uptake of plants is called phytodegradation or phytotransformation [42].

Rhizofiltration is the process, where plant roots can absorb the pollutants. Separation of pollutants, such as radionuclide contamination in groundwater, HM pollutants and surface water pollutants by the activity of microbes, which is increased by the plant roots, is called rhizodegradation. For the nutrition and energy generation, microbes deplete the pollutants and use them as their nutritive ingredients. Plant roots release natural sugar, acids, and carbon, which is a nutrient for the soil. Condensed and abridged roots can hold a large amount of water; therefore, plantation is highly encouraged [42].

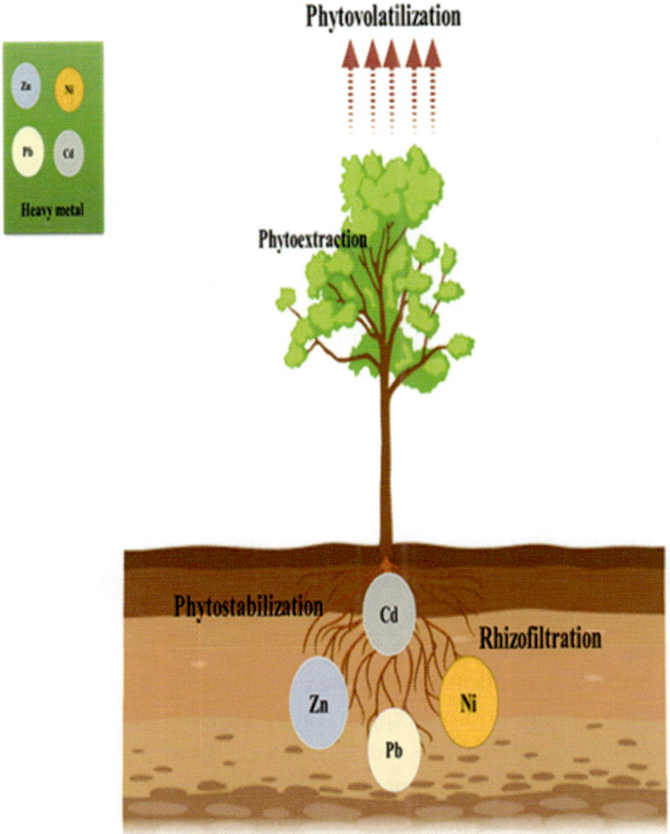

FIGURE 10.3 Overview of phytoremediation process.

Usually, the biomedical waste from the hospitals is treated through incineration and autoclaving, which are not the eco-friendly tools. However, phytoremediation can be the best tool for the biological management and remediation of hospital wastes involving the use of plants. However, the use of bacteria for the remediation of these wastes has also proven successful, which results in the restoration of the previous natural soil environs without harming it further. The hydrolytic bacteria are known to produce different kinds of enzymes, which are capable of degrading the hospital-based wastes in the affected soils [2, 4, 8].

10.4.2 MECHANISMS OF HEAVY METAL (HM) UPTAKE

Plants can act as accumulators and excluders. Even though the pollutants get absorbed by the plant parts and accumulated in the aerial tissue of plants, does not pose a direct effect on the floral endurance. The pollutants are biodegraded and are transformed into a less toxic state [7, 12]. The excluders confine uptake of pollutants [37]. Plants are evolved in such a way that they can absorb micronutrients even if in a very less quantities of parts per million. Plants have the property of translocation and storing up the nutrients. Therefore, phytoremediation is a unique feature to uptake the pollutants and to transform it [17, 62].

For the translocation and uptake of ions, some proteins are laden in the plasma membrane of a plant cell. It includes proton pump where ATPase can devour energy and an electrochemical gradient is formed (Figures 10.4 and 10.6). This electrochemical gradient is being used by proteins for the uptake of ions; co and anti-transporters are involved in this process. For the transport of ions inside the cell, channel proteins are involved (Figure 10.5). Every transport mechanism has its unique way for transportation of the ions. Presence of HM pollutants proves an inordinate hindrance for the uptake of ions. The uptake of ions by the roots is followed by the process of translocation whereby the ions are moved upwards into the shoots. The exact mechanism through which ions move from roots to shoots is less known [17].

Uptake and translocation mechanism in plants is closely related to each other. Typically, plants do not acquire trace elements unless it is required to perform a certain metabolic process. The necessity of trace elements is 10 to 15 parts per million [39, 41]. Hyperaccumulator plants are an exception to it. Metal ions with high toxicity generally accumulate in the hyperaccumulating plants. Several mechanisms are involved in it and typically stored in the vacuoles [9, 17].

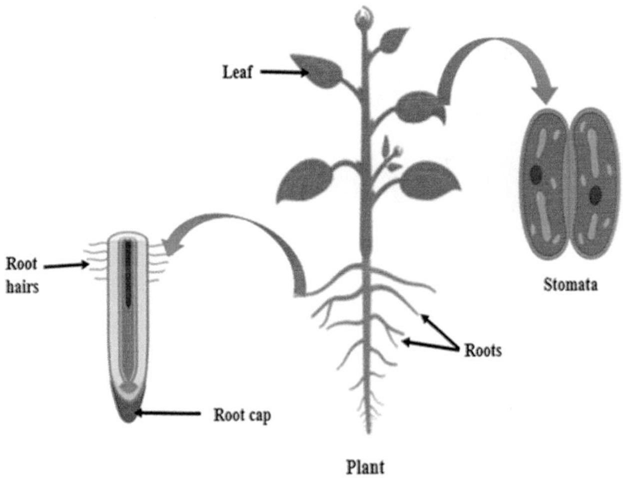

FIGURE 10.4 Some parts of the plant, where the proton pump (ATPase) is found.

Evaporation of water from leaves act as a pump which makes the plant roots to absorb the nutrients from the soil and is called evapotranspiration. Consequently, the pollutants and contaminants are also moved from the root sites to the shoot through the evapotranspiration pull. Likewise, toxic components and biomedical based elements are wiped out, and the soil quality remains intact [4].

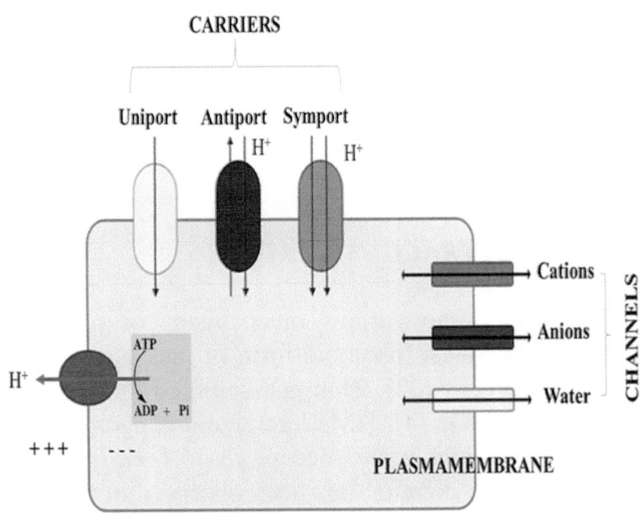

FIGURE 10.5 Transporters across plasma membrane.

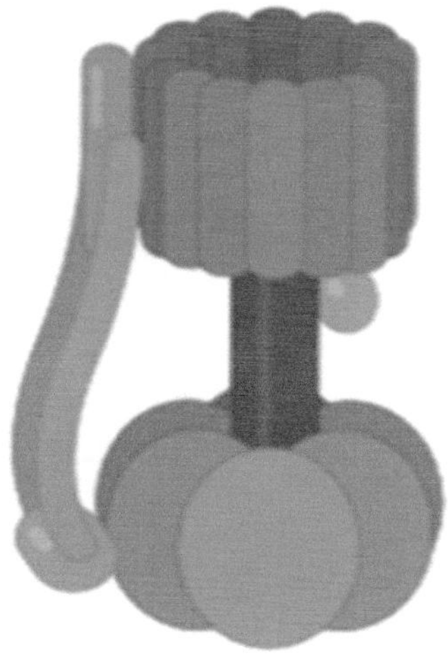

FIGURE 10.6 ATPase.

Hyperaccumulator plants can be used in the phytoremediation process to reduce the concentration of cadmium, zinc, cobalt, nickel, and lead [51, 59]. Hyperaccumulators can concentrate up to 1,000 times more than the excluders, hence have proved to be most effective towards the restoration of the soils contaminated with biomedical and hospital-based wastes. Microorganisms found in the rhizosphere (like bacteria and fungi) can help in the uptake process, which is more effective in removing organic pollutants rather than the inorganic ones [19, 20]. The phytoremediation process involves processes, such as [8, 9, 18–20, 24, 26, 41–43, 54]: phytoextraction, phytostabilization, rhizofiltration (Figures 10.7 and 10.8) and phytovolatilization for the HM uptake.

10.4.3 ROLE OF HYPERACCUMULATORS

Researchers have found the various mechanisms of hyperaccumulating plant species to uptake HMs. Insoluble form of zinc is accumulated in the vacuoles of *T. caerulescens* [22]. Zinc is assembled in the mesophyll cells of *A. halleri* leaves [33, 53, 74]. HMs like zinc and cadmium are taken up via compartmentation of ions in the mesophyll of *A. halleri* and *T. caerulescens* [14]. Chemical modification has been observed in the rhizosphere of *T. caerulescens* and *T. goesingense* in response to the hyperaccumulation process. Reduction of zinc is detected in the rhizosphere of *T. goesingense*. *T. goesingense* only accumulates nickel [46], when it is planted on non-polluted soil (Table 10.1).

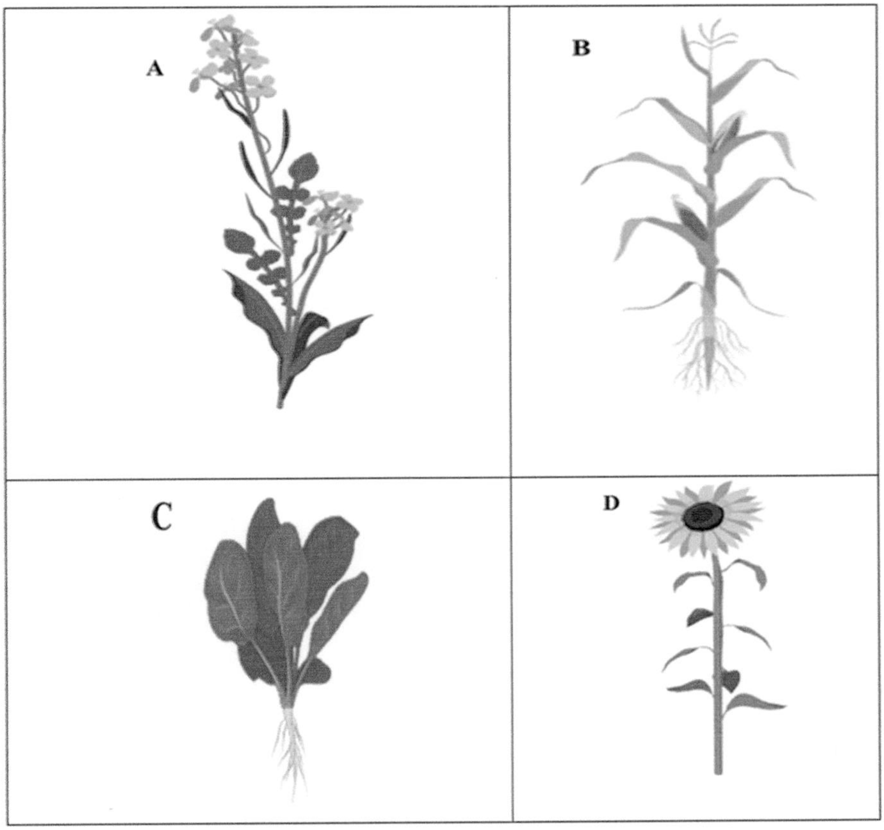

FIGURE 10.7 Plants used in rhizofiltration process: (a) rapeseed; (b) corn; (c) spinach; (d) sunflower .

Ueno et al. [61] studied the synergy between Cd and zinc in pot soil and envisaged that the aggregation of cadmium in the shoot is not afflicted if zinc is supplied in huge quantity, however, increase in the concentration of zinc in solution decreases the cadmium amount in the roots. *T. caerulescens* plantation in the Ganges showed that cadmium struggles with zinc for the uptake. However, zinc does not compete with cadmium for the uptake [61].

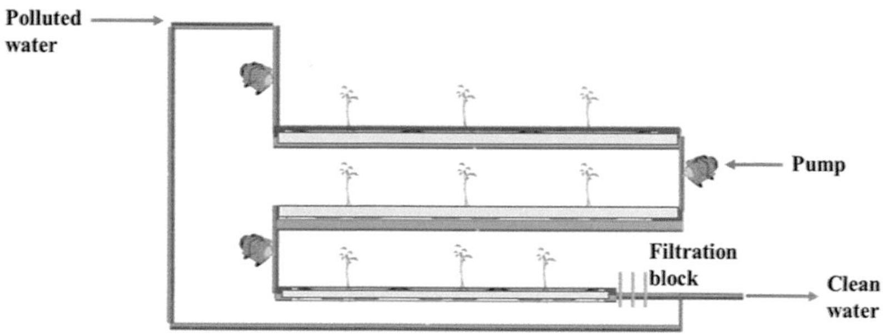

FIGURE 10.8 Representation of rhizofiltration.

TABLE 10.1 List of Plant Species Used in Hyperaccumulation

Plant Species	References
A. bertolonii	[34]
A. halleri	[33]
A. lesbiacum	[34]
H. vulgare	[23, 66]
L. heterophyllum	[23, 66]
P. cretica	[10]
P. longifolia	[10]
P. tremula	[10]
P. umbrosa	[10]
P. vitta	[10]
S. alfredii	[35, 68–71]
T. caerulescens	[14, 22, 23, 46, 60, 61, 66]
T. goesingense	[46]
Z. mays	[68]

The amount of cadmium concentration in the soil is in the order of: $CdSO_4$ > $CdCO_3$ > CdS. *Thlaspi caerulescens* accumulates 6,900 mg/kg of zinc. *T. caerulescens* is capable of using insoluble cadmium and zinc elements in the soils [61].

Root biomass and its length increases, when cadmium is accumulated in *T. carerulescens*. An increase in root biomass is not observed when cadmium is not absorbed by *T. carerulescens. T. carerulescens, H. vulgare* and *L. heterophyllum*, when grown together to recognize the uptake of cadmium, it was found that cadmium uptake is increased by *H. vulgare.* However, the concentration of zinc accumulation has decreased. *T. carerulescens* can promote the uptake amount if grown along with other plants [23, 66].

When *S. alfredii* and *Z. mays* are planted together, the uptake of copper and zinc is diminished compared to monoculture. However, the plant growth is seen to be enhanced [68]. The root of *T. geosingense* Halacsy secretes organic compounds that lead to nickel hyperaccumulation [65]. In hydroponics study, it was found that *Sedum alfredii* Hance can accumulate zinc and cadmium [35, 36, 68–71]. More zinc and cadmium absorption lead to their further accumulation in the leaves. The decreasing order of zinc and cadmium accumulation is: stem > leaf > root and leaf > stem > root. *S. alfredii* can sustain zinc and cadmium pollutants and the sustaining capacity of cadmium in mined ecotypes is more than non-mined ecotypes [69].

Using radioactive tracer flux method, it was found that *S. alfredii* shoots concentrate zinc more than 2% of its weight. The concentration of zinc in leaves and stems is 28 times higher than shoots. In the root vacuoles also, zinc is stored and not available for stowing into xylem. Tonoplast is involved in hyperaccumulation of zinc in *S. alfredii* H [70]. Higher V_{max} of zinc indicates its higher transport value through the per unit membrane of the root cells [35].

In a study by Li et al. [36], it was determined that the amount of zinc is higher in leaves and stems of *S. alfridii* H. The sustaining capacity of *S. alfridii* H towards zinc and lead is related to the physiology of the plant and growth of roots. Secretion by roots increases the availability of zinc and lead in the mining areas [36].

For the hyperaccumulation of nickel, *Alyssum* species was studied by Kidd and Monterroso [30]. *A. bertolonii* (Desv), *A. lesbiacum* (Candargy) and *T. goesingense* (Halacsy) were examined for the nickel hyperaccumulation. The sustaining ability of *T. goesingense*is was found to be less towards nickel compared to the other species. X-ray microanalysis proved that nickel is dispersed in epidermal cells and vacuoles of leaves and stems [34]. *Alyssum serpyllifolium* sp. were planted in different mining contaminated sites where one part of the land was contaminated with chromium and the other part contaminated with chromium, lead, copper, and zinc. *A. serpyllifolium* was found to be an appropriate and suitable option for multi-HM polluted soil [30]. Arsenic can be hyperaccumulated by *P. vitta, P. longifolia,*

P. umbrosa and *P. Cretica* [73]. Most of the arsenic is recovered from fronds and rhizomes [72]. *P. vitta* concentrates more arsenic than *P. Tremula* [10].

10.5 SUMMARY

Phytoremediation is induced by solar energy. Some invasive plant species have been used for the phytoremediation process. While selecting a plant for phytoremediation, it is necessary to check the biomass and uptake capacity of plant species to accumulate pollutants. There are certain plant species, which grow in the contaminated area and are generally not considered for consumption. The benefit of using those plant species is that it will not enter into the food chain. For the removal of pollutants, plants disintegrate the toxic materials and thereby decontaminate the sites. Microbes associated with soil can also help in phytoremediation and work in symbiosis with the plants. Chromatography method can be used to determine the accumulation of toxic elements. Exact recognition, separation, and disposition of biomedical waste is a crucial prerequisite.

ACKNOWLEDGMENTS

Pimpi Sahu is an awardee of the Chinese Government Scholarship bilateral program with the Ministry of Human Resource Development, Government of India, in category A. Kumar Kritartha Kaushik is a DST-INSPIRE Fellow, Government of India. We are grateful to the Key Laboratory for Space Bioscience and Biotechnology, School of Life Sciences, Northwestern Polytechnical University in China for providing the necessary facilities.

KEYWORDS

- **biomedical waste**
- **heavy metal**
- **hospital-based waste**
- **hyperaccumulation**
- **microbiological waste**
- **phytoextraction**

REFERENCES

1. Ali, H., Khan, E., & Sajad, M. A., (2013). Phytoremediation of heavy metals—concepts and applications. *Chemosphere, 91*(7), 869–881.
2. Ali, S. M., Pervaiz, A., Afzal, B., Hamid, N., & Yasmin, A., (2014). Open dumping of municipal solid waste and its hazardous impacts on soil and vegetation diversity at waste dumping sites of Islamabad city. *Journal of King Saud University - Science, 26*(1), 59–65.
3. Alkorta, I., Hernández-Allica, J., Becerril, J. M., Amezaga, I., Albizu, I., & Garbisu, C., (2004). Recent findings on the phytoremediation of soils contaminated with environmentally toxic heavy metals and metalloids such as zinc, cadmium, lead, and arsenic. *Reviews in Environmental Science and Biotechnology, 3*(1), 71–90.
4. Alok, R. D. S., & Nirmala, N., (2017). Analytical studies on the impact of bio-medical waste on garden and landfill soil. *International Journal of ChemTech Research, 10*(3), 656–665.
5. Antoniadis, V., Levizou, E., & Shaheen, S. M., (2017). Trace elements in the soil-plant interface: Phytoavailability, translocation, and phytoremediation: A review. *Earth-Science Reviews, 171*, 621–645.
6. Ashokkumar, B., Jothiramalingam, S., & Thiyagarajan, S., (2014). Removal of heavy metals from contaminated soil using phytoremediation. *International Journal of Chemical and Physical Sciences, 3*(5), 88–94.
7. Bharti, R. P., Vastava, A. S., Soni, K., Tiwari, A., & More, S., (2014). Phytoremediation of heavy metal toxicity and role of soil in rhizobacteria. *International Journal of Scientific and Research Publications, 4*(1), 1–5.
8. Brennan, M. A., & Shelley, M. L., (1999). A model of the uptake, translocation, and accumulation of lead (Pb) by maize for the purpose of phytoextraction. *Ecological Engineering, 12*(3), 271–297.
9. Burken, J. G., & Schnoor, J. L., (1996). Phytoremediation: Plant uptake of atrazine and role of root exudates. *Journal of Environmental Engineering, 122*(11), 958–963.
10. Caille, N., Vauleon, C., Leyval, C., & Morel, J. L., (2005). Metal transfer to plants grown on a dredged sediment: Use of radioactive isotope 203 Hg and titanium. *Science of the Total Environment, 341*(1), 227–239.
11. Capuana, M., (2011). Heavy metals and woody plants - biotechnologies for phytoremediation. *iForest - Biogeosciences and Forestry, 4*(1), 7–15.
12. Chandra, H. J. C., (1999). Hospital waste: An environmental hazard and its management. *Newsletter of ISEB India, 5*(3), 80–82.
13. Chaudhry, T., Hayes, W., Khan, A., & Khoo, C., (1998). Phytoremediation-focusing on accumulator plants that remediate metal contaminated soils. *Australasian Journal of Ecotoxicology, 4*, 37–51.
14. Cosio, C., Martinoia, E., & Keller, C., (2004). Hyperaccumulation of cadmium and zinc in *Thlaspi caerulescens*; *Arabidopsis halleri* at the leaf cellular level. *Plant Physiology, 134*(2), 716–720.
15. Cunningham, S. D., & Ow, D. W., (1996). Promises and prospects of phytoremediation. *Plant Physiology, 110*(3), 715–719.
16. De Campos, F. V., De Oliveira, J. A., Da Silva, A. A., Ribeiro, C., & Dos, S. F. F., (2019). Phytoremediation of arsenite-contaminated environments: Is *Pistia stratiotes* L. a useful tool? *Ecological Indicators, 104*, 794–801.

17. ENERGY-U.S.D.O., (1994). Plume focus area. *Mechanisms of Plant Uptake, Translocation, and Storage of Toxic Elements*. Summary Report of a workshop on phytoremediation research needs. https://www.osti.gov/servlets/purl/10109412 (accessed on 5 August 2021).
18. Erakhrumen, A. A., & Agbontalor, A. J., (2007). Phytoremediation: An environmentally sound technology for pollution prevention, control and remediation in developing countries. *Educational Research and Reviews, 2*, 151–156.
19. Erdei, L. J., Mezôsi, G., Mécs, I., Vass, I., & Fôglein, F., (2005). Phytoremediation as a program for decontamination of heavy-metal polluted environment. *Proceedings of the 8th Hungarian Congress on Plant Physiology and the 6th Hungarian Conference on Photosynthesis, 2005*, 75–76.
20. Erdei, L. J., (2005). Phytoremediation as a program for decontamination of heavy-metal polluted environment. *Acta Biologica Szegediensis, 49*(1, 2), 75–76.
21. Favas, P. J. C., Pratas, J., Rodrigues, N., D'souza, R., Varun, M., & Paul, M. S., (2018). Metal(loid) accumulation in aquatic plants of a mining area: Potential for water quality biomonitoring and biogeochemical prospecting. *Chemosphere, 194*, 158–170.
22. Frey, B., Keller, C., & Zierold, K., (2000). Distribution of Zn in functionally different leaf epidermal cells of the hyperaccumulator *Thlaspi caerulescens*. *Plant, Cell & Environment, 23*(7), 675–687.
23. Gove, B., Hutchinson, J. J., Young, S. D., Craigon, J., & McGrath, S. P., (2002). Uptake of metals by plants sharing a rhizosphere with the hyperaccumulator *Thlaspi caerulescens*. *International Journal of Phytoremediation, 4*(4), 267–281.
24. Gupta, P., & Kumar, V., (2016). Value-added phytoremediation of metal stressed soils using phosphate solubilizing microbial consortium. *World Journal of Microbiology and Biotechnology, 33*(1), 9–15.
25. Heera, S., Kunal, & Rajor, A., (2014). Bacterial treatment and metal characterization of biomedical waste ash. *Journal of Waste Management, 9563*16.
26. Ibeanusi, V. M., Grab, D. A., Jensen, L., & Ostrodka, S., (2004). *Radionuclide Biological Remediation Resource Guide* (p. 65). Washington – DC: US Environmental Protection Agency, (EPA) Region 5, Superfund Division.
27. Jiang, Y., Lei, M., Duan, L., & Longhurst, P., (2015). Integrating phytoremediation with biomass valorization and critical element recovery: A UK contaminated land perspective. *Biomass and Bioenergy, 83*, 328–339.
28. Joshi, S. C., Diwan, V., & Tamhankar, A. J., (2015). Staff perception on biomedical or health care waste management: A qualitative study in a rural tertiary care hospital in India. *PLOS One, 10*. Article ID: e0128383.
29. Karthikeyan, S., Balasubramanian, R., & Iouri, K., (2006). Particulate air pollution from bushfires: Human exposure and possible health effects. *Journal of Toxicology and Environmental Health - Part A, 69*(21), 1895–1908.
30. Kidd, P. S., & Monterroso, C., (2005). Metal extraction by *Alyssum serpyllifolium* ssp. *lusitanicum* on mine-spoil soils from Spain. *Science of the Total Environment, 336*(1), 1–11.
31. Kumar, A., Duggal, S., & Gur, R., (2015). Safe transportation of biomedical waste in a health care institution. *Indian Journal of Medical Microbiology, 33*(3), 383–386.
32. Kumar-Yadav, K., Gupta, N., & Kumar, A., (2018). Mechanistic understanding and holistic approach of phytoremediation: A review on application and future prospects. *Ecological Engineering, 120*, 274–298.

33. Küpper, H., Lombi, E., Zhao, F. J., & McGrath, S. P., (2002). Cellular compartmentation of cadmium and zinc in relation to other elements in the hyperaccumulator *Arabidopsis halleri*. *Planta, 212*(1), 75–84.
34. Küpper, H., Lombi, E., Zhao, F. J., Wieshammer, G., & McGrath, S. P., (2001). Cellular compartmentation of nickel in the hyperaccumulators *Alyssum lesbiacum*, *Alyssum bertolonii* and *Thlaspi goesingense*. *Journal of Experimental Botany, 52*(365), 2291–2300.
35. Li, H., Cheng, F., Wang, A., & Wu, T., (2005). Cadmium removal from water by hydrophytes and its toxic effects. *Unpublished Article in Proc. of the International Symposium of Phytoremediation and Ecosystem Health* (p. 8). Hangzhou – China.
36. Li, T. Q., Yang, X. E., He, Z. L., & Yang, J. Y., (2005). Root morphology and Zn^{2+} uptake kinetics of the Zn hyperaccumulator of *Sedum alfredii* hance. *Journal of Integrative Plant Biology, 47*(8), 927–934. https://www.researchgate.net/journal/1672-9072_Journal_of_Integrative_Plant_Biology (accessed on 2 August 2021).
37. Lone, M. I., He, Z. L., Stoffella, P. J., & Yang, X. E., (2008). Phytoremediation of heavy metal polluted soils and water: Progresses and perspectives. *Journal of Zhejiang University Science B, 9*(3), 210–220.
38. Mastorakis, N. E., Bulucea, C. A., Oprea, T. A., Bulucea, C. A., & Dondon, P. J., (2010). Environmental and health risks associated with biomedical waste management. *Development, Energy, Environment, Economics, 7*(8), 288–294.
39. Mathur, P., Patan, S., & Shobhawat, A. S. J., (2012). Need of biomedical waste management system in hospitals-an emerging issue: A review. *Current World Environment, 7*(1), 117–120.
40. Mench, M., Bussière, S., & Boisson, J., (2003). Progress in remediation and revegetation of the barren jales gold mine spoil after *in situ* treatments. *Plant and Soil, 249*(1), 187–202.
41. Merkl, N., Schultze-Kraft, R., & Infante, C., (2005). Phytoremediation in the tropics - influence of heavy crude oil on root morphological characteristics of graminoids. *Environmental Pollution, 138*(1), 86–91.
42. National Risk Management Research Laboratory (US), (2000). *Introduction to Phytoremediation* (p. 9). National Risk Management Research Laboratory, Office of Research and Development, US Environmental Protection Agency.
43. Negri, M. C., Hinchman, R. R., & Gatliff, E. G., (1996). *Phytoremediation: Using Green Plants to Clean Up Contaminate Soil, Groundwater, and Wastewater*. United States. Research Org.: Argonne National Lab., IL (United States). https://www.osti.gov/biblio/274159 (accessed on 2 August 2021).
44. Phieler, R., Merten, D., Roth, M., Büchel, G., & Kothe, E., (2015). Phytoremediation using microbially mediated metal accumulation in *Sorghum bicolor*. *Environmental Science and Pollution Research, 22*(24), 19408–19416.
45. Program, U. N. E. (2002). *Phytoremediation: An Environmentally Sound Technology for Pollution Prevention, Control and Remediation: An Introductory Guide to Decision-Makers*. https://www.unenvironment.org/ (accessed on 2 August 2021).
46. Puschenreiter, M., Wieczorek, S., Horak, O., & Wenzel, W. W., (2003). Chemical changes in the rhizosphere of metal hyperaccumulator and excluder *Thlaspi* species. *Journal of Plant Nutrition and Soil Science, 166*(5), 579–584.

47. Rajor, A., Xaxa, M., Mehta, R., & Kunal, (2012). An overview on characterization, utilization and leachate analysis of biomedical waste incinerator ash. *Journal of Environmental Management, 108*, 36–41.
48. Ranjan, R., Robin, D. T., & Vandanarani, M., (2019). Biomedical waste management in Ayurveda hospitals - current practices and future prospective. *Journal of Ayurveda and Integrative Medicine, 10*(3), 214–221.
49. Raskin, I., & Ensley, B. (2000). *Recent Developments for In Situ Treatment of Metal Contaminated Soils*. Phytoremediation of toxic metals: Using plants to clean up the environment. John Wiley & Sons Inc., New York. https://www.wiley.com/en-in/Phytoremediation+of+Toxic+Metals%3A+Using+Plants+to+Clean+Up+the+Environment-p-9780471192541 (accessed on 5 August 2021).
50. Sabiha, J., Tufail, M., & Khalid, S., (2008). Heavy metal pollution from medical waste incineration at Islamabad and Rawalpindi. *Pakistan. Microchemical Journal, 90*(1), 77–81.
51. Salido, A. L., Hasty, K. L., Lim, J. M., & Butcher, D. J., (2003). Phytoremediation of arsenic and lead in contaminated soil using Chinese brake ferns (*Pteris vittata*) and Indian mustard (*Brassica juncea*). *International Journal of Phytoremediation, 5*(2), 89–103.
52. Sarma, H., (2011). Metal hyperaccumulation in plants: A review focusing on phytoremediation technology. *Journal of Environmental Science and Technology, 4*(2), 118–138.
53. Sarret, G., Saumitou-Laprade, P., Manceau, A., Panfili, F., & Geoffroy, N. J. P., (2002). Mechanized hyperaccumulation of zinc in *Arabidopsis halleri*. *Plant Physiology, 130*(4), 1815–1826.
54. Sas-Nowosielska, A., Galimska-Stypa, R., & Kucharski, R., (2008). Remediation aspect of microbial changes of plant rhizosphere in mercury-contaminated soil. *Environmental Monitoring and Assessment, 137*(1), 101–109.
55. Sasmita, B., (2013). Liquid biomedical waste management: An emerging concern for physicians. *Muller Journal of Medical Sciences and Research, 4*(2), 99–106.
56. Sikdar, S., & Kundu, M., (2018). A review on detection and abatement of heavy metals. *Chem. Bio. Eng. Reviews, 5*(1), 18–29.
57. Silva, A. C. D. N., Bernardes, R. S., Moraes, L. R. S., & Reis, J. D., (2002). *Critérios Adotados Para Seleção de Indicadores de Contaminação ambiental relacionados aos resíduos sólidos de Serviços de Saúde: Uma proposta de Avaliação (Criteria Adopted for Selection of Environmental Contamination Indicators Related to Solid Waste from Health Services: An Evaluation Proposal)* (Vol. 18, No. 5, pp. 1401–1409). Cadernos de Saúde Pública (Public Health Notebooks).
58. Townsend, W. K., & Pruss, A., (1998). *Teacher's Guide: Management of Wastes from Health-Care Activities*. https://apps.who.int/iris/handle/10665/64002 (accessed on 2 August 2021).
59. Traunfeld, J. H., & Clement, D. L., (2001). *Lead in Garden Soils* (p. 5). Home and Garden series. College Park, MD: Maryland Cooperative Extension, University of Maryland.
60. Ueno, D., Zhao, F. J., Shen, R., & Ma, J. F., (2004). Cadmium and zinc accumulation by the hyperaccumulator *Thlaspi caerulescens* from soils enriched with insoluble metal compounds. *Soil Science and Plant Nutrition, 50*(4), 511–515.

61. Ueno, D., Zhao, F. L., & Ma, J. F., (2004). Interactions between Cd and Zn in relation to their hyperaccumulation in *Thlaspi caerulescens*. *Soil Science and Plant Nutrition, 50*(4), 591–597.
62. Vara, P. M. N., & De Oliveira, F. H. M., (2003). Metal hyperaccumulation in plants: Biodiversity prospecting for phytoremediation technology. *Electronic Journal of Biotechnology, 6*(3), 285–321.
63. Varun, M., Jaggi, D., D'souza, R., Paul, M. S., & Kumar, B., (2015). *Abutilon Indicum* L.: A prospective weed for phytoremediation. *Environmental Monitoring and Assessment, 187*(8), 527–532.
64. Vithanage, M., Dabrowska, B. B., Mukherjee, A. B., Sandhi, A., & Bhattacharya, P., (2012). Arsenic uptake by plants and possible phytoremediation applications: A brief overview. *Environmental Chemistry Letters, 10*(3), 217–224.
65. Wenzel, W. W., Bunkowski, M., Puschenreiter, M., & Horak, O., (2003). Rhizosphere characteristics of indigenously growing nickel hyperaccumulator and excluder plants on serpentine soil. *Environmental Pollution, 123*(1), 131–138.
66. Whiting, S. N., Leake, J. R., Mcgrath, S. P., & Baker, A. J. M., (2000). Positive responses to Zn and Cd by roots of the Zn and Cd hyperaccumulator *Thlaspi caerulescens*. *New Phytologist, 145*(2), 199–210.
67. WHO. (2018). *Health Care Waste*. https://www.who.int/news-room/fact-sheets/detail/health-care-waste (accessed on 5 August 2021).
68. Xiaomei, L., Qitang, W., & Banks, M. K., (2005). Effect of simultaneous establishment of *Sedum alfredii* and *Zea mays* on heavy metal accumulation in plants. *International Journal of Phytoremediation, 7*(1), 43–53.
69. Xiong, Y. H., Yang, X. E., Ye, Z. Q., & He, Z. L., (2004). Characteristics of cadmium uptake and accumulation by two contrasting ecotypes of *Sedum alfredii* hance. *Journal of Environmental Science and Health, Part A*, *39*(11, 12), 2925–2940.
70. Yang, X., Li, T., Yang, J., He, Z., Lu, L., & Meng, F., (2006). Zinc compartmentation in root, transport into xylem, and absorption into leaf cells in the hyperaccumulating species of *Sedum alfredii* hance. *Planta, 224*(1), 185–195.
71. Yang, X. E., Ye, H. B., Long, X. X., He, B., He, Z. L., Stoffella, P. J., & Calvert, D. V., (2005). Uptake and accumulation of cadmium and zinc by *Sedum Alfredii* hance at different Cd/Zn supply levels. *Journal of Plant Nutrition, 27*(11), 1963–1977.
72. Zhang, W., Cai, Y., Tu, C., & Ma, L. Q., (2002). Arsenic speciation and distribution in an arsenic hyperaccumulating plant. *Science of the Total Environment, 300*(1), 167–177.
73. Zhao, F. J., Dunham, S. J., & Mcgrath, S. P., (2002). Arsenic hyperaccumulation by different fern species. *New Phytologist, 156*(1), 27–31.
74. Zhao, F. J., Lombi, E., & Breedon, T. M S. P., (2000). Zinc hyperaccumulation and cellular distribution in *Arabidopsis halleri*. *Plant, Cell & Environment, 23*(5), 507–514.

CHAPTER 11

BIOREMEDIATION OF WASTE DUMPING SITES

TRINATH BISWAL

ABSTRACT

Different methods have been followed to degrade the waste-contaminated materials, but the technique of bioremediation is more preferred. The bioremediation of hydrocarbons, chlorinated compounds, aromatic hydrocarbons, toxic heavy metals (HMs), and plastic wastes have been observed at landfill sites. Most countries prefer landfilling as the leading method for the reduction and management of solid wastes. Since the landfill is the popular method of waste reduction and disposal, therefore proper monitoring and identification of hazards are necessary. The accumulation of plastic wastes in the landfill is an important component. Till now three different methods (recycling, incineration, and landfill) are adopted for proper management and disposal of plastic wastes. The bioremediation of plastic wastes and toxic metals in a landfill is limited as few secondary pollutants may be emitted into the environment. Although bioremediation is an attractive selection in landfill due to cost and sustainability towards the environment, yet it has many drawbacks; therefore, substantial research is still needed for effective bioremediation of all kinds of wastes available in landfills.

11.1 INTRODUCTION

Due to increase in population, industrial sector, mining activity, urbanization, living standard of human beings, generation of more hazardous materials as wastes, our natural and mining resources are rapidly degrading, which results in the generation of huge amounts of wastes (such as plastic waste, rubber waste, garbage waste, industrial, urban, and agricultural wastes) thus causing ecological imbalance and the change in habitats and creation of adverse effects to the living organisms. Solid wastes of huge amounts deposited in the landfill sites are a major problem to our biosphere [40, 48, 61].

The process of bioremediation mainly involves biodegradation of toxic substances by converting these into harmless species, particularly CO_2, CH_4 and H_2O. The process is carried out either through *in-situ* or *ex-situ* bioremediation. The technology of bioremediation is the method of degradation of organic wastes of ecological contaminants in the biological process by using various kinds of microorganisms producing non-toxic or less toxic simple substances [6, 30, 63].

In this technique, local natural occurring fungi and bacteria are used to decontaminate the hazardous materials. The bioremediation is more popular because of many advantages, such as, low depletion of energy, higher efficiency of biodegradation, no evolution of toxic secondary pollutants, simple technical method, long-term viability, and no need for additional infrastructure. Hence, this technique is most successfully and extensively being applied in many countries of our globe [25, 26, 35, 36].

The microorganisms (bacteria, fungi, and archaea) are bioremediators, which are the biological agents applied in the process of bioremediation in the form of a biotechnological method involving microorganisms to reduce pollutants through the process of biodegradation from our natural environment [4, 10, 28, 37, 50].

This chapter describes key features of biological processing of the waste dumping sites and discusses the techniques of applying the biological methods to remediate the sites contaminated with hazardous wastes.

11.2 THE ROLE OF MICROBES IN BIOREMEDIATION

The major aim of bioremediation is to activate the microbial communities along with nutrients with other chemical constituents present at the landfill sites to destroy the pollutants by breaking these into simple compounds. Bioremediation also encompasses the degradation of organic matter by using bacteria, fungi, yeast, and some other organisms producing energy and non-toxic simple products. The process of biotransformation and bioremediation are natural processes that occur in the presence of microbial community to degrade, accumulate, or transform polychlorinated biphenyls (PCBs), aliphatic hydrocarbons, polyaromatic hydrocarbons, metals, and radionuclides into less harmful substances.

Microorganisms obtain their energy by catalyzing the chemical reactions, which in turn liberate energy due to the cleavage of chemical bonds present in the contaminant molecules, and then electrons are moved away from the pollutants. Hence, the chemical reaction carried out by the microorganisms in bioremediation is called redox reaction or oxidation-reduction reaction. In this reaction, the organic pollutants are oxidized and lose electrons, whereas some other chemical compounds simultaneously gain electrons and are reduced. Hence, the organic waste substances are termed electron donor, whereas, the chemical substances are termed electron acceptors. Due to the transfer of these electrons, energy is produced, and this energy is responsible for the cell growth of microorganisms and termed primary substrates. The microorganisms break the organic wastes through their metabolism and aerobic digestion by using molecular O_2. The microorganisms utilize O_2 to oxidize a part of the carbon into CO_2 from the contaminated wastes at the landfill sites, and the remaining part of the carbon is utilized for creating new cell-mass [25].

Although, the system of bioremediation is restricted by the capacity of the native microbial communities, but the researchers are recently investigating a path that supplement the contaminated landfill sites with the non-native microbial community such as genetically engineered microorganisms (GEMs) which are suitable for biodegradation of pollutants at specific sites [1, 10, 17, 22, 41].

The microorganisms use the waste organic materials for two purposes:

- The organic waste is available as a source of carbon, which is considered as the basic building block for the formation of new cells [70].
- Microorganisms gain energy from the organic waste by breaking the chemical bonds and transferring the electrons from the organic matter to an electron acceptor and the energy produced along with the electrons and carbon is sufficient to produce new cells [31, 70].

11.3 BRIEF REVIEW ON LANDFILL SITES

The landfill site is termed as rubbish dump or dumping yard or tip garbage dumping site, which is the area of disposal of waste materials. Landfill is considered the oldest and widely used process of waste disposal. The sanitary landfill means a planned bioreactor, where the microbial community breaks the chemical bonds of the complex organic matter and converts it into low toxic substances [51].

The reactor for this purpose is designed and controlled on the basis of the guidelines of regulatory standards. The small organic species formed in anaerobic biodegradation normally dissolve and diffuse at the liquid phase in the landfill area due to rain; and the organic matter from waste undergoes hydrolysis and the compounds obtained after hydrolysis undergo conversion and volatilization to form CO_2 and CH_4 and the remaining part of the wastes are present as liquid and partly in solid phase [54].

The total life cycle of a sanitary landfill generally is arranged in the following five separate phases [15, 51, 54]:

1. **Initial Adjustment:** Initially, the void spaces present in the landfill area contain a high concentration of O_2 but gradually with the addition of more and more wastes, it becomes compacted with a reduction in volume. The oxygen concentration in the bioreactor layers of the landfill sites progressively decreases. Consequently, the density of the wastes increases with an increase in the population of microorganisms, where the aerobic process of biodegradation dominates in which O_2 is considered as the electron acceptor.
2. **Transition:** Due to the process of aerobic biodegradation, the concentration of O_2 quickly decreases with the present population of microorganisms. Due to continuous decrease in the level of O_2, the process gradually changes from aerobic to anaerobic. Therefore, O_2 is continuously substituted by CO_2 from the landfill gases (LFGs) and the key electron acceptors in the process of transition are SO_4^{3-} and NO_3^-.
3. **Generation of Acids:** The biodegradable part of the solid waste present at the landfill area on hydrolysis will form acids, which are converted quickly as volatile fatty acids (VFAs) and mixed with the leachate generated at the landfill sites. The increase in the concentration of the acid decreases the pH of the leachate from about 7.5 to 5.6. In this phase of the degradation process, the transitional compounds formed like VFAs adds more COD to the landfill wastes [15, 54]. The volatile organic acids (VOAs) of long-chain are transformed into CH_3COOH, H_2, CO_2. The presence of VFAs in high concentration contributes both BOD and VOAs, which stimulates the production of H_2 due to the activity of fermentative bacteria and supports the growth and existence of H_2-oxidizing bacteria. The phase of H_2 production is

normally small because it ends after the acid creation phase. Hence the population of acidogenic bacteria gradually increases results in an increase in the rate of biodegradation by consuming more nutrients. The water-soluble metals and their salts having less pH can easily disperse within that phase and increases the concentration of metals in the leachate [15].

4. **Methane Production:** The methanogenic bacteria convert the acids, such as butyric acids, acetic acid and propanoic acid to CO_2 and CH_4. Hence, methanogenic bacteria can metabolize VFAs with neutralizing the pH of the effluent liquid generated at the landfill sites. The organic strength of leachate is expressed in the form of oxygen demand, which gradually decreases at a quicker rate with an increase in the production of CO_2 and CH_4 [54].
5. **Final Stabilization and Maturation:** In the last phase of the bioremediation, the function of the microbial community diminishes due to the supply of a smaller amount of nutrients, which restricts the chemical reactions and the bioavailable phosphorus. The CH_4 generation totally vanishes, but the oxidized products and oxygen gradually appear and O_2 infuses in a downward direction from the troposphere. The residual byproducts of organic materials are then converted into a gaseous phase, whereas the remaining organic substances form composts and finally are converted into substances of humic categories [15, 51].

The proper management of solid waste is now an important issue in most countries. The common methods used for the management of wastes are incineration, landfilling, composting, pyrolysis, etc. However, it is not applied in modern technology because of high cost. The sanitary landfill process is used in many countries because of the requirement of less technological assistance and cost. According to ASCE (American Society of Civil Engineers), a sanitary landfill is defined as a controlled regulatory operation of MSW deposited in layer like form, the layers are compressed and covered by a layer of soil before adding the next layer. The basic distinction between landfill and dumping is that: in dumping, no effort is taken to separate the waste materials from the soil or layers of the rocks, the gap created spreads towards the groundwater aquifer, and the dumped wastes are directly mixed with groundwater [15, 54].

The sanitary landfill is made in a proper engineering plan with bottom lining with leachate collection, separation systems and proper covering. The landfill is designed in such a manner that it can store and treat the dumped wastes, simultaneously. The major components generated in the landfill are biogas and leachates, which could affect our biological system. Biogas is the cause of air pollution, whereas leachates are the cause of groundwater pollution. Hence, more attention is given on pre-treatment of MSW before dumping to the landfill sites, and the landfill sites should be properly covered [51].

The collected biogas may be used as a source of energy, whereas the leachates generated must have to be again treated before mixed into water sources. The LFG or biogas is generated due to the anaerobic biodegradation of biological materials. The rate of production of biogas and its chemical composition mostly depends on the service life of the landfill [15].

The LFG produced in the landfill sites contains an average of 40–60% of CH_4 with an average value of heating capacity 17,765 kJ/N-m^3 and the efficiency of energy conversion is about 34%, which is able to generate the electric energy of about 2.5 kW-h/N-m^3. LFG is a potential source of greenhouse gas production because of the presence of CO_2 and CH_4 are the major components [51, 54]. The key factors that influence the production of LFG are: chemical composition, structure of wastes, percentage of moisture, age of the landfill, temperature, and pH, etc.

11.3.1 LANDFILL MINING (LFM)

It is a new advanced process to lessen environmental hazards and risks related to landfills and is able to get the energy and secondary materials from the wastes and permits the land uses in the command area of the landfill. Although landfill mining is able to recover the resource materials from the waste, but it has a number of drawbacks, such as [2, 7, 11, 29]: (i) The LFM process requires an average of 103 MJ diesel and 1.9 MJ of electricity per ton of mined wastes, which in turn produces 12 kg of CO_2 or an equivalent. (ii) The transportation and processing during LFM emitted a massive amount of greenhouse gas in the troposphere, therefore, causes negative impact on climate. (iii) Hence, in most of the cases, the current scenario of LFM illustrates negative result. (iv) It is a great challenge to develop and design LFM for maximum recovery of useful materials with the least environmental risks and hazards.

11.4 SETTLEMENT OF MSW IN LANDFILL

The bioremediation of biological matter contained in the MSW present in the landfill sites is the potential cause of enhancing the gap ratio that results in the decrease in the density and structural strength of the waste materials present in the landfill area with subsequent reduction of volume due to settlement [25]. The settlement of wastes of MSW always obstructs the design and expansion of ideal landfill sites, which is highly desirable in the landfills, where the land area is limited or not enough.

The mechanism of settlement of MSW in landfills is the correct prediction of the magnitude and rate of settlement, which is the most required factor for proper design of a landfill with perfect operation. The settlement forecasting helps to determine the lifespan and design technology of landfill. Although the differential settlement of MSW is more fruitful, yet its application is critical in comparison to the total settlement [51]. It should be unavoidable because of non-homogeneity of MSW in the landfill sites.

The major limitation of the differential treatment process is the production of large quantities of water on the covering area and accumulation of water at the drainage layers; therefore, it is the cause of increase in the rate of infiltration of water into the leachate and waste itself.

If the biodegradation of MSW occurs at the landfill sites for a longer time period, the amount of the settlement of MSW develops more. Now different mathematical modeling has been formulated to improve the mechanism of settlement in the landfill sites [37]. The models developed are normally used in Geotechnical engineering, which can estimate the magnitude and rate of the settlement depending on the nature of the soil and MSW [21, 51, 68].

Although a number of models are being developed to identify the role of bioremediation on settlement of MSW, yet very low number of these have a direct relationship associated quantitatively to predict the different parameters of the model for describing the biological processes of settlement within the landfill. However, all these developed models are successful in providing information to a limited degree, but more is estimated at the time of settlement of the waste in correlation with physicochemical and biological processes occurring at the landfill sites.

The three different stages of settlement of MSW are given below [20, 47]:

1. **Initial Compression:** This is the settlement, which occurs directly by the application of external force or load at the landfill. This stage

is the compaction of layers immediately with the decrease of void space and volume of the wastes.

2. **Primary Compression:** This is owing to the debauchery of water of different pore spaces and the gaseous substances from the void spaces. This compression generally happens after 30 days of dumping the wastes into the landfill, but however the initiation of the processes of dissipation is not accountable for the primary settlement of MSW in landfill. The wastes to dump at the landfill are rarely saturated because the conventional management of MSW in a landfill does not allow the addition of water into the filling area. The property of permeability of the wastes is of similar type (magnitude and order) like gravel and sand. Since the water formed can easily escape out from the pores, therefore no water pressure is generated at the pore space.
3. **Secondary Compression:** This is because of the biological disintegration and development of creep at the skeleton of the waste present at the landfilling area. Generally, we can say that the stage of secondary compression comprises a noticeable fraction of the whole landfill sites of settlement and happens over many years. The long-term settlements at the stage of primary compression results in the biodegradation of wastes, and the volume of the waste reduces theoretically to about 60% of the total waste in the landfill. After few years the waste is completely degraded depending on the process of stabilization.

Normally it was predicted that almost 15% thickness of the original waste in the landfill is settled due to decay of the waste materials. It is believed now that the secondary compression in settlement of wastes within the landfill comprises of three factors, such as mechanical secondary compression, biological disintegration, and physicochemical actions. Hence, the secondary-compression is directly proportional to the condition of favorable degradation and gap ratio [20, 47].

11.4.1 MECHANISMS OF SETTLEMENT IN LANDFILL

There are five mechanisms leading to the settlement of MSW in landfills, such as mechanical change, physicochemical change, raveling, biochemical decay, and interaction between these mechanisms. Due to dumping stress or applied load to the waste particles it crushes, bend, and also move them to settle in a better compressed region creating mechanical deformation. The waste particles are of different shapes and sizes; the small size particles have greater diffusion capacity because of lightweight and escape from the voids more rapidly (raveling) [60, 65]. This results in more settling during the period of compaction. Normally, it is very tough to differentiate raveling from other different kinds of mechanisms. The physicochemical transformation (such as corrosion, combustion, and oxidation) in combination with biological degradation leads to the reduction of waste mass residues resulting in more vertical deformations. The process of combustion is stimulated due to the production of heat and CH_4 and on merging it activates raveling [64].

The MSW is biodegradable and is also compressible, therefore when it is placed in the landfill sites, which can be easily settled and the settlement continues for a longer time period and the height of the landfill is reduced up to 75%–50% in comparison to the volume at the time of dumping. The more is the settlement of the wastes, more is the capacity of storage at the landfill sites, which is the indication of good operation and proper management of the landfill. After the settlement procedure is closed, the cover or caps of the leachate or gas storage systems may hamper reclamation [65].

Hence, correct predictions of the rate of settlement are always necessary both during and after the filling of the landfill, so that design of landfill area, leachate collection system, gas storage system and expansion of the landfill areas have been properly estimated. For MSW settlement in the landfill areas, various kinds of mechanisms have to be adopted on the basis of timescale. The most accepted mechanisms of settlement are described below:

- The matrix or base phase of the solid wastes should be rearranged through reorientation, sliding or distortion because of increase in vertical force either in the period of compaction or due to further dumping of the waste materials at the top of the deposit. This mechanism is analogous to the compression mechanism used normally in soil mechanics, where it was assumed that the voids where water is present and the particles of the solid waste both are incompressible in nature and the changes in whole volume are constantly related with the change in the void ratio because the waste particles are distorted and maybe rearranged among these. The settlement mechanism of the solid wastes is more effective immediately after loading, when the waste dumped is highly permeable and unsaturated in nature. But otherwise, if the dumped wastes are saturated and having low permeability, then the settlement rate is less owing to the reduced volume of the gaps, which leads to the poor water escaping capacity [60, 67].
- There is another mechanism, which is the compression of the fluid present at pore space or voids. This is not usually similar to the

mechanism of saturated soil because the water present in the voids is incompressible in nature. But this mechanism is more significant in the case of unsaturated soil, where the air, water or both cannot judiciously be observed as incompressible [65].

- Crushing and compression of the particles of the wastes is usually treated as the conventional mechanism of soil mechanics.
- The breaking of the waste particles is caused by an increase in pressure, and due to soaking of the waste, the particles in the waste are softened and losing their structure and shape. These mechanisms usually are not considered as conventional soil mechanics, even though the importance of the breakdown of the particles recognized is continuously increasing [60, 63].
- The processes of settlement, such as physicochemical method and biological processes such as oxidation and corrosion are the cause of biodegradation of the wastes. The MSW present at the landfill will be converted into leachate and gaseous components. Then the wastes remain at the matrix phase and cannot evade the force applied and the garbage materials undergo settlement.

The normal mechanical creep developed in the landfill areas are due to the following causes and undergoes settlement:

- Erosion and transfer of fine particles into the void space in between larger sized particles;
- The fine particles of the waste materials are diffused from the void space and slowly degraded;
- Sustained plastic deformation.

The continued migration of tiny particles into the gaps, which is commonly called raveling, is detected as a new mechanism. It is the mechanisms that involves in the settlement of degraded products and mechanical creep produced at the site of the landfill [60, 64, 65].

11.4.2 FACTORS AFFECTING SETTLEMENT IN THE LANDFILL

The rate of the settlement at the landfill primarily depends upon the composition and structure of the wastes, operational procedure, microbial activity, moisture content, pH, and temperature.

11.4.2.1 MECHANISMS RELATED TO SETTLEMENT AND TEMPORAL PHASES

1. **The Immediate Compression:** It is mainly owing to the overloading of the dumped wastes and the additional stress produced due to excessive dumping at the landfill. This process is completed within few hours leading to the formation of excessive strain.
2. **The Primary Consolidation:** This only occurs in the saturated soil, but the condition of saturation is seldom true in case of MSW because of the unsaturated condition and more hydraulic conductivity of the wastes after the settlement. Henceforth the primary consolidation is not well-thought-out.
3. **The Long-Term Settlement:** It is associated with a number of mechanisms such as creep, biodegradation, raveling, and physico-chemical degradation [32, 45].

11.4.2.2 METHODOLOGY OF SETTLEMENT

Again, the rate of settlement depends upon the methodology, which is explained by the vertical strain at the landfill sites. The vertical strain depends upon the original height of MSW and is determined after the settlement because of immediate compression [32, 49].

11.4.2.3 EFFECT OF EXTERNAL VERTICAL STRESS

The external stress at the landfill is due to covering of MSW by the soil layers. The compaction of the wastes in the landfill also brings together the immediate compression to the waste products. The influence of the external vertical stress is calculated by proper analysis of the data found in the anaerobic tests without and with the application of vertical stress [45, 49].

11.4.2.4 EFFECT OF AERATION

The capturing of oxygen by the matrix phase of the waste is quickly depleted after dumping the waste at the landfill site. Hence, the anaerobic bioremediation conditions are usually more common both in the process of bioreactor and conventional landfills. The process of aeration generally does not change the volume of the gap space with the waste at the landfill, but can be able to speed up the rate of macroscopic biodegradation of MSW and enhances the biodegradability of the wastes.

Papers and organic waste materials degrade at a faster rate in aerobic than in anaerobic conditions because of the production of high energy. However, the bioremediation of wood and wood products are highly hostile and unfinished in the anaerobic process yet facilitated under the aerobic environment. The distribution and infusion of air are affected by method of aeration, void ratio, connectivity in between voids, simulator configurations and hydraulic conductivity [32, 49].

11.4.2.5 EFFECT OF COMPOSITION OF THE WASTES

The composition of MSW in the landfill spatially varies regionally and temporally. The constituents present in the waste materials from landfill are divided into a number of groups on the basis of kind, size, and geometrical shape. For carrying out experiments in the laboratory, the waste materials are always shredded in order to accommodate in small simulator. The fraction of the wastes having size >20 mm normally includes organic materials, paper, waste plastic products, waste textile products, wood, rubber after use and gravel pieces, which produces more strain in the process of settlement. These wastes are undergoing deformation and compression in both bio-inaccessible and bio-accessible gaps [32, 45].

11.4.2.6 EFFECT OF SIMULATOR AND SPECIMEN SIZE

The specimen size used for testing of MSW is usually more important. For the biodegradation studies of the MSW of crushed and uniform specimens, small simulators are normally appropriate. These specimens used are not illustrative of the dumped wastes of the landfill, whereas, the distinctive MSW are never processed. The small sized biodegradable wastes are used in small stimulators in order to enhance bioaccessibility and the rate of biodegradation resulting in an increase in the compression ratios and strain rate. The wastes having at least a radius of 0.15 m gives best result in larger simulators. The settlement is quicker and larger if the specimen used is of radius more than 0.5 m, whereas, almost no difference is observed in the waste particles having radius in the range of 0.05–0.5 m [32, 45, 49].

11.5 ENVIRONMENTAL FACTORS AFFECTING BIOREMEDIATION

11.5.1 GEOLOGIC AND HYDROGEOLOGICAL FACTORS

11.5.1.1 ADSORPTION AND ABSORPTION

The binding of organic substances over the surface of a solid is termed as adsorption. The degree of adsorption is calculated and depends upon the affinity of the adsorbent materials on the surface of the solid matrix phase, the surface area of the solid matrix in the landfill and the solubility and volatility of the waste particles in water. If the chemicals are hydrophobic in nature, then it can adsorb the hydrophobic particles and insoluble organic particles. Similarly, if the chemicals are hydrophilic in nature, then it mostly binds minerals. Hence, the pollutants which are adsorbed are always in equilibrium with the volatilized and dissolved pollutants until the concentration in the gaseous or aqueous phases is decreased.

The constituent particles, which are adsorbed strongly with the solid matrix phase, are less transported towards the gaseous and aqueous phases, and hence the mobility of these particles within the system is inhibited. However, the constituent particles, which are not adsorbed strongly, can easily be moved through the aquatic systems, atmosphere, and soil [14]. The rate of adoption totally depends upon the organic content present in the soil and kind of soil. Generally, in the superficial soils, the percentage of organic carbon varies in the range of 1 to 3.5% [5, 14]. Hence, the concentration of organic matter is less in the subsurface soils because of the fact that a major fraction of the organic residue is deposited on the surface soils and not easily migrated toward the subsurface.

Under some special situations, co-solvents or surfactants are added in the aqueous phase in order to enhance the ionic mobility and relative solubility of the polluted particles leading to the decrease in adsorption. In the process of absorption, the microscopic polluted particles are infiltrated into the bulk soil matrix. Both the process of absorption and adsorption are able to decrease the accessibility of the pollutants to various kinds of microorganisms.

11.5.1.2 CONTAMINANT MIGRATION IN GROUNDWATER

The movement of the contaminants within the aquifer can be regulated and controlled due to various physical and chemical properties of the pollutants and also the hydrogeological setting of the waste materials at the landfill. The hydraulic conductivity is considered as one of the significant features of the aquifer, which can properly and correctly forecast the movement of the pollutants and estimated the probability of adding the substance in order to increase the rate of bioremediation. The size of the particles and classes of the soil influences the hydraulic conductivity: the smaller is the size of the particles, the higher is the density of the soil, and lower is the rate of the hydraulic conductivity [5, 14].

The low value of hydraulic conductivity is accepted if the other factors are favorable. The change in the porosity of both saturated and unsaturated zones of the aquifer affects significantly the diffusion of fluids with water. The capability of the base phase after settlement to diffuse gases such as O_2, CO_2, CH_4 is reduced in the sediments containing fine grain sized particles, and the soil becomes comparatively better saturated in water [14]. This fact may influence the kind and rate of bioremediation that takes place at the landfill. However, fine grain size of silt and clay particles continuously retards the movement of the pollutants. There are also many other factors, which may affect the dispersal and distribution of the pollutants, that includes concentration of the pollutants, temperature, and distribution pattern. Dispersal is a process of mixing the pollutants mechanically and scattering the pollutants within the aquifer because of their kinetic energy [5, 14].

11.5.2 BIOAVAILABILITY

The portions of the pollutants, which are available for bioremediation by the microorganisms, are entitled as bioavailable. The amount of the bioavailable pollutants is always less than the total quantity calculated by using the process of solvent extraction. Hence, the rate and degree of pollutant undergoing bioremediation is directly proportional to the amount of bioavailable pollutants, but not the total quantity [51].

11.5.3 SOIL MATRIC POTENTIAL

It is defined as the determination of energy that is necessary to overcome the force of adsorption and capillary action and hence reflects the "work" of the microbial cells to extract water from the soil. The matric potential is an important characteristic of the soil and highly depends on the soil type, soil texture, field capacity, and soil moisture. If the soil moisture is in the range of

50–80%, it approximately corresponds to the matric potentials in the range of –0.01 to –0.015 MPa (sands), and in the case of fine-textured soil, it is in the range of –0.03 to –0.05 MPa. It was observed that the activity of the microorganisms is less in the soil having the metric potential approximately –0.1 MPa and it totally ceases at a matric potential of approximately –8 MPa. Hence, the matric potential of the soil in the range of –0.5 to –1.5 MPa is normally more appropriate and if the matric potential is very low, the rate of degradation of the pollutants in the landfill is accordingly low [51, 52, 65].

11.5.4 REDOX POTENTIAL

The redox potential determines the electron density of the soil. The biological energy produced is due to the oxidation of substances; in which the electrons move to the different substances called electron acceptors. The low electron density is the indication of oxidation under aerobic environment, whereas the high electron density is the indication of reduction in an anaerobic environment. Hence redox potential is always contingent from other soil parameters, which determines the kind of biodegradation.

11.5.5 NUTRIENTS

Normally microorganisms are already present in the contaminated soil, and for the process of bioremediation, there is no need of addition of microorganisms. However, to quicken the bioremediation process, the function and growth of the microorganisms must have to be stimulated by the addition of oxygen and nutrients in the waste, which supports the growth of native microorganisms and is known as biostimulation [13]. Nutrients are considered as the basic building blocks in the existence of microbial community, which is required by the enzymes in order to cleave the pollutants.

The activity and growth of the microorganisms normally depends on pH, moisture, and temperature. Although the microorganisms are normally isolated in extreme environment, but are able to grow comfortably at optimum conditions in a narrow range, so for proper growth of the microorganisms, optimum conditions are maintained. If the soil in the landfill is acidic in nature, then it should be first neutralized by the addition of some lime. The temperature at the site of the landfill influences the rate of biochemical reactions, and it almost doubles at every 10°C temperature increment, but up to a certain extent; and after a specific temperature, the microorganisms die [56].

Hence, the quantity of available oxygen determines the anaerobic or aerobic condition of the biodegraded system.

The compounds of hydrocarbons and toxic chlorinated compounds can be easily degraded in an anaerobic environment. The amount of oxygen can be increased by sparging air, and in some situations, H_2O_2 and manganese peroxide can be added within the system. The structure of the soil regulates and controls the actual distribution of air, nutrients, and water. The addition of organic materials or gypsum may improve the structure of the soil. The low soil permeability can obstruct the diffusion of nutrients, oxygen, and water, therefore soil having low permeability is not suitable in this process [13, 56].

11.6 BIOREMEDIATION OF LEACHATE GENERATED FROM LANDFILL SITES

Leachate is a thick liquid, which leaches or drains from the landfill sites and usually contains both suspended and dissolved impurities, and its composition varies on the type of wastes, age of the landfill sites and rainfall rate. It is the most popular term used in the background of landfill in industrial or putrescible waste materials. Leachate is actually a black colored thick liquid, toxic material that leaches from any stored or accumulated materials containing potentially rich quantities of hazardous substances.

The rainwater is the key reason for leachate formation, which is mainly caused by the precipitation infiltrating towards the accumulated wastes and interacts either physically or chemically with the constituent particles of the non-dissolved and dissolved materials deposited at the landfill. The biological decomposition of solid wastes combines with soil moisture and rainwater and forms a black color viscous liquid containing toxic substance. This sometimes flows out from the landfill and mixed with nearest sources of groundwater and surface water. The biological degradation of carbonaceous waste in landfill produce leachate in addition with some gaseous substances such as CH_4, CO_2 and some complex form of organic acids, carbonyl compounds, alcohols, and simple carbohydrate units [27].

The moisture found at the landfill can be removed by consuming water in the formation of LFG because the water vapor evaporates from the LFG or leachate may be drained out from the landfill. Hence, the drainage of the leachates is normally closely associated with surface runoff, rainfall pattern and rate of infiltration into the earth's crust from the landfill. The requirement of the landfill facility such as water preventing cover, insulating layer of materials, such as, clay or cohesive soil, geotextiles, and presence of plastic wastes is highly vital to the movement of water in the upper layer of the wastes of the landfill, and therefore it influences the risk of the migration of the toxic pollutants [55].

The quantity of leachate formed vastly depends on the environmental conditions because of the precipitation, which is the input of the leachate and evaporation leading to the drastic reduction of leachate and it also depends on the climatic factors. Normally in most of the countries in the world, landfills are not properly covered and insulated, this is the cause of leakage of leachate effluents into the surrounding, widely, and ultimately mixed with both surface water and groundwater [27, 55].

The quality of leachate generated from the landfill depends on the composition of the waste, water budget or water balance, physical, chemical, biological characteristics, and age of the landfill. In the newly formed landfill, the composition of the leachate generated cannot be easily measured, and it is difficult to predict the leachate composition in the future of the existing landfill. The presence of a high percentage of organic waste is due to uncontrolled anaerobic oxidation, which is generally operating within the first 10 years of the landfill formation. The leachate formation and composition are influenced by physical, chemical, biological reactions carried out at the landfill and also other factors such as pH, relative velocities of the different metal ions such as Fe, Ca, Mg, Al, etc., and other simple species [69].

When the landfill operation is completed, the pollution due to leachate reaches to the highest level and remains for a longer period of time with very slow decreasing rate. Since the prediction of interaction in between the different constituents of the landfill is very complex, therefore the determination of the composition and rate of flow of the leachate is very difficult.

The impact of leachate towards the environment varies with the variation of high or low concentrations of hazardous waste in different situations. Hence, there is a little quantity of hazardous materials found in the landfill sites such as expired medicine, battery case after use, detergents, some biomedical wastes, etc., whose presence is not much affecting the concentration of the leachate formed even for a longer period of time [53, 69].

11.6.1 COMPOSITION OF LEACHATE GENERATED IN LANDFILL

In the landfill site, the volume of the solid waste gradually increases because of the continuous addition on a daily basis. During rain when water is percolated inside the waste, the biodegradation of the organic or biological waste is initiated by the native fungi or bacteria present in the landfill. In this process, at first, the oxygen present in the landfill is used rapidly by the microorganisms leading to the formation of some by-products with the formation of an anoxic environment. In the process of waste biodegradation, the temperature increases with the quick decrease in pH value and the metal ions, which are originally insoluble at neutral condition is now dissolved in the leachate [19].

This biodegradation process also generates some water, which is mixed with the leachate to increases its volume. The leachate generated can also react with the substances, which are susceptible to biodegradation, such as fly ash, building materials made of cement, refractories, and materials containing gypsum may change the chemical composition of the generated leachate. If the huge volume of gypsum-based building waste is impending to the landfill site, it reacts with the generated leachate and forms a huge volume of H_2S, mercaptans, and becomes a major component of LFG and the leachate shows strong displeasing odor [68]. The color of the leachate changes to black or yellow and is transformed to a cloudy liquid with high turbidity. The smell of the leachate is more aggressive and acidic because of the presence of H_2S and mercaptans, which are the compounds rich in N, S, and simple organic species [19].

The wastes added in the landfill are the combination of four different categories of wastes, including urban waste, commercial wastes, biomedical wastes, and industrial wastes. These wastes in aqueous solution consist of three different classes of pollutants:

- Dissolved organic wastes, including alcohols, carboxylic acids, aromatic acids, carbonyl compounds, simple carbohydrates [19].
- Some inorganic components such as anions and cations like Cl^-, SO_4^{2-}, Fe, Al, Zn, Mg, Ca, and NH_3 [19, 68].
- Some toxic heavy metals (HMs) such as Pb, Hg, Ni, Cu, and xenobiotic hazardous organic substances like PCBs, halogenated organic compounds, dioxins, etc. In addition to that, some other complex organic compounds such as 2-OH-benzothiazole, perfluorooctanoic acid and Bisphenol-A may also be present in the landfill leachate [19, 68].

11.7 BIOREMEDIATION OF SOLID WASTES OF LANDFILL SITES

The aim and purpose of bioremediation is to make a clean green environment, reducing the pollution load with the help of eco-friendly microbial communities. MSW is predominantly the domestic wastes obtained from the local community of the municipality or cities. In most of the countries of the world, MSW are mainly generated from three different sources as follows [38, 66];

- The waste materials from households and public places such as residential buildings, streets, litter bins, marine coastal areas, and parking lots, which are commonly termed as domestic solid waste;
- The waste materials generated from shops, fish markets, hotels, chicken shops, restaurants, offices, vegetable markets and some other market areas, which is commonly termed as commercial solid waste; and
- The waste materials ejected from the industries without the hazardous waste which are commonly termed as industrial solid waste.

It was estimated that about 1.3 billion tons of MSW are obtained at every year globally, and the volume of the wastes is continuously increasing very quickly due to rapid development and growth of urbanization, more stock resource consumption and modification in the lifestyle. The total generation of MSW is assumed to be almost doubling in 2025, especially in developed and developing countries. In many countries worldwide, open dumping of MSW is a common practice, and its disposal in the landfill does not follow any scientific procedure and no care is taken related to the pollution of the surrounding environment.

MSW contains toxic HMs, organic pollutants, plastic wastes, biomedical wastes, which on bioremediation releases some simple gaseous components such as VOCs, benzene, ethyl-benzene, toluene, and isomers of xylene [38]. The wastes on oxidation forms CO_2, CH_4, and some other gaseous compounds of their derivatives. The rapid increase in population, modernization, and decrease in the natural resources, the generation of solid waste and e-waste in the landfill increases rapidly, therefore it is an obligation, to adopt a safe and proper technique for disposal of such a huge volume of solid wastes [38, 66].

The MSW present in the landfill are more commonly and popularly used as an important tool for waste management, and the microbial community are the core components of bioremediation. Bioprocessing of MSW in the landfill normally originates due to the metabolism of the substrate and co-metabolism by different syntrophic microorganisms leading to the partly conversion of complex substances into simple materials and undergoes mineralization into simple inorganic salts, gaseous components, water, and CH_4 (used as biofuel). The bio-decomposition of substrate is a result of the interaction among the various bacteria, fungi, and archaeal domains because of the prevalent environmental conditions of both abiotic and biotic substances in landfill sites, which facilities the processes like hydrolysis, acetogenesis, fermentation, and methanogenesis successfully [3, 37].

11.8 BIOREMEDIATION OF HEAVY METALS (HMS)

Although there are 20 different kinds of HMs found in nature, yet only a few metals including Cd, Ar, Ni, Cu, Ag, Ur, As, Sb, Hg, Zn, Pb, Ra, etc., exhibit the property of toxicity [16, 44]. Not only soil but also groundwater is specially contaminated through the process of leaching the leachates formed at the landfill [17, 62]. The removal of these toxic metals is essential because these metals can easily enter the food chain and food web and finally get deposited in the tissues of the human and animal bodies creating severe health hazard effects. The process of biosorption of the cations of the metal is strongly affected by the pH of the landfill [8, 14, 44, 55].

Experimentally it was found that the presence of metals in the biosorption is highly influenced by various species of fungi and is very much sensitive to pH. It was also found that in many cases, the removal efficiency of the metal ions like Cd^{2+}, Zn^{2+} and Cr^{6+} ranges in between 85% and 60% predominantly due to the mechanism of intracellular accumulation. There are two important species of microorganisms (*Aspergillus niger* and *Pseudomonas aeruginosa*), which are able to remove almost all the toxic HMs through the process of bioremediation. The HMs Pb, Cr, Cd, Hg, Cu, Zn, and Selenite (SeO_3^{2-}) are found to be more toxic towards human health [16, 17, 44, 62].

The species of microorganisms, which are effective towards the bioremediation of HMs, are: *Flavobacterium, Pseudomonas, Arthrobacter, Bacillus, Corynebacterium, Rhodococcus, Methosinus, Mycobacterium, Stereum hirsutum, Methanogens, Nocardia, Aspergillus niger, Rhizopus arrhizus, Pleurotus ostreatus, Azotobacter, Phormidium valderium, Alcaligenes, and Ganoderma applanatum* [44, 62].

Generally, it was observed that industrial effluents contain HMs beyond the acceptable limit fixed by different regulating agencies, and in most of the countries, they are used for cultivation purposes. Indirectly, the HMs enter into river water and groundwater, which are the major sources of freshwater for human consumption. The chance of increase in HM in soil, landfill waste and water are due to the use of advanced technology, industrial exposure, and agricultural development without sustainable manner. Hence, it is a great challenge for us to remove the toxic metallic compounds from the environment [16, 17, 44, 62].

11.9 BIOREMEDIATION OF RUBBER WASTE

The solid waste generated from various sources contains about 12% of rubber, which is non-degradable and non-recyclable because of its unique physical properties and chemical composition. Rubber is either natural or synthetic polymer, which on combustion emits huge quantities of toxic fumes and CO causing various environmental issues. Although, the utilization of rubber is increasing day by day, yet the maximum use of rubber is in the automobile sector, i.e., 65% mainly as tires and tubes [33].

The presence of ZnO in rubber prevents the growth and survival of sulfur-oxidizing bacteria and some other naturally found native bacteria, which may be the reason for the slow rate of degradation of rubber waste. Hence, for easy degradation of rubber, there is a need to eliminate the various toxic components from rubber by the fungal microorganisms, such as, *Recinicium bicolor*. The rubber wastes are devulcanized by using sulfur-oxidizing or reducing bacteria commonly *Pyrococcus furiosus* and *Thiobacillus ferroxidans*; and the rubber obtained after devulcanization can also be recycled [21]. Hence, the combustion of waste rubber products in a controlled manner is considered as the best management practice.

The natural biodegradation of rubber wastes is very difficult and extremely slow process, as the polymer chain segments of the rubber are cross-linked with each other. During the manufacturing process, it requires additives, stabilizers, initiators, and other chemicals for prevention of biodegradation. Most of the recycling process of the rubber involves conversion of unused rubber into tiny particles of rubber, termed as, rubber crumb [33]. The crumb produced in combination with newly prepared rubber forms water emulsion of rubber particles known as latex adhesives. In this process, the original rubber loses its physical properties, which inhibits the performance of the prepared rubber, therefore in addition to recycling, the waste from unused rubber is broken into small pieces and utilized for production of cementing materials for road surface construction and or used as fuel for the production of electricity. The tiny particles of rubber on mixing with asphalt forms a new composite material, which is economical, durable, and successfully used for road construction purposes. The bioremediation of waste tires and tube rubbers can be carried out by devulcanization, rubber metabolism and detoxification [21, 33].

11.9.1 MECHANISM OF RUBBER BIOREMEDIATION

The waste scrap rubber product materials are devulcanized, degraded, and detoxified by using different kinds of fungi and prokaryotes. Among these, the fungi *Recinicium* is more important and successful species of bacteria for the rubber degradation. The other bacteria, such as *Rhodococcus rhodochrous, Pseudomonas* spp., *Corynebacteria* spp., and *Escherichia coli*, can also be effectively used for the detoxification of waste vulcanized rubber products [22, 41, 56].

The sulfur-oxidizing bacteria, particularly *Thiobacillus* spp., is effective towards the oxidation of disulfide bonds and therefore can be easily devulcanized. The sulfur-reducing archaea, such as, *Pyrococcus furiosus* is also effective for devulcanization of waste rubber. In the process of bioremediation, the poly (cis-1,4-isoprene) chain segments present in the rubber can easily undergo degradation by the bacteria. The enzymes identified for metabolism of rubber waste products include: rubber oxygenase (RoxA) and latex-clearing protein (LCP) to initiate the breaking of poly (cis-1,4-isoprene) bond [44, 66, 69].

11.9.1.1 BIOLOGICAL DETOXIFICATION

During the process of manufacturing of the rubber products, many chemical compounds are added, such as zinc salts and ZnO. These compounds prevent microbial growth; therefore, it is essential to remove these products before metabolism and devulcanization.

11.9.1.2 FUNGAL DETOXIFICATION

There are many fungal species, which have the capability of degrading many toxic aromatic compounds present in rubber; therefore, these fungal species can be effectively used for the degradation of rubber materials. There are a number of species of white-rot fungi (WRF), which are more effective enzymes towards the degradation of lignin, and this class of enzyme is also fruitful in the process of soil bioremediation. Since this kind of fungi is effective for degradation of xenobiotics; therefore, these are recommended for using degradation of rubber because rubber contains many structurally similar kinds of aromatic substances. The brown rot fungi similarly are also effective towards the degradation process of rubber.

11.9.1.3 BACTERIAL DETOXIFICATION

MBT (2-methyl-benzothiazole) is considered as the most toxic material added in the rubber, which serves as an accelerator during the process of vulcanization. It was observed that the bacteria *Rhodococcus rhodochrous* has a great capability of cleaving or biotransformation of MBT. The aromatic compound MBT is a particular cause of bladder cancer in human beings. The simple products produced due to the bioremediation of MBT are 2-methl-thio benzothiazole (MTBT) and benzothiazolyl sulfonate, both of these are comparatively less toxic. The other kinds of bacteria having the capability of metabolizing MBT include *Corynebacteria, Pseudomonas* genera and *Escherichia coli* [5, 67].

11.9.1.4 BIOLOGICAL DEVULCANIZATION

Before being subjected to the process of recycling, the rubber products have to be devulcanized, and during the process of devulcanization, oxidation of sulfur present in the rubber produces an active surface, and the outer layer of the rubber is changed from elastic to viscoelastic state. The viscous materials of the rubber combine with the newly formed rubber and finally are revulcanized [35]. The sulfur atoms, which are cross-linked with the rubber chain molecules, are removed due to the sulfur-oxidizing action of enzymes; and in this process, although sulfur is removed, yet the hydrocarbon chains are never broken down and remain intact.

Both the anaerobic sulfur-reducing archaeon and aerobic sulfur-oxidizing bacteria can actively desulfurize the rubber wastes. The anaerobic sulfur-reducing bacteria resist to the oxidation of the parent chain of hydrocarbons and form a high-grade quality of rubber than the aerobic oxidation carried out by sulfur-oxidizing bacteria [29].

11.9.1.4.1 Sulfur-Oxidizing Bacteria

The number of sulfur-oxidizing bacteria comfortably grow on the surface of the waste rubber products. The disulfide bonds present in the rubber undergo cleavage by sulfur-oxidizing bacteria. The major bacterial species used in this process of bioremediation are three bacteria of Thiobacillus class, such as *T. ferrooxidans, T. thiooxidans and T. thioparus*. These bacteria have the capability of producing strains to desulfurize the rubber products containing sulfur and degrade DBT, which is the indicator of the desulfurizing action of these bacteria [58].

11.9.1.4.2 Sulfur-Reducing Archaeon

The sulfur-reducing archaeon (*Pyrococcus furiosus*) in the anaerobic process is more effective towards devulcanization of rubber products. This *Pyrococcus furiosus* bacterium is of thermophile heterotroph class and its rate of growth and reproduction is inhibited due to the production of hydrogen. This prevention can also be protected by the addition of sulfur. Some compounds such as ZnO, zinc-based salts and some therapeutic chemicals like N-oxydiethene-2-benzothiazyl sulfenamide are toxic towards *Pyrococcus furiosus* bacteria. It was also observed that the new rubber on mixing with waste rubber can be effectively re-vulcanized efficiently by using *P. furiosus* bacteria. The bacteria *P. furiosus* helps in cleaving the sulfide bridge bonds in between the chain segments of the polyisoprene unit. This reaction sets the reactive sides free, which are used for cross-linking in the process of vulcanization [23, 42, 58].

11.10 BIOREMEDIATION OF POLYMER AND PLASTIC WASTES

The traditional polymer and plastic materials are very stable and cannot be easily degraded under normal environmental conditions and now is a major problem throughout the globe. The bioremediation of polymeric materials and their composites by the microbes are mainly governed by its structure, crystal-ability, molecular mass, chemical, and physical properties of the polymeric materials. The enzymes, which are accountable for the bioremediation of polymers, and plastics are divided into two kinds: intracellular depolymerase and extracellular depolymerase. These enzymes degrade complex form of polymeric materials into simple monomers, dimers, and trimmers with the release of energy. The degradation of polymeric materials forms simple products, such as simple organic acids, CH_4, CO_2, and H_2O at the end of the process [22, 59].

The natural polymers, such as chitin, chitosan, alginate, pectin, cellulose, etc., and the biocomposites are more susceptible towards bioremediation by microorganisms and the process of degradation is influenced by the accessibility of O_2. The polyolefins and their plastic derivatives are normally non-biodegradable in natural environmental conditions, however can be easily degraded by adopting chemical or photo-chemical degradation. Polyester, polyurethane, and aliphatic polyester are more prone towards microorganisms. The disruption of the polymer chain initiates the breakdown of the polymeric materials called aging [52].

More focus is given globally towards the research work on the bioremediation of plastic materials. The waste plastic materials generally come from industrial and household solid wastes. Both prokaryotic and eukaryotic bacteria are effective towards the bioremediation of plastic materials [22].

Microorganisms, such as, S*treptomyces, Arthrobacter, Micrococcus, Pseudomonas, Rhodococcus,* and *Corynebacterium* are the most effective microbial community being used in the process of bioremediation of plastic materials. The microorganisms can cleavage the polymer compounds to more simple, non-toxic, or less toxic materials with change in molecular mass, decrease in mechanical strength and some surface properties. The process of bioremediation is nothing but enzymatic cleavage of covalent bonds of the plastic and polymeric materials. The degradation occurs in the following consecutive steps:

1. **Bio-Deterioration:** The change in physical, structural, and chemical characteristics of the plastic and polymer.
2. **Bio-Fragmentation:** The plastic and polymeric materials are broken down into simple species by enzymatic activity.
3. **Assimilation:** The microorganism consumes the molecules as a source of food, and it grows.
4. **Mineralization:** The generation of oxidized and simple non-toxic species such as CO_2, H_2S, H_2O, CH_4 after the degradation. It can be carried out in both anaerobic and aerobic environments. In aerobic state, H_2O and CO_2 are produced, whereas in anaerobic state CH_4, H_2S, H_2O and CO_2 are produced.

The bacteria *Pseudomonas citronellolis* and *Streptomyces coelicolor* are effective towards the degradation of poly (cis-1, 4-isoprene) and vulcanized rubber. Styrene can be degraded by bacterial enzymes into simple products, such as, Pyruvate, 2-vinylmuconate and acetaldehyde 2-phenylethanol [22]. The solubilized natural biopolymers, such as lipids, carbohydrates, proteins, produce H_2O and CO_2 in the aerobic degradation process, whereas it produces CO_2, H_2S, CH_4, and NH_3 under anaerobic situation by sulfate-reducing bacteria or methanogens [52, 65].

It was explored that PEGs (polyethylene glycols) undergo degradation by sludge microorganisms both in the anaerobic and aerobic environment. The degradation in aerobic situation is observed to be more effective compared to anaerobic state. In the extreme weathered conditions, only surface of the polymeric materials undergo biodegradation, but in normal weathering, the structure of the whole polymer is affected mechanically, which can facilitate to permit microbes to enter inside the material resulting in the biodegradation of whole plastic materials [22, 52, 59].

11.11 BIOREMEDIATION STRATEGIES

The strategies of bioremediation of polymer and plastic wastes mean adopting suitable degradation techniques for proper management of plastic wastes, which includes disposal, structural composition, and toxicity. The presence of various kinds of waste synthetic plastics, microplastic, and nano-plastic are the cause of potential health hazards. Therefore, its proper sustainable management is required.

Although various kinds of techniques are adopted both in anaerobic and aerobic environments, but the most important strategies of bioremediation include *ex situ* and *in situ* techniques. The process of bioaugmentation is the process of addition of the microbes for degradation of the contaminants [9, 46].

11.11.1 IN SITU BIOREMEDIATION

The *in-situ* bioremediation process is applied for the treatment of contaminated soil, landfill, and groundwater and is classified into two categories: enhanced and intrinsic. The extent and rate of removal of the pollutants are based on the variability of the distribution of the contaminants, concentration of the pollutants, reaction kinetics, populations of microbial community and the environmental factors, such as pH, temperature, moisture content, supply of nutrient. During the *in-situ* bioremediation for the treatment of solid wastes, proper monitoring of landfill sites must be established to remove the primary pollutants by microbial activity. In situ technology is more popular because of cost-effective, less disruption, easy operational condition and adopts the process of treatment instead of the distribution and transfer of the pollutants. The treatment of the soil by *in situ* bioremediation process is restricted up to the length of 30 cm inside the surface of the earth's crust, but in some cases, it is effective up to 60 cm or more [24, 67].

The process of bioventing is the common *in situ* process, in which nutrients and air are supplied to the site of contamination to stimulate the native bacteria to degrade the organic contaminants present in the landfill. In this process, the rate of airflow is less; therefore, the limited flow of oxygen is available for biodegradation with the decrease in volatilization. This process is helpful for purification of groundwater and soil. In the treatment of groundwater, the water solution of the nutrients along with O_2 are injected into the groundwater [67].

Biosparging is the process of injecting air under pressure into the water table of the aquifer in order to enhance the concentration of O_2 in groundwater and increase the biological degradation of the pollutants with the help of local naturally found bacteria. This process enhances the combination of saturated zone with native bacteria leading to the increase in interaction between groundwater and soil. Bioaugmentation is an *in-situ* process of bioremediation in which microorganisms are continuously added either in the exogenous or indigenous process to the landfill [8, 24, 67].

11.11.2 EX SITU BIOREMEDIATION

This technology mostly includes excavation or separation of pollutants from the land surface. The process of excavation is generally carried out by separating the pollutants from the landfill areas and subsequently transferring these for one side to another for treatment. This technique of bioremediation is selected totally on the basis of cost, pollution level, kind of pollutants, nature of pollution, geographical location, meteorological factors, geological mapping of the contaminated area. The criteria of performance, which determines the nature of *ex-situ* bioremediation process, are described in subsections.

11.11.2.1 LAND FARMING

In this process, the polluted soil is properly excavated and uniformly distributed over the manufactured bed frequently until the whole contaminants are degraded. It is the simplest process among the bioremediation techniques because of its cost-effectiveness and simple operation. The land farming in most of the situation favors ex-situ process of bioremediation but rarely in few cases *in situ* process of bioremediation is favorable. When the contaminated soil undergoes treatment at the site, it is known as in situ, otherwise it is commonly termed as ex-situ technique of bioremediation.

The excavated contaminated soils are applied judiciously on a fixed layer above the surface to allow and facilitate the aerobic degradation of the pollutants by using autochthonous microorganisms. Cultivation or farming, which is the cause of aeration by mixing the extra nutrients such as N, K, P, and irrigation is considered as the key operational factors for stimulating the function of the autochthonous microorganisms, which leads to increased bioremediation during the period of land farming. The process of land farming is normally applied in the bioremediation of hydrocarbon products specially PAHs. Overall, land farming is very simple for designing and implementation and less initial cost is required. By this process, a huge volume of soil can be treated with less impact on the environment [8, 34].

11.11.2.2 COMPOSTING

This process involves the mixing of the polluted soil with the natural non-toxic organic substances, mainly agricultural waste products or manure. The presence of the natural organic substances facilitates the growth of microbial population and increases the temperature in that area, which is the characteristic property of composting.

11.11.2.3 BIOPILES

It is usually a hybrid technique of composting and land farming. The engineered cells are normally built in the form of piles of aerated compost, which is used for the treatment of usual surface contamination and the petroleum product of hydrocarbons. This is considered as a modified form of the land farming, which is able to control and prevent the physical loss of the pollutants due to volatilization, leaching, and delivers a promising environment for the growth of indigenous anaerobic and aerobic microbial population.

The process of bioremediation is based on biopile generally occurring above the ground. The piling of the polluted soil is done with the help of nutrients and extra aeration, which increase the function of the microbes leading to the increase in the rate of bioremediation. The major steps of this process are irrigating, leachate collection schemes, aeration, nutrient addition, and treatment bed. This procedure can be applied efficiently to remediate the pollutants even in the extreme condition of the environment.

11.11.2.4 WINDROWS

This is the method, where the piled contaminated soil is periodically changed to increase the efficiency of bioremediation due to an increase in the efficiency of degradation by indigenous or transient hydrocarbonoclastic microorganism or both, which are already present within the polluted site during the process of bioremediation. The periodical change of the polluted soil with the sprinkling of water enhances the aeration, uniformity in the distribution of contaminants, increase in nutrients and degradative activities of the local microorganisms, thus leading to the increase in efficiency of bioremediation. This process has a higher capacity for the removal of hydrocarbons as compared to the biopile process of treatment [18].

11.11.2.5 BIOREACTORS

The aqueous or slurry bioreactors are generally applied in the *ex-situ* technique for the treatment of the polluted soil and water and is driven out from the dirty plume. In this kind of reactor, the bioremediation is normally carried out by the processing of polluted sludge, soil, and sediments or water in an industrially developed system. The waste bioreactor can be defined as a polluted container or equipment which is used to improve the efficiency of the bioremediation in the form of slurry of the biomass to degrade the target species of contaminants. Normally the efficiency and extent of bioremediation is more in the bioreactor method in comparison to *in situ* or any other solid-phase schemes because the environmental condition can be properly managed, controlled, and forecasted.

In bioreactor, the polluted materials are converted into some simple non-polluted species through a number of biological reactions. There are various modes of operation in the process of bioreactor such as sequencing batch, batch, fed-batch, multistage, and continuous process of operation [39]. The samples of the polluted substances that are normally introduced into the bioreactor are either slurry or in a dry state. This method controls various parameters including temperature, agitation, pH, substrate, and rate of aeration, which is the cause of effective decrease in time of the bioremediation. The different limiting factors such as controlling bioaugmentation, addition of nutrients, an increase in the bioavailability of the contaminants, mass transfer in between the microorganisms and contaminants are the influencing factors for the process of bioremediation [18, 34, 39, 43].

11.12 SUMMARY

The process of bioremediation is greatly affected by the presence of local microbial population, bioavailability of the pollutants, type of pollutant, chemical structure, and concentration of the pollutants. In addition to these many environmental factors, including the nature of the soil, pH, nutrient availability, temperature, and presence of electron acceptors including O_2 are greatly influencing bioremediation. In bioremediation process, the microorganisms consume complex form of organic matter and degrade these into simpler less toxic pollutants with release of energy and some gaseous components, such as, CO_2, H_2O, and CH_4. The process of bioremediation cannot fully remove the toxic inorganic pollutants, metallic contaminants, and radionuclides, but it can change the oxidation state of the pollutants by the process of adsorption, acceptance, and accumulation. Finally, bioremediation normally involves the application of local microorganisms in a controlled manner for achieving a substantial reduction of the volume of wastes. The leachates produced in the process of bioremediation in landfill should be properly managed.

KEYWORDS

- **bioremediation**
- **contaminates**
- **landfill**
- **microorganisms**
- **toxic materials**
- **toxic substances**
- **volatile fatty acids**

REFERENCES

1. Abatenh, E., Gizaw, B., Tsegaye, Z., & Wassie, M., (2017). The role of microorganisms in bioremediation: A review. *Open Journal of Environmental Biology, 2*(1), 38–46.
2. Abdel-Shafy, H. I., & Mansour, M. S. M., (2018). Solid waste issue: Sources, composition, disposal, recycling and valorization. *Egyptian Journal of Petroleum, 27*(4), 1275–1290.
3. Abdullah, J. J., El-Imam, A. A., Greetham, D., Du, C., & Tucker, G. A., (2017). The application of fungi for bioleaching of municipal solid wastes for the production of environmental acceptable compost production. *Journal of Environmental Science and Public Health, 1*(3), 167–194.
4. Adhikari, B., Dahal, K. R., & Khanal, S. N., (2014). Review of factors affecting the composition of municipal solid waste landfill leachate. *International Journal of Engineering Science and Innovative Technology, 3*(5), 237–281.
5. Alkorta, I., Epelde, L., & Garbisu, C., (2017). Environmental parameters altered by climate change affect the activity of soil microorganisms involved in bioremediation. *FEMS Microbiology Letters, 364*(19), 1–9.
6. Amritha, P. K., & Anilkumar, P. P., (2016). Development of landscaped landfills using organic waste for sustainable urban waste management. *Procedia Environmental Sciences, 35*, 368–376.
7. Benjamin, S. R., (2019). Current trends in nanotechnology for bioremediation. *International Journal of Environment and Pollution, 66*, 19–40.
8. Beškoski, V. P., Cvijovic, G. G., & Milic, J., (2011). *Ex Situ* bioremediation of a soil contaminated by mazut (heavy residual fuel oil) - field experiment. *Chemosphere, 83*, 34–40.
9. Davoodi, S. M., Miri, S., Taheran, M., Brar, S. K., & Galvez, R., (2020). Bioremediation of unconventional oil-contaminated ecosystems under natural and assisted conditions: A review. *Environmental Science & Technology, 54*(4), 2054–2067.
10. Devi, L. H. S., & Arjuna, A., (2017). Bioremediation. *Biosciences Biotechnology Research, 14*(1), 367–372.
11. Dubey, A., Chakraborty, M., & Pandit, D., (2016). Landfill mining as a remediation technique for open dumpsites in India. *Procedia Environmental Sciences, 35*, 319–327.
12. Dzionek, A., Wojcieszyńska, D., & Guzik, U., (2016). Natural carriers in bioremediation: A review. *Electronic Journal of Biotechnology, 23*, 28–36.
13. Enim, A. E., (2013). Factors that determine bioremediation of organic compounds in the soil. *Academic Journal of Interdisciplinary Studies, 2*(13), 125–128.
14. Fei, X., & Zekkos, D., (2013). Factors influencing long-term settlement of municipal solid waste in laboratory bioreactor landfill simulators. *Journal of Hazardous, Toxic and Radioactive Waste, 17*(4), 259–271.
15. Ferronato, N., & Torretta, V., (2019). Waste mismanagement in developing countries: A review of global issues. *International Journal of Environmental Research and Public Health, 16*(6), 1060–1088.
16. Garima, T., & Singh, S. P., (2014). Application of bioremediation on solid waste management: A review. *Journal of Bioremediation & Biodegradation, 5*, 248–256.
17. Gaur, N., Flora, G., Yadav, M., & Tiwaria, A., (2014). Review with recent advancements on bioremediation-based abolition of heavy metals. *Environmental Science Processes and Impact, 16*, 180–193.

18. Gomes, H. I., Ferreira, C. D., & Ribeiro, A. B., (2013). Overview of *in situ* and *ex situ* remediation technologies for PCB-Contaminated soils and sediments and obstacles for full-scale application. *Science of the Total Environment, 445*, 237–260.
19. Gupta, A., & Paulraj, R., (2017). Leachate composition and toxicity assessment: An integrated approach correlating physicochemical parameters and toxicity of leachates from MSW landfill in Delhi. *Environmental Technology, 38*(13, 14), 1599–1605.
20. Hanc, A., Novak, P., Dvorak, M., Habart, J., & Svehla, P., (2011). Composition and parameters of household bio-waste in four seasons. *Waste Management, 31*, 1450–1460.
21. Hejna, A., Korol, J., Romatowska, M. P., Zedler, L., Chmielnicki, B., & Formela, K., (2020). Waste tire rubber as low-cost and environmentally-friendly modifier in thermoset polymers: A review. *Waste Management, 108*, 106–118.
22. Jang, J. H., Abu-Omar, M. M., Scott, S. L., & Suh, S., (2020). Degradation rates of plastics in the environment. *ACS Sustainable Chemistry and Engineering, 8*(9), 3494–3511.
23. Janssen, D. B., & Stucki, G., (2020). Perspectives of genetically engineered microbes for groundwater bioremediation. *Environmental Science: Processes & Impacts, 22*, 487–499.
24. Jørgensen, K. S., (2007). *In situ* bioremediation. *Advances in Applied Microbiology, 61*, 285–305.
25. Kapahi, M., & Sachdeva, S., (2019). Bioremediation options for heavy metal pollution. *Journal of Health and Pollution, 9*(24), 1–20.
26. Khalid, S., Shahid, M., Niazi, N. K., Murtaza, B., & Bibi, I., (2016). A Comparison of technologies for remediation of heavy metal contaminated soils. *Journal of Geochemical Exploration, 182*, 247–268.
27. Koshlaf, E., & Ball, A. S., (2017). Soil bioremediation approaches for petroleum hydrocarbon polluted environments. *AIMS Microbiology, 3*(1), 25–49.
28. Kotwal, D. R., Shewale, N. B., Tambat, U. S., Thakare, M. J., & Bholay, A. D., (2018). Bioremediation of mercury using mercury resistant bacteria. *Research Journal of Life Science, Bioinformatics, Pharmaceutical and Clinical Sciences, 4*(2), 146–156.
29. Krook, J., Svensson, N., & Eklund, M., (2012). Landfill mining: A critical review of two decades of research. *Waste Management, 32*(3), 513–520.
30. Kumar, S. S., Shantkriti, S., Muruganandham, T., Niraj-Rane, E. M., & Govindwar, S. P., (2016). Bioinformatics aided microbial approach for bioremediation of wastewater containing textile dyes. *Ecological Informatics, 31*, 112–121.
31. Leal, A. J., Rodrigues, E. M., & Leal, P. L., (2017). Changes in the microbial community during bioremediation of gasoline-contaminated soil. *Brazilian Journal of Microbiology, 48*, 342–351.
32. Lee, U., & Wang, J. M., (2017). Evaluation of landfill gas emissions from municipal solid waste landfills for the life-cycle analysis of waste-to-energy pathways. *Journal of Cleaner Production, 166*, 335–342.
33. Leong, Y. K., & Chang, J. S., (2020). Bioremediation of heavy metals using microalgae: Recent advances and mechanisms. *Bioresource Technology, 303*, 1–48.
34. Lina, T. C., Panb, P. T., & Cheng, S. S., (2010). *Ex-situ* bioremediation of oil-contaminated soil. *Journal of Hazardous Materials, 176*, 27–34.
35. Litchfield, C., (2005). Thirty years and counting: Bioremediation in its prime. *Bioscience, 55*(3), 273–279.

36. Martínez, A. G., Martín, M. S., López, R., Fernández, R. T., & Sánchez, A. B., (2019). Remediation of potential toxic elements from wastes and soils: Analysis and energy prospects. *Sustainability, 11*, 3307–3334.
37. Massoud, R., Hadiani, M. R., Hamzehlou, P., & Khosravi-Darani, K., (2019). Bioremediation of heavy metals in food industry: Application of *Saccharomyces cerevisiae*. *Electronic Journal of Biotechnology, 37*, 56–60.
38. Maximova, A., & Koumanova, B., (2006). Study on the content of chemicals in landfill leachate. In: Simeonov, L., & Chirila, E., (eds.), *Chemicals as Intentional and Accidental Global Environmental Threats* (pp. 345–356). Dordrecht, Springer.
39. Moldes, A. B., Paradelo, R., Rubinos, D., Rey, R. D., & Cruz, J. M., (2011). *Ex-situ* treatment of hydrocarbon-contaminated soil using biosurfactants from *Lactobacillus pentosus*. *Journal of Agricultural and Food Chemistry, 59*, 9443–9447.
40. Morris, S., Guiomar, C., Ryan, D., & Enright, A. M., (2018). Bioremediation of landfill leachate using isolated bacterial strains. *International Journal of Environmental Bioremediation & Biodegradation, 6*(1), 26–35.
41. Naik, G. M., & Duraphe, M. D., (2012). Review paper on - parameters affecting bioremediation. *International Journal of Life Science and Pharma Research, 2*(3), 77–80.
42. Nawong, C., Umsakul, K., & Sermwittayawong, N., (2018). Rubber gloves biodegradation by a consortium, mixed culture and pure culture isolated from soil samples. *Brazilian Journal of Microbiology, 49*(3), 481–488.
43. Ni, Z., Wu, X., Li, L., Lv, Z., & Zhang, Z., (2018). Pollution control and in situ bioremediation for lake aquaculture using an ecological dam. *Journal of Cleaner Production, 172*, 2256–2265.
44. Ojuederie, O. B., & Babalola, O. O., (2017). Microbial and plant-assisted bioremediation of heavy metal polluted environments: A review. *International Journal of Environmental Research and Public Health, 14*(12), 1504–1530.
45. Park, H., Park, B., & Lee, S. R., (2007). Analysis of long-term settlement of municipal solid waste landfills as determined by various settlement estimation methods. *Journal of the Air & Waste Management Association, 57*(2), 243–251.
46. Pathak, V. M., & Navneet, (2017). Review on the current status of polymer degradation: A microbial approach. *Bioresources and Bioprocessing, 4*, 15–46.
47. Pietzsch, N., Ribeiro, J. L. D., & Medeiros, F., (2017). Benefits, challenges and critical factors of success for zero waste: A systematic literature review. *Waste Management, 67*, 324–353.
48. Raghab, S. M., Maggid, A. M. A., & Hegazi, H. A., (2013). Treatment of leachate from municipal solid waste landfill. *HBRC Journal, 9*, 187–192.
49. Ronga, F., Zhaogui, G., & Tugen, F., (2011). Analysis of stability and control in landfill sites expansion. *International Conference on Advances in Engineering, 24*, 667–671.
50. Schindler, F., Merbold, L., Karlsson, S., Sprocati, A. R., & Kothe, E., (2016). Seasonal change of microbial activity in microbially aided bioremediation. *Journal of Geochemical Exploration, 174*, 4–9.
51. Sekhohola, D. L., & Tekere, M., (2020). Microbiology of municipal solid waste landfills: A review of microbial dynamics and ecological influences in waste bioprocessing. *Biodegradation, 31*, 1–21.
52. Shah, A. A., Hasan, F., Hameed, A., & Ahmed, S., (2008). Biological degradation of plastics: A comprehensive review. *Biotechnology Advances, 26*, 246–265.

53. Singh, C. K., Kumar, A., & Roy, S. S., (2018). Quantitative analysis of the methane gas emissions from municipal solid waste in India. *Science Report, 8*, 2913–2939.
54. Singh, R., Singh, P., & Sharma, R., (2014). Microorganism as a tool of bioremediation technology for cleaning environment: A review. *Proceedings of the International Academy of Ecology and Environmental Sciences, 4*(1), 1–6.
55. Spina, F., Tigini, V., Romagnolo, A., & Varese, G. C., (2018). Bioremediation of landfill leachate with fungi: Autochthonous vs. allochthonous strains. *Life* (*Basel*), *8*(3), 27–42.
56. Srivastava, J. R., Naraian, S. J., Kalra, S., & Chandra, H., (2014). Advances in microbial bioremediation and the factors influencing the process. *International Journal of Environmental Science and Technology, 11*, 1787–1800.
57. Stamps, B. W., Lyles, C. N., & Suflita, J. M., (2016). Municipal solid waste landfills harbor distinct microbiomes. *Front. Microbiology, 7*, 534–545.
58. Stevenson, K., Stallwood, B., & Hart, A. G., (2008). Tire rubber recycling and bioremediation: A review. *Bioremediation Journal, 12*(1), 1–11.
59. Sutar, H., & Das, C. K., (2012). A review on: Bioremediation. *International Journal of Research in Chemistry and Environment, 2*(1), 13–21.
60. Tahmoorian, F., & Khabbaz, H., (2020). Performance comparison of a MSW settlement prediction model in Tehran landfill. *Journal of Environmental Management, 15*, 254–266.
61. Vaverková, D. M., (2019). Landfill impacts on the environment: Review. *Geosciences, 9*, 431–447.
62. Verma, S., & Kuila, A., (2019). Bioremediation of heavy metals by microbial process. *Environmental Technology & Innovation, 14*, 1–11.
63. Vidali, M., (2011). Bioremediation: An overview. *Pure Applied Chemistry, 73*(7), 1163–1172.
64. White, J. K., & Beaven, R. P., (2013). Developments to a landfill processes model following its application to two landfill modelling challenges. *Waste Management, 33*(10), 1969–1981.
65. William, P., Bing, X. X., David, R., & Zhan, L. T., (*2019*). Mechanisms of settlement in municipal solid waste landfills. *Journal of Zhejiang University - Science A* (*Applied Physics & Engineering*), *20*, 927–947.
66. Yasmin, C., Mariam, K., Lobna, E., Abdelwaheb, C., & Ismail, T., (2019). Simultaneous removal of organic compounds and heavy metals from Tunisian landfill leachate using dairy processing wastes. *Water Practice & Technology, 14*(2), 278–288.
67. Ying, T., & Wei, C., (2019). Soil microbiomes- a promising strategy for contaminated soil remediation: A review. *Pedosphere, 29*(3), 283–297.
68. Zegzouti, Y., Boutafda, A., & Ezzariai, A., (2020). Bioremediation of landfill leachate by *Aspergillus flavus* in submerged culture: Evaluation of the process efficiency by physicochemical methods and 3-D fluorescence spectroscopy. *Journal of Environmental Management, 255*, 1–11.
69. Zegzouti, Y., Boutafda, A., & Fels, L., (2020). Screening and selection of autochthonous fungi from leachate contaminated soil for bioremediation of different types of leachate. *Environmental Engineering Research, 25*(5), 722–734.
70. Zhou, Q., & Hua, T., (2004). Bioremediation: A review of applications and problems to be resolved. *Progress in Natural Science, 14*(11), 937–944.

INDEX

B

C

D

E

I

N

R

T

For Product Safety Concerns and Information please contact our
EU representative GPSR@taylorandfrancis.com Taylor & Francis
Verlag GmbH, Kaufingerstraße 24, 80331 München, Germany